Social-Ecological Diversity and Traditional Food Systems

Social-Ecological Diversity and Traditional Food Systems
Opportunities from the Biocultural World

Edited by

Ranjay Kumar Singh
Central Soil Salinity Research Institute
Karnal, Haryana, India

Nancy J. Turner
School of Environmental Studies
University of Victoria, Victoria
B.C., Canada V8W 2Y2

Victoria Reyes-Garcia
ICREA and Institut
de Ciencia i Technologia Ambientals
Universitat Autonoma de
Barcelona, Spain

and

Jules Pretty
Essex Sustainability Institute and
Department of Biological Sciences
University of Essex, Colchester CO4 3SQ, UK

CRC Press
Taylor & Francis Group
Boca Raton London New York

CRC Press is an imprint of the
Taylor & Francis Group, an **informa** business

NEW INDIA PUBLISHING AGENCY
New Delhi – 110 034

First published 2022
by CRC Press
2 Park Square, Milton Park, Abingdon, Oxon, OX14 4RN

and by CRC Press
6000 Broken Sound Parkway NW, Suite 300, Boca Raton, FL 33487-2742

© 2022 selection and editorial matter, NIPA.; individual chapters, the contributors

CRC Press is an imprint of Informa UK Limited

The right of Ranjay Kumar Singh et.al. to be identified as the authors of the editorial material, and of the authors for their individual chapters, has been asserted in accordance with sections 77 and 78 of the Copyright, Designs and Patents Act 1988.

All rights reserved. No part of this book may be reprinted or reproduced or utilised in any form or by any electronic, mechanical, or other means, now known or hereafter invented, including photocopying and recording, or in any information storage or retrieval system, without permission in writing from the publishers.

For permission to photocopy or use material electronically from this work, access www.copyright.com or contact the Copyright Clearance Center, Inc. (CCC), 222 Rosewood Drive, Danvers, MA 01923, 978-750-8400. For works that are not available on CCC please contact mpkbookspermissions@tandf.co.uk

Trademark notice: Product or corporate names may be trademarks or registered trademarks, and are used only for identification and explanation without intent to infringe.

Print edition not for sale in South Asia (India, Sri Lanka, Nepal, Bangladesh, Pakistan or Bhutan).

British Library Cataloguing-in-Publication Data
A catalogue record for this book is available from the British Library

Library of Congress Cataloging-in-Publication Data
A catalog record has been requested

ISBN: 978-1-032-15903-4 (hbk)
ISBN: 978-1-003-24622-0 (ebk)

DOI: 10.1201/9781003246220

Dedicated to
Indigenous People and
Native Communities of the World

Contents

Contributors .... *ix*
Preface .... *xi*
Acknowledgements .... *xv*
Introduction
Social-Ecological Diversity and Traditional Food Systems:
Opportunities from the Biocultural World .... *xvii*
Ranjay K. Singh, Nancy J Turner, Victoria Reyes-García and Jules Pretty

1. **The Food/Medicine/Poison Triangle: Implications for Traditional Ecological Knowledge Systems of Indigenous Peoples of British Columbia, Canada** 1
   Nancy J. Turner

2. **Integration into the Market Economy and Dietary Change: An Empirical Study of Dietary Transition in the Amazon** 33
   Elizabeth Byron and Victoria Reyes-García

3. **The Loss of Local Livelihoods and Local Knowledge: Implications for Local Food Systems** 65
   Sarah Pilgrim-Morrison and Jules Pretty

4. **The Seasonal Migration of Thai Berry Pickers in Finland: Non-wood Forest Products for Poverty Alleviation or Source of Imminent Conflict?** 91
   Celeste Lacuna-Richman

5. Sustainable Management of Natural Resources and Biocultural Diversity for Subsistence Livelihoods: A Cross Cultural Study 107
   Ranjay K. Singh, Anamika Singh, Anshuman Singh and B.S. Dwivedi

6. Status and Contribution of Non-cultivated Food Plants Used by Dawro People in Loma District, South Ethiopia 137
   Kebu Balemie

7. Biocultural Resources and Traditional Food Systems of *Nyishi* Tribe of Arunachal Pradesh (India): An Empirical Learning on the Role of Mythology and Folklore in Conservation 155
   Hui Tag, P. Kalita, Ranjay, K. Singh and A.K. Das

8. New Shoots, Old Roots — the Incorporation of Alien Weeds into Traditional Food Systems 199
   Michelle Cocks, Tony Dold and Madeleen Husselman

9. Edible Fungi in Mesoamerican Lowlands: A Barely Studied Resource 217
   Felipe Ruan-Soto and Joaquín Cifuentes

10. Menu for Survival: Plants, Architecture, and Stories of the Nisga'a Oolichan Fishery 237
    Nancy Mackin and Deanna Nyce

11. Salmon Food Webs: SAANICH First Nation Peoples' Intrinsic Interconnectedness to Salmon Fishing and Conservation on Southern Vancouver Island, British Columbia, Canada 261
    Roxanne Paul

12. *Tsampa* of Ladakh: Adaptation of a Traditional Food at Higher Altitude and Emergent Changes 293
    Konchok Targais, Dorjey Angchok, Tsering Stobdan, R.B. Srivastava and Ranjay Singh

13. Bioculturally Important Indigenous Fruit Tree *Mahua* (*Madhuca* spp.; Sapotaceae): It's' Role in Community -Based Adaptive Management 307
    Anshuman Singh, Ranjay K. Singh, Sarvesh Tripathy and BS Dwivedi

Contributors

A.K. Das
Plant Systematic and Ethnobotanical
Research Laboratory
Department of Botany
Faculty of Life Science
Rajiv Gandhi University
Rono Hills, Itanagar – 791112
Arunachal Pradesh, India

Deanna Nyce
University of Victoria and Wilp
Wilxo'oskwhl Nisga'a
Canada, 2919 Tower Hill
West Vancouver, BC
Canada V7V4W6

Dorjey Angchok
Defence Institute of High Altitude Research
Leh 194 101, India

Elizabeth Byron
Independent Researcher
International Health
Washington DC, USA

Felipe Ruan-Soto
Sección de Micología Herbario Eizi Matuda
Facultad de Ciencias Biológicas
Universidad de Ciencias y Artes de Chiapas
Libramiento Norte Poniente S/N Col. Lajas
Maciel
Tuxtla Gutiérrez. Chiapas
México. CP 29039

Hui Tag
Plant Systematic and Ethnobotanical
Research Laboratory
Department of Botany
Faculty of Life Science
Rajiv Gandhi University, Rono Hills
Itanagar – 791112, Arunachal Pradesh, India

Joaquín Cifuentes
Laboratorio de biodiversidad y biogeografía
ecológica de hongos
Facultad de Ciencias, Universidad Nacional
Autónoma de México
A.P. 70-181, Coyoacán, C.P. 04510. México
D.F. México

Jules Pretty
Essex Sustainability Institute and
Department of Biological Sciences
University of Essex, Wivenhoe Park
Colchester CO4 3SQ, UK

Kebu Balemie
Institute of Biodiversity Conservation (IBC)
Addis Ababa, Ethiopia; P.O. Box 30726

Konchok Targais
Defence Institute of High Altitude Research
Leh 194 101, India

Nancy J. Turner
School of Environmental Studies
University of Victoria, Victoria
British Columbia, Canada V8W 2Y2

Nancy Mackin
University of Victoria and Wilp
Wilxo'oskwhl Nisga'a
Canada, 2919 Tower Hill
West Vancouver, BC
Canada V7V4W6

Madeleen Husselman
Center for International Forestry Research
(CIFOR)
P.O. Box 0113, BOCBD
Bogor 16000 Indonesia

Michelle Cocks
Institute of Social and Economic Research
(ISER)
Rhodes University
7 Prince Alfred Street
P.O. Box 94, Grahamstown 6140
South Africa

P. Kalita
Plant Systematic and Ethnobotanical
Research Laboratory
Department of Botany
Faculty of Life Science
Rajiv Gandhi University
Rono Hills, Itanagar 791112
Arunachal Pradesh, India

Ranjay K. Singh
Central Soil Salinity Research Institute
Karnal 132 001, Haryana, India

R.B. Srivastava
Defence Institute of High Altitude Research
Leh 194 101, India

Roxanne Paul
School of Environmental Studies
University of Victoria, Victoria
British Columbia
Canada V8W 2Y2

Sarah Pilgrim-Morrison
Essex Sustainability Institute and
Department of Biological Sciences
University of Essex
Wivenhoe Park
Colchester CO4 3SQ, UK

Tony Dold
Selmar Schonland Herbarium
Rhodes University Botany Department
P.O. Box 94, Grahamstown 6140
South Africa

Tsering Stobdan
Defence Institute of High Altitude Research
Leh 194 101, India

Victoria Reyes-García
ICREA and Institut de Ciència i Tecnologia
Ambientals
Universitat Autònoma de Barcelona, 08193
Bellatera, Barcelona, Spain

Preface

Indigenous and local peoples throughout the world have from time immemorial been sustainably interacting with nature to maintain their life support systems. This process has resulted in long-term accumulation of rich biocultural diversity. Many indigenous and long-resident communities, including those living in fragile and stressful environments, have developed adaptive practices, sustainable regimes, and location-specific traditional food systems that maintain the biocultural diversity of their ecosystems. Today this diversity is considered an essential component in maintaining the earth's life-sustaining processes. However, global environmental and societal change is threatening this biocultural capital. Unprecedented changes in social-ecological systems due to anthropogenic and climatic factors have raised concerns over the sustainability of the food systems of many indigenous and local communities and, ultimately, of their cultures and the ecosystems they inhabit. Scholars, policymakers and governments of many countries have started to recognize the impacts of biocultural loss on indigenous and local peoples' food systems, their health and well-being, and even their own identities.

Today, many location specific food production systems and sustainable natural resources practices of indigenous and local communities are overlooked in mainstream society, although there is a growing appreciation of the value of such knowledge and resources in sustainable development. Extensive work with some of these communities, documenting their knowledge, cultures and institutions relating to conservation and adaptation, especially linked to their food systems, has shown some pathways that might enable societies to mitigate and adapt to global change. The diversity of social-ecological systems and biocultural resources – sustained by indigenous and local communities through their foods, medicines and diverse practices – provides many opportunities for adaptation, and indigenous knowledge has demonstrated extensive potential for generating insights valuable for global conservation science and policy. This integration offers opportunities for reciprocal learning and co-production of knowledge between different stakeholders (indigenous peoples, scientists, civil society) with potential multiple outcomes (promoting conservation, ecosystem services, food behaviours and cultural identity of indigenous and local communities) in order to support overall food security, conservation and sustainability.

There is an urgent need to consider these precious resources of knowledge systems in indigenous and native communities, and to promote effective partnerships. Traditional knowledge related to food, nutrition, materials and medicines of indigenous and local communities has developed from centuries of experience, experimentation and observation. It has been orally transmitted from generation to generation, and adapted to local situations. This knowledge tends to be collectively owned, and transmitted through stories, songs, proverbs, cultural values and beliefs, community laws and institutions, and local languages. Many indigenous and local peoples, who are among the most knowledgeable in relation to plants and environments, live in areas where the majority of the world's plant genetic resources are found. The overlap is not surprising, given the practices of cultivation and sustainable use of biological diversity by these peoples over thousands of years. Furthermore, the adaptive approaches and rich ecological ethics of indigenous and local communities to the conservation and sustainable use of biological diversity go far beyond their role as natural resource managers. Their wisdom, skills and techniques provide valuable information to the global community and serve as models for biodiversity policy development.

As we consider the breadth and diversity of work reported in this volume, we are struck by the comprehensive knowledge relating to foods resources, their management and related dynamics. The book is designed to share some of the more interesting lessons arising from the empirical research. It documents a large diversity of knowledge relating to traditional foods and associated social, cultural, health, nutritional, ecological and livelihood perspectives. The researchers' efforts in learning about biocultural diversity are an interesting aspect of the book, representing a new gathering of information on the dynamics of culture, biodiversity, traditional food systems, livelihoods, sustainability. These dimensions can, in the near future, make a larger impact in the field of social-ecological systems and in studying community-based conservation and adaptations.

The book targets different audiences including food researchers, scholars working on human-environment interactions and climate adaptation science, biodiversity conservators, non-government organizations and institutions involved in revitalizing traditional foods and community-based conservation and adaptation. In particular, we hope that it will serve as a useful resource for those in local communities who aim to document, retain and promote their own cultural approaches, knowledge and language in relation to the world's ecosystems. It will also be, we hope, a source book for educators who are advocating for and collaborating with indigenous and local peoples to promote location specific adaptations to the impacts of multiple stressors. Finally, we also hope this volume will provide insights and lessons to policy makers who are advocating participatory conservation of biocultural resources.

We sincerely hope that this publication will have a long-lasting impact in promoting biocultural diversity, location specific natural and historical sustainable adaptations, conservation and adaptive co-management in a changing world.

Ranjay K. Singh **Nancy J Turner** **Victoria Reyes-García** **Jules Pretty**

Editors

Acknowledgements

The editors wish to acknowledge the traditional knowledge holders who kindly shared their precious knowledge and practices for this book. We also thank all the authors for their valuable contributions. The photographs, art and paintings used in this book, taken by the respective authors from the studied communities, are gratefully acknowledged. All the funders who provided financial support to the respective authors and editors of this book are appreciated and gratefully acknowledged. The logistical supports provided to authors and editors of this volume by their respective parent institutions are appreciated. The editors thank New India Publishing Agency, New Delhi, India; and the team, especially Mr Sumit and his colleagues, who devoted their time and energy for production of this volume.

Introduction

Social-Ecological Diversity and Traditional Food Systems: Opportunities from the Biocultural World

Ranjay K. Singh*, Nancy J Turner, Victoria Reyes-García and Jules Pretty

Abstract

This introduction describes the role of biocultural diversity in food systems of native and indigenous peoples across the world: an issue which has attracted global attention in the recent past. It has been recognized that culture, language, worldviews and livelihood of a particular community are intricately linked with the ecosystem they inhabit. The types and scales of interaction between local communities with their surrounding environment shape both social-ecological systems and biocultural diversity. Ecological and environmental considerations compel local communities to decide 'what to eat', 'when to eat', 'how to eat' and 'how to produce the food stuffs'. Such decisions are governed by social characteristics, including local and traditional knowledge systems and their nurturing institutions. Biocultural diversity has a key role to play, not only in assuring the food security of local communities but also in determining the sustainability of natural resources in their territories. Socioeconomic and climatic changes at the global level have threatened the diversity of both cultures and biological resources. The industrialization of the process of food production, by negating and overlooking the traditional ways and means of

*ranjaysingh_jbp@rediffmail.com

food security, has threatened long term biocultural sustainability. In this introduction, in addition to providing the framework for the rest of this volume, we provide a synthesis of the contributions made by authors on biocultural resources from a variety of locations across Asia, Africa, Latin America and Europe. The diversity of knowledge systems and institutions developed by native and indigenous peoples worldwide illustrates the breadth of information on how best biocultural diversity can provide a foundation for facing challenges and creating a sustainable future in the scenario of global environmental change. What opportunities are available in focusing on biocultural resources of native and indigenous peoples? and what do these peoples need external helps in terms of inputs, scientific knowledge and policy support from the formal systems? These questions are discussed in order to learn about biocultural sustainability at a global scale.

Kewords: Biocultural diversity, conservation, opportunity, adaptation, global challenges.

The importance of traditional ecological knowledge systems in restoring microecosystems, in sustaining plant and animal species, and in providing returns to individuals and societies are well documented in many parts of the world (Anderson 2006; Turner and Turner 2007; McDade et al. 2007; Reyes-Garcia et al. 2008; Bharucha and Pretty 2010; Singh et al. 2013a). Traditional communities all around the world have managed their land, water, and natural resources collectively in a sustainable manner while retaining their cultural heritage (Singh 2013). These communities hold a deep understanding of their environments and they connect their ecosystems with world views and livelihood resources closely and meticulously (Pretty 2007). Indigenous and tribal peoples, living in varying social-ecological systems, have a rich collective wisdom allowing them to cope with food shortage, hunger and malnutrition (Turner and Turner 2007; Kwik 2008).

In the recent past, the sustainability of many local food production systems has been increasingly recognized as the processes and interactions of social and ecological systems (Berkes et al. 2003; Folke 2004; Singh and *Adi* Women 2010). The systems, learning processes and diversity of institutions have provided ample opportunities for local and traditional communities to interact with nature, and thus to evolve their bioculturally-rich food resources (Kwik 2008; Jeanine and Robert 2008). Such communities have developed location specific adaptive practices and sociocultural processes in sustaining their food systems and livelihoods (Belay 2012). They have been able to achieve food and nutritional security by ensuring access to healthy and diverse traditional foods from a variety of aquatic and terrestrial ecosystems (Singh et al. 2013a).

Furthermore, they have strengthened their food security and resilience to stressors by developing adaptive societal interactions manifested by creative harvesting of food resources, processing technologies and resource management strategies that provided them predictability with greater quantities and higher quality of food year round (Turner and Turner 2007), as well as during contingency times (Gomez-Baggethun et al. 2012; Singh et al. 2013a). Such location-specific cultural knowledge and approaches have played a pivotal role, not only in food and nutritional security, but also in conserving biodiversity, including agricultural diversity, in the areas managed by native peoples (Krystyna 2006; Singh et al. 2013b).

The transition from locally produced food resources to the large-scale commercial economy and agricultural processes has changed the ways and systems of managing food and related ecocultural heritage almost everywhere. The global restructuring of food systems, together with the globalization process, has greatly impacted ecosystems and human health (Kuhnlein et al. 2004; Pretty 2007, 2013; Kuhnlein et al. 2013). This problem is further aggravated by ongoing climatic changes (Adger et al. 2012), posing serious threats to the biocultural diversity of native communities in particular regions (Christopher et al. 2007; Singh et al. 2011a). In fact, a "nutrition transition" is underway in most parts of the world, with food produced through industrial processes replacing traditional and local sourced diets. Such industrially processed food, procured from distant parts of the globe, is often of lower nutritional quality than traditionally produced local food (Samson and Pretty 2007; Kuhnlein et al. 2013). Such process is also associated to a decrease of the diversity of foods cultivated and gathered (Menendez-Baceta et al. 2012). Changing life styles and changing food systems coupled with a shift from subsistence to commercial economies have posed threats to social-ecological resilience (Adger et al. 2012). The *top-down* policies of land use and associated political processes have greatly affected food diversity and conservation of crop varieties and native food species (Shiva 2011).

In recent years, wild and traditional food plants have increasingly became a focus of scholarly attention. Turner and Turner (2007) reported that renewed interest in local traditional foods and neglected plant food sources, intangible cultural heritage, and nutraceutical and therapeutic properties of the traditional diet are some of the driving forces behind this attention. Recently, many national and international organizations and movements have emerged with the aim to redefine the values assigned to biocultural diversity and emphasize its role in the conservation and sustainability of biodiversity and other natural resources (Maffi 2005; Maffi and Woodley 2010). These movements, supporting sustainable food production systems led by collective approaches, are also designed to enhance social-ecological resilience (Shiva 2005; Preston-Pile 2007).

In many countries, there are still diverse ethnic, racial and religious populations which continue to grow and conserve bioculturally important indigenous agro-biodiversity (Figure 1). In most cases, native and traditional communities do not separate traditional food from traditional medicine, therefore simultaneously contributing to the conservation of both food and ethnomedicinal resources (Reyes-García 2006). Such a combined approach of conservation and use of indigenous plant resources additionally provides sustainable ecological services in the form of significantly enhanced biodiversity. This has further provided an opportunity to scientists and policy makers to apply these resources in science and technology for enhancing food production (Singh et al. 2011b), and nutritional and medicinal security. These community-based management practices, developed over generations, are, however, witnessing some fundamental changes as a result of adverse effects of various socioeconomic and environmental factors (Christopher et al. 2007). On the one hand, some of the ethnic groups living in isolation from rest of the population continue to select and access their local, bioresources for food and medicinal requirements keeping their customs and traditions alive. On the other hand, some other groups tend to adapt some of their customs and traditions (Turner 2005) due to global changes (Turner and Clifton 2009; Adger et al. 2012). Despite their history of heavy cultural losses and social injustices suffered by many indigenous peoples from colonialization and industrialization, many traditional communities today are more interested in looking forward, in renewing their connections with their lands and cultural heritage (Pretty 2003), and in reviving their traditional food systems to regain eco-cultural strength and personal and community health (Pretty et al. 2005; Samson and Pretty 2006; Kuhnlein et al. 2013).

Figure 1: Mixed cropping of finger millet, amaranthus and field pea with mosaic agricultural system in a micro-ecosystem of Tawang (Arunachal Pradesh, India) sustained by *Monpa* tribe to meet out their multifarious needs. A classical example of culturally rich, biodynamic and energy efficient agriculture.
Photo: Ranjay K. Singh

From a social-ecological and biodiversity conservation perspective, the changing landscape of biocultural diversity (Maffi and Woodley 2010) provides new and exciting opportunities for enhancing sustainability. The linguistic and cultural diversity that gives different shapes to biodiversity is intricately linked with human biodiversity use and conservation of natural resources (Maffi 2005; Maffi and Woodley 2005) (Figure 2). Different groups use and conserve different species and related resources for location specific traditional food products (fresh, fermented, processed, etc.) to meet their food, medicine, and other livelihood needs. Over centuries, such groups have devised and evolved culturally mediated practices to sustain effective agroecological practices, and related food chain (Figure 3 &4). These groups have developed networks of services, socio-cultural institutions and unique cost effective approaches to identify, select, use and process traditional foods, as well as to conserve related biodiversity and cultural heritage. However, there are a number of challenges and opportunities arising from such systems among diverse cultural and ethnic groups worldwide.

Figure 2: Traditional painting of *Monpa* hanged in house indicating dynamics of culture, language and their inter-relation with ecology.
Photo: Ranjay K. Singh

Figure 3: Burning of bong (barley) crop in field itself to collect roasted ear-heads. Barley grains processed with this method, and flour prepared from it is exchanged among pastoral *Brokpa* people.
Photo: Ranjay K. Singh

The key aim of this volume is to show the range of interactions relating to biocultural diversity, local and traditional systems of ecological knowledge, traditional food systems, conservation strategies, and mitigation and adaptation to climate change. There are 13 chapters presenting research from various parts of the world on these relations, and including different issues that define, challenge, and open up new avenues associated with culturally important food diversity and conservation.

In her chapter, Nancy J. Turner discusses the dynamics of traditional foods, medicinal and potentially toxic plants among indigenous peoples from Northwestern North America who have utilized several hundred local plant species, as well as some algae, lichens and fungi in sustaining themselves over millennia. She further correlates the linguistic diversity of First Nations groups and diversity of foods and medicines originated from plant and fungi species. Ultimately, she stresses the importance of people's traditional knowledge about the details of plants' edibility, medicinal properties and potentially harmful traits, and how this knowledge supports group's resilience and ability to survive using local resources. She concludes by stressing that these aspects of a people's cultural and environmental heritage must be recognized and taken into account in efforts for cultural renewal and ethnoecological restoration.

Figure 4: A traditional painting on *Brokpa* (herder from *Monpa* tribe who lives at higher altitude) indicating harmony between man, yak and other natural resources. *Brokpa* exchange their yak based products in exchange of grains and flour of barley, buckwheat and maize from *Monpa* tribe (lives at lower altitude).
Painting courtsey: State Tourism Department (Dirang), Government of Arunachal Pradesh.
Photo: Ranjay K. Singh

Elizabeth Byron and Victoria Reyes-García contribute a chapter on changes on the nutritional system of the Tsimane', a native Amazonian group, as they integrate into the market economy. They report that greater integration into the market economy is associated with a higher percentage of market-procured foods in the household diet, although market-produced foods still do not surpass farm foods among the studied group, The authors further found that households that are more integrated into the market economy show signs of turning away from local forest foods to incorporate more market foods in the daily diet. These findings thus could be used by policy-makers in creating programs to mitigate the potential detrimental consequences of integration into the market economy for native cultures and livelihoods.

The work of Sarah Pilgrim-Morrison and Jules Pretty shows that our collective local knowledge is eroding due to several factors such as livelihood diversification towards non-resource dependent strategies, the emergence of local markets as a consequence of globalization process, and externally driven economic development. With the loss of local livelihoods and knowledge comes a departure from traditional food systems. In that process, hunters, fishers, gatherers and cultivators lose the skills needed to locate, collect, preserve, prepare, consume and manage indigenous foods. This shift has negative consequences for human health, as well as for cultural diversity, as it has for ecosystem sustainability in both industrialized and developing countries.

Celeste Lacuna-Richman reports that in Finland, the decrease in berry consumption has led to under-harvesting of berries. This situation has generated a problem in providing economic benefits to local berry harvesters and pickers, and to seasonal migrants who arrived to the area in berry harvesting season. The author illustrates the issue with the case study of Thai people working in Finland, who are the main harvesters of underutilized wild berries in Lapland. The future of these Thai berry pickers in Finland depends partly on how work authorities accommodate their needs in the future, such as ensuring their social well-being, and predicting the abundance of berries for the coming season.

The work of Ranjay K Singh and colleagues demonstrates how local farmers (both men and women) of traditional communities have been conserving and using a variety of indigenous biological resources in agriculture and food systems. This study, conducted across the eastern Himalaya and Indo-Gangetic plains, reveals location specific farmers' creativity in developing and reviving indigenous plant varieties, and the role of women in domestication of wild germplasm in order to sustain their subsistence livelihoods as well as the ecosystems on which they rely. The study further indicates that efforts in domestication and conservation of plant biodiversity were higher among elders than among young people, as they were among women than among men, irrespective of the ecosystems. This may reflect survival strategies evolved by the traditional communities situated in harsh and marginal ecosystems.

The study of Kebu Balemie' documents the role and importance of non-cultivated food plants for the indigenous Dawro people in the Loma District of South Ethiopia. This study provides an overview on local knowledge, and on the range of indigenous plant biodiversity use and conservation by the Dawro people. It also highlights various challenges in sustaining biocultural knowledge systems among the Darwo, and suggest that some promising non-cultivated food plant species can be domesticated for enhancing local food security and sustain cultural resources.

The study of Hui Tag and colleagues reports on the significance and role of cultural knowledge in food resource conservation and sustaining ecosystem services by the *Nyishi* tribe of Arunachal Pradesh, India. This study shows that each of the plant and animal species used by the *Nyshi* has a deeply rooted cultural history and that many of the important species for this group are now very rare and threatened in their natural habitats. Folklore, myths and legends play a significant role in the *in situ* conservation of traditional food bioresources. The study suggests that impacts from outside cultures and modern education play a significant role in changing attitudes regarding traditional beliefs and food habits among younger *Nyishi* generations.

Michelle Cocks and colleagues' work presents evidence to show that the consumption of wild leafy vegetables (pot-herbs) represents more than just an important source of nutrition and/or a safety net function for economically poor households. The consumption of such vegetables also fulfills an important cultural function for both poor and wealthy Xhosa speaking people in the Eastern Cape, South Africa. The study further shows that the consumption of food is best studied as a biocultural phenomenon, one which includes both nutritional and anthropological understanding. The study concludes with an emphasis on a need for those studying agricultural biodiversity systems to give attention to alien weeds and plant species which may not be actively managed but which may have a significant role in food systems.

Felipe Ruan-Soto and Joaquín Cifuentes elaborate on the utility and values of edible fungi resources in the Mesoamerican lowlands. In their chapter, the authors emphasize that edible indigenous fungi have been important in the survival patterns of both highland and lowland societies of Mesoamerican during the different historic periods. The chapter highlights the social, food, economic and ecological issues related to edible fungi consumption and discusses further policy implications of these issues.

Nancy Mackin and Deanna Nyce report on the dynamics of plants, architecture, and folk stories relating to Nisga'a oolichan fishery technology on the north coast of British Columbia, Canada. The authors emphasize the role of plant technologies in processing and preserving of oolichans and oolichan "grease," a nutritious oil rendered from these small fish. The study reports that technological knowledge, gathered and adapted over countless generations, continues to be important in the Nass Valley oolichan fishery, which remains strongly rooted in traditions that retain their value within an increasingly technological world. The authors conclude with an emphasis on the implications of such technology at a global level.

Roxanne Paul reports on her research on Pacific salmon food webs carried out in collaboration with the Saanich First Nation peoples of Southern Vancouver Island, British Columbia, Canada. The nature of traditional ecological knowledge in fishing and processing of fish, along with the various factors influencing the social-ecological systems of Saanich indigenous people, are discussed in critical ways. The work ends with an emphasis on conservation practices of Saanich fishers and on the need for integrating traditional ecological knowledge with formal scientific knowledge and policy relating to the salmon fishery.

The primary research work of Targais and his associates depicts the unique preparatory method and use of a barley-based (*Hordeum vulgare*) food, locally called *tsampa*, from the Ladakh region, India. Along with methods of *tsampa* preparation and use, and methods for the conservation of related indigenous varieties, the authors describe factors that have caused changes in use and conservation mechanisms of food related resource at Ladakh's higher altitudes. This study also reports that global changes have impacted the Ladakh region (trans-Himalaya), thereby reducing the availability of *tsampa*, as well as its use and related knowledge among the local inhabitants. Nevertheless, *tsampa* prepared using traditional methods is still preferred by the local inhabitants. The authors stress that there is a need for preservation of traditional *tsampa* preparation methods and the variety of foods prepared from it.

Anshuman Singh and his colleagues discuss the biocultural significance of Indian fruit tree *mahua* (*Madhuca* spp.). This study, carried out with diverse traditional communities of India, highlights the importance of *mahua* from the socio-economic, cultural and environmental perspectives. Study also reveals that *Madhuca* spp. widely distributed in the north and central Indian plains, has huge biocultural significance as it satisfies the multifarious livelihood requirements of the poor and marginal people. *Mahua* has been vital to the survival of tribal communities from the time immemorial, and has been a major source of energy in the form of sugar, seed oil and alcohol. Different plant parts of *mahua* tree constitute an integral part of the ethnomedicines. The high regard paid to the *mahua* tree is reflected by its sustainable and regulated use. Taking insights from key results of this study, authors present an in-depth analysis regarding the conservation, management and the sustainable use of *mahua* in diverse social-ecological systems.

Reviewing the works of all the contributors to this volume leads us to focus on the basic concern as to whether or not the diversity of biocultural resources developed by native and indigenous peoples can be used to address the global challenges of conservation, sustainability and adaptation (e.g. Figure 5a & 5b) to climatic change for a healthy future. Although the answer is relatively complex, the strength and opportunity available through the existing resources

Figure 5a: A *monpa* community forest grove of *paisang* (*Quercus griffithii*).
Photo: Ranjay K Singh

sustained by native and indigenous peoples can provide a better lead for options on food security and sustainability. A main challenging issue is how best synergy and policies might be framed to meaningfully bring native and indigenous peoples into policy, research and developmental frameworks. In most of the cases, the voices and knowledge systems together with supportive innovations available from local and indigenous communities (Future Earth 2013) are unheard and unnoticed. There is a need for a radical change and shift from the old-style way of thinking with the imposition of *top-down* policy, conservation and developmental processes to new *bottom-up* approaches incorporating a more dynamic perspective of food and livelihood systems and sustainability. The question arises as to whether the knowledge systems, energy and institutional strength available from native and indigenous peoples would be enough to contend with the current challenges of globalization, climate change and sustaining biocultural diversity. The need is there for broader perspectives in light of current practices, and identifying intervention points where biocultural diversity and traditional practices of local communities seem to be vulnerable. Such an analysis would enable us to determine more precisely the level and types of interventions and policy support needed.

Figure 5b: The leaves of *paisang* tree (Fig. 5a) are used by *Monpa* tribe to conserve more than 32 landraces.
Photo: Ranjay K Singh

Acknowledgements

The editors of this book are grateful to all the authors who contributed to this book generously with their research and synthesis-based articles. The first editor (RKS) is thankful to all the native and tribal communities with whom he has been associated since last over 12 years and from whom he has learned the knowledge and wisdom on biocultural diversity. The photographs (by Ranjay K. Singh) used in the introduction section from *Monpa* tribe, Arunachal Pradesh, were taken with their consent. Financial and logistical support obtained from the National Innovation Foundation-India, Ahmedabad and College of Horticulture and Forestry, Central Agricultural University, Pasighat, Arunachal Pradesh for some of the data collected from Arunachal Pradesh and reported in this book are thankfully acknowledged. The logistical support from Central Soil Salinity Research Institute, Karnal, Haryana is appreciated. Editorial inputs to the introduction section from Dr. Anshuman Singh are thankfully acknowledged.

References

Adger WN, Barnett J, Brown K, Marshall N, O'Brien K (2012) Cultural dimensions of climate change impacts and adaptation. *Nature Climate Change*, DOI:1038/NCLIMATE1666

Adger WN, Quinn T, Lorenzoni I, Murphy C, Sweeney J (2013) Changing social contracts in climate-change adaptation. *Nature Climate Change*, 3: 330-333

Anderson MK (2006) Tending the wild: native American knowledge and the management of California's natural resources. University of California Press, Berkeley, USA

Belay M (2012) Participatory mapping, learning and change in the context of biocultural diversity and resilience. Ph. D. thesis, Rhodes University, UK. P. 43

Berkes F, Colding J Folke (2003) Navigating social–ecological systems: building resilience for complexity and change. Cambridge University Press, Cambridge, UK

Bharucha Z and Pretty J (2010). The role and importance of wild foods in agricultural systems. *Philliosophical Transection of Royal Society of London B* 365:2913-2926

Christopher CP, Turner NJ, Shirley M, Solberg SM (2007) Resetting the kitchen table: food security, culture, health and resilience in coastal communities. Nova Science Pub Inc. Canada

Folke C (2004) Traditional knowledge in social–ecological systems. *Ecology and Society,* 9(3): 7. www.ecologyandsociety.org/vol9/iss3/art7/

Future Earth (2013) Future earth: research for global sustainability. Draft of initial design report, 17th April 2013. http://www.icsu.org/future-earth/media-centre/relevant_publications/FutureEarthdraftinitialdesignreport.pdf. Accessed on 08-12-2013

Gomez-Baggethun E, Reyes Garcia V, Olsson P, Montes C (2012) Traditional ecological knowledge and community resilience to environmental extremes: A case study in Doñana, SW Spain. *Global Environmental Change,* 22:640-650

Jeanine MP, Robert AV (2008) Biological invasions and biocultural diversity: linking ecological and cultural systems. *Environment Conservation,* doi:10.1017/S0376892908005146

Krystyna S (2012) Banishing the biopirates: A new approach to protecting traditional knowledge. Gatekeeper Series 129, IIED, London, UK.pp. 2-19

Kuhnlein HV, Erasmus B, Spigelski D, Burlingame B (2013) Indigenous Peoples' food systems & well-being interventions & policies for healthy communities. Food and Agriculture Organization of the United Nations Centre for Indigenous Peoples' Nutrition and Environment, FAO, Italy.

Kuhnlein HV, Receveur O, Soueida R, Egeland GM (2013) Arctic indigenous peoples experience the nutrition transition with changing dietary patterns and obesity. *Journal of Nutrition,* 34(6):1447-53

Kwik J (2008) Traditional food knowledge: A case study of an Immigrant Canadian "foodscape". *Environments,* 36(1):59-74

Maffi L (2005) Linguistic, cultural and biological diversity. *Annual Review of Anthropology,* 34:599-618

Maffi L, Woodley E (2010) Biocultural diversity conservation: a global sourcebook. Earthscan, London, UK

McDade T, Reyes-Garcia V, Leonard W, Tanner S, Huanca T (2007) Maternal ethnobotanical knowledge is associated with multiple measures of child health in the Bolivian Amazon. *Proceedings of National Academy Sciences, United States of America,* 104:6134-6139

Preston-Pile K (2007) A return to the earth: Vandana Shiva's campaign for sustainable agriculture in India. http://www.calpeacepower.org/0302/pdf/08_09_10_A_Return_to_the_earth.pdf. Accessed on 01-12-2013

Pretty J (2007) The earth only endures: on reconnecting with nature and our place in it. Earthscan Publication, UK

Pretty J, Peacock J, Sellens M, Griffin M (2005) The mental and physical health outcomes of green exercise. *International Journal Environmental Health Researh,* 15(5):319-337

Pretty J (2013) The consumption of a finite planet: well-being, convergence, divergence, and the nascent green economy. *Environmental and Resource Economics,* 55(4):475-499

Reyes-García V (2006) Eating and healing: Traditional food as medicine. *Economic Botany* 60(4):389-389.

Reyes-Garcia V, McDade T, Vadez V, Huanca T, Leonard WR, Tanner S, Godoy R (2008) Non-market returns to traditional human capital: Nutritional status and traditional knowledge in a native Amazonian society. *Journal of Development Studies,* 44:217-232

Samson C, Pretty J (2006) Environmental and health benefits of hunting lifestyles and diets for the Innu of Labrador. *Food Policy,* 31:528–553

Shiva V (2005) Earth democracy: Justice, sustainability and peace. South End Press, New York, USA

Singh RK (2013) Eco-culture and subsistence living of *Monpa* community in the eastern Himalayas: An ethnoecological study in Arunachal Pradesh. *Indian Journal of Traditional Knowledge,* 12(3):441-453

Singh RK, Adi Women (2010) Biocultural knowledge systems of tribes of eastern Himalayas. NISCAIR, CSIR, New Delhi

Singh RK, Bhowmik SN, Pandey CB (2011a) Biocultural diversity, climate change and livelihood security of the *Adi* community: grassroots conservators of eastern Himalaya Arunachal Pradesh. *Indian Journal Traditional Knowledge,* 10(1):39-56

Singh RK, Rallen O, Padung E (2013a) Elderly *Adi* women of Arunachal Pradesh: 'living encyclopedias' and cultural refugia in biodiversity conservation of the eastern Himalaya, India. *Environmental Management,* 52(3):712-735

Singh RK, Srivastava RC, Pandey CB, Singh A (2013b) Tribal institutions and conservation of the bioculturally valuable '*tasat*' tree (*Arenga obtusifolia*) in the Eastern Himalaya. *Journal of Environmental Management and Planning.* DOI:10.1080/09640568.2013.847821

Singh RK, Turner NJ, Pandey CB (2011b) '*Tinni*' rice (*Oryza rufipogon* Griff.) production: An integrated sociocultural agroecosystem in eastern Uttar Pradesh of India. *Environmental Management,* 49:26–43

Chapter – 1

The Food/Medicine/Poison Triangle: Implications for Traditional Ecological Knowledge Systems of Indigenous Peoples of British Columbia, Canada

*Nancy J. Turner**

Abstract

Indigenous Peoples of northwestern North America have utilized several hundred local plant species, as well as some algae, lichens and fungi, in their traditional diets. Many of these are also used in traditional healing, and some have potentially harmful or poisonous properties that must be taken into account in harvesting and preparation. Other species, known primarily for their medicinal use, also have noted toxicity if used improperly or without constraints. In this study, a total of 375 traditional food and medicine plants, fungi, lichen and algal species, as well as those species reported to be toxic by indigenous knowledge holders, were surveyed across 12 First Nations groups, each with distinct languages, for overlaps across the categories of food, medicine and toxic species. A complex of species use and avoidance emerged, modulated by processing, dosage, and other factors, with significant numbers of food species with reported healing properties, food species also used as medicine, and both food and medicinal species known to be harmful or toxic under some circumstances. Even most of those species considered extremely poisonous have been taken internally as medicines in some way by some groups.

---

*nturner@uvic.ca

In traditional use, people have taken the potential for harm of these species into account when consuming them as food or medicine. In view of the inextricable relationships among food, medicine and toxic species, detailed knowledge about their selection, harvesting, processing and consumption is critically important for survival. As such knowledge is lost or eroded due to the forces of acculturation and globalization, the continued use of these traditional species takes on higher risks. Ultimately, people's resilience and ability to survive using local resources is threatened by the loss of this knowledge. Its importance as a living part of peoples' cultural and environmental heritage must be recognized and taken into account in efforts for cultural renewal and ecological restoration.

Keywords: British Columbia, Indigenous peoples, food plants, medicinal plants, poisonous plants.

Introduction

Knowledge of food and medicine always goes hand in hand with knowledge of harmful properties of species utilized, as well as those to be avoided altogether. There is a notable three-way relationship across the categories of food, medicine and toxins, which can be framed as a food/medicine/poison triangle. In any consideration of one of these three categories, the other two must invariably be linked.

To start with, many commonly used foods and beverages around the world are known to be potentially toxic in some way, either if taken in large quantities or without some type of processing to reduce their toxic properties (Pieroni 1999; Etkin 2006a). A classic example is the potato (*Solanum tuberosum* L.), that delicious and globally treasured vegetable tuber of the nightshade family (Solanaceae). Even the edible tubers of potato, however contain traces of the bitter alkaloid solanine and related compounds, which are found in higher concentrations throughout the potato plant, rendering the green leaves, sprouts, and green, light-exposed tubers very poisonous (Turner and von Aderkas 2009). Relatives of potato, including the nightshades (*S. dulcamara* L., *S. nigrum* L.), are similarly poisonous due to high concentrations of these alkaloids. The wild ancestral forms of cultivated potato, even the tubers, are also toxic (Johns and Kubo 1988; Johns 1996). Domestication, involving generations of careful and continued selection and breeding of less toxic potato varieties by indigenous Andean farmers, is thus responsible not only for producing tubers of larger size and greater diversity of colour and shape compared with the wild progenitors, but also for developing tubers with minimal levels of bitterness and toxicity (reflecting reduced alkaloid content).

Knowing how to distinguish less poisonous from more poisonous varieties and species, and how to harvest and process potentially toxic species for safe consumption are critical components of any group's traditional ecological knowledge system. In some cases, animals can provide clues to which foods and parts are safe to eat. Indeed, humans and animals share many of the same foods (Martin et al. 1989). However, the physiological and digestive capacities of animals are in many cases quite different from those of humans. (For example, deer have been observed to eat potato leaves, which would be toxic to humans.) Therefore traditional knowledge systems must incorporate understandings of animals' use of food (and medicine) and what similarities and differences may exist between humans and animals in this regard (Turner 1997).

In a different but related context, many food plants are also known as sources of medicine or as providing health benefits in some way (Moerman 1996; Etkin 2006a). Moerman (1996), for example, documented this food-medicine congruence for plant species and families of plants used by North American Indigenous Peoples. While, in a broadly based continent-wide ethnobotanical survey, he noted a substantial overlapping of medicinal and food floras, he also found significant differences, in that food and medicine tend to involve different plant parts, plant habits, and plant characters. In other words, in many cases the same species were used for both food and medicine, but for species with both uses, different parts, different modes of preparation or different life cycle stages of the plant were sought for use as food and medicine, respectively.

The third component of the food/medicine/poison triangle – the intersection between medicinal plants and potentially toxic plants – is perhaps most widely recognized across the world's cultures. Since medicinal species are commonly selected because of traits and effects that reflect physiologically active compounds – for their effects on the human digestive, circulatory, or nervous systems, or their antibiotic properties for example – many medicinal species can have the capacity for harm if consumed in higher concentrations (Duke 1985; Tyler 1987; Blumenthal et al. 2000; Turner and von Aderkas 2009). Medicinal plants can regulate the heartbeat, stimulate the nervous system, bring healing relaxation, control internal bleeding, heal a range of infections, and reduce fever or numb pain. However, some of the same species can cause irreparable damage to the liver or kidneys, lower blood sugar, interfere with normal blood clotting, prevent cell division, or affect immune systems.

Foxglove, or digitalis (*Digitalis purpurea* L.) is one of the best known and most widely used poisonous medicinal plants. Foxglove has been well known as a poison through history, but was nevertheless used for centuries by farmers and herbalists in England and Europe for treating "dropsy," a condition of massive fluid retention in the body. A prominent British physician, William

Withering, first proposed in the late 1700s a link between foxglove, folk medicine, dropsy (edema), and poor heart function. We now know that foxglove contains a mixture of cardiac glycosides, including digitoxin, gitoxin, and gitaloxin, which have a strong effect on the heart muscles. When used in specific dosages they change the rhythm of the heartbeat, lengthening the time between heart contractions and thus allowing the ventricle to be emptied more completely. These compounds also improve general circulation, relieve edema, and help kidney secretion. However, the difference between an effective therapeutic dose and a toxic dose is small, in the order of only 30 per cent; this factor and the need for careful control of dosage was well understood long before Withering's time. Today, millions of people take some form of digitalis to help regulate and strengthen their heartbeat – proof of the close linkage between medicines and poisons (Balick and Cox 1996).

There are many other examples worldwide of toxic plants that are also medicinal. Opium poppy (*Papaver somniferum* L.) yields, among other compounds, the potent but habit-forming painkiller, morphine, as well as the milder, usually non-habit-forming pain relief medicine, codeine. Both are alkaloids, isolated from opium, which has been used for at least 5,000 years in the Middle East and Mediterranean regions as a sedative, and to alleviate pain and insomnia. A third opium alkaloid, papaverine, is used mainly in the treatment of internal spasms, particularly of the intestines. Yet, opium poppy is also the source of the highly addictive refined drug heroin, which can easily cause death if taken in too high a dosage. Another example from Papaveraceae is bloodroot (*Sanguinaria canadensis* L.), an herbaceous perennial of eastern North American hardwood forests, which contains numerous benzophenanthridine alkaloids, most notably sanguinarine, which have antimicrobial activity. Extracts of bloodroot were taken orally by Native Americans as emetics, expectorants and respiratory aids and were applied topically for ailments from ringworm to skin ulcers and eczema (Campbell et al. 2007). Yet, bloodroot can also be considered poisonous, capable of causing vomiting, diarrhea, fainting and a whole host of other negative consequences.

Mayapple (*Podophyllum peltatum* L.), an herbaceous perennial growing in similar habitats to bloodroot, contains lignans having anti-cancer and anti-viral properties, and today provides a drug of choice for treating human venereal warts. However, the entire plant, except the ripe fruits, is highly toxic and potentially fatal if ingested. Autumn crocus (*Colchicum autumnale* L.), another extremely toxic plant, has been investigated for use in cancer chemotherapy because one of its alkaloids, colchicine, interferes with cell division and hence with the proliferation of rapidly growing cancer cells.

Often, individual plants, local populations of plants, or related species within a genus vary in the relative concentrations of medicinal and/or toxic compounds they contain. The level of phytochemicals also varies with the part of plant used and its development stage and season. Bloodroot, mentioned previously, was found in a recent study to contain dramatically different concentrations of the alkaloid sanguinarine in different parts of the plant; concentrations in leaves, flower and fruit were one-thousandth of the rhizome concentration (Campbell et al. 2007). In another study, Rocky Mountain juniper (*Juniperus scopulorum*) was found to exhibit significant seasonal changes in volatile oil composition (Powell and Adams 1973). Location, too, can influence the nature and concentration of plant compounds. Elevation and level of shading, for example, can affect the level of ultraviolet light to which a plant is exposed and in turn influence the concentrations of compounds like furanocoumarins in some plants (e.g., cow parsnip, *Heracleum maximum,* Bartram-Kuhnlein and Turner 1987). Processing and ways of administering a medicinal plant are also significant variables in terms of the concentrations of toxic or chemically active compounds a person might consume as part of a medical treatment. Knowledge of all of these factors is, again, an important component of any cultural group's traditional ecological knowledge system.

In agricultural societies, some medicinal plants, as with food plants, have been domesticated and grown in more controlled conditions, and their medicinal qualities standardized. Indigenous peoples of western Canada, and many others of North America, did not practice agriculture or obvious genetic manipulation of species, either food or medicine (with the exception of two species of ceremonially used tobacco, *Nicotiana* spp.). Rather, these peoples developed other practices of maintaining, enhancing and intensifying the various species on which they relied (Deur and Turner 2005; Turner et al. 2013). An intimate knowledge of the edible qualities and physiological effects of all foods and medicines was, and is, as critically important for them as for agrarian peoples.

Methods

In order to better understand the relationships across the domains of food, medicine and poisons for the First Peoples of western Canada – and of the traditional ecological knowledge embodying these relationships – I undertook a survey across 12 linguistically and geographically distinct Indigenous groups of British Columbia. These were groups representative of the major language groups (Na-dené, Salishan, Wakashan, Ts′msyenic, Haida) of the province, as well as the major ecological regions, for which substantive, detailed ethnobotanical data are available as per the references cited (Table 1). In the original research, most of the interviews were conducted in English, but in collaboration with linguists who recorded the Indigenous names of the plants in the Indigenous languages. Standard methods in collaborative ethnobotanical documentation were used in these studies, and are described in greater detail in the publications themselves. This paper shows the relative extent of the food/medicine/poison triangle as reflected in the ethnobotanical knowledge systems of these peoples.

A total of 375 species of plants, algae, lichens and fungi were identified in this survey as having known edible and/or (internally administered) medicinal and/or poisonous properties, as reported from available elders and recognized plant specialists within these groups (see references cited, Table 1). Species reported as being a specific food of one or more types of animals were also included, as a means of investigating the similarities and differences perceived between human and animal foods.

Some of the food and medicine species are widely known and used across virtually all the groups, whereas others are documented as used by only one or two groups. Some species reflect an array of diverse applications as food, animal food and/or medicine, while others are documented as being used in more restricted ways only. A "best fit," conservative approach was used in determining assignment of categories beyond the obvious two or three that might fit a given species and range of uses. Only those categories that were most consistent and prevalent were counted in the summaries. In some cases there were difficulties with conflicting information or applications, with a species said to be inedible for people in one area but eaten by people in another (e.g. the berries of *Streptopus amplexifolius* (L.) Desf. are widely considered inedible or poisonous, but reported as edible in one or two cases). In this case, the predominant view was used in assigning the use categories.

The body of empirical ethnobotanical data is derived from interviews with knowledgeable elders and plant specialists, undertaken with prior informed consent, and approval for publication according to standard ethnobiological principles as outlined originally in the Declaration of Belem (International Society of Ethnobiology 1988; see also International Society of Ethnobiology 2006). The knowledge and use of these species has changed over time. Some food and medicine species are still commonly harvested and used today, whereas knowledge of others has diminished, and in some cases only members of the oldest generation, or those who have recently passed away, recalled their use. The depth of knowledge varies depending on the sources, the plant species available, and opportunities to ask focused questions, and hence the inventory reflects some gaps or potential missing elements. Nevertheless, though not completely equivalent, the data derived from the diverse groups shows the

Table 1: Indigenous language groups surveyed for knowledge of edible, medicinal and poisonous plants and fungal species, with traditional territories and associated references.

Indigenous group and language family	General location of territory (BC = British Columbia)	Ethnobotanical reference(s)
Ukatcho Dakelh (Na-dené Family)	West central interior of BC, vicinity of Anahim Lake	Hebda et al. 1996
Secwepemc (Shuswap) (Salish Family - Interior)	East central interior of BC, vicinity of Kamloops, Salmon Arm, Williams Lake	Turner et al. n.d. unpubl. ms.
Okanagan-Colville (Salish Family - Interior)	Southern interior of BC, vicinity of Vernon, Kelowna, Penticton, across US border to C Washington	Turner et al. (1980)
Nlaka'pamux (Thompson) (Salish Family - Interior)	West southern interior of BC, vicinity of Spences Bridge, Merritt, Spuzzum	Turner et al. (1990)
Saanich (Straits) (Salish Family - Coast)	Southern Vancouver Island, Saanich Peninsula and adjacent Gulf Islands	Turner and Hebda 1990; 2012; Turner and Bell (1971)
Ditidaht (Nitinaht) (Wakashan Family)	Southwest coast of VI, vicinity of Port Renfrew, North to Nitnat Lake	Turner et al. (1983)
Hesquiaht (Nuu-chah-nulth)	Clayoquot Sound, VI, Hesquiat Peninsula (Wakashan Family)	Turner and Efrat (1982)
Kwakwaka'wakw (Southern Kwakiutl) (Wakashan Family)	NE VI and adjacent Mainland, vicinity of Campbell River, Alert Bay, Kingcome Inlet	Boas (1921); Turner and Bell (1973)
Haisla and Hanaksiala (Wakashan Family)	N Central BC coast, vicinity of Kitimat and Kitlope	Compton (1993)
Nuxalk (Bella Coola) (Salish Family isolate)	Central BC coast, vicinity of Bella Coola	Smith (1928); Turner (1973)
Gitga'at (Ts'msyen)	North BC coast, entrance to Douglas Channel and surrounding Islands and mainland	Turner and Thompson (2006)
Haida (Haida Family)	Haida Gwaii (Queen Charlotte Islands)	Turner (2004)

existence of important congruencies across the spectrum of food, medicine and poison for the peoples of this region.

Results

Table 2 shows a summary from ethnobotanical lexicons and knowledge systems of 12 indigenous language groups of British Columbia, of assessments of species of plants (as well as some algae, fungi, and lichens) used as food, medicine, and/or known as food for animals, or as having poisonous properties, including various overlapping categories across the food/medicine/poison triangle. The table includes the total number of species documented for each category and examples for each. Since the goal of this research was to examine the link between medicine and food, only medicines taken internally were included in the survey; species used solely externally, as salves, skin washes or in some other capacity were not counted. There are almost twice as many species entries (661) as the total number of species identified in the study, since most of them fell into more than one of the designated categories. For example, 56 of the species were known to be food for particular animals as well as for people, and all of these are included in at least one of the various other food categories.

A significant proportion of the species (91 in total – just under 25%) were reported as being used solely as food, with no reported medicinal use or known toxic properties. Ten species (less than 3%) were known only as poisons within one or more groups, and were not known to be used as food or medicine. A total of 68 species (18%) used as internal medicines had no reported use as food and no reported toxicity. All of the other species (over half, at 55%) represent some degree of overlap in use and knowledge across the food/medicine/poison continuum.

Of the identified food and beverage plants 39 (10.4%) were recognized as having some type of poisonous or harmful property, requiring special preparation or consideration around harvesting or processing, and a further 10 (>3%) had other parts, not eaten, reported to be toxic. Nearly one third of the food and beverage species (118, or 31.5%) were also used in some way as internally consumed medicines. Of these, 55 had the same part used as both food and medicine, and 63 had different parts used as food and medicine (e.g. fruit eaten, and bark or leaves used in a medicinal preparation). Eighty medicinal species (over 21%) were also reported to have some potential toxic or harmful properties if not prepared correctly or if taken in too high a dosage.

Table 2: Summary and categorizations of food, medicine and poisonous plant species (including algae, fungi and lichens), from ethnobotanical lexicons of 12 indigenous language groups from British Columbia, showing examples of plants assigned to various categories.

Categories of food, medicine, poisonous or harmful to eat species (including plants, marine algae, lichens and fungi)	No. of species documented*	Selected examples (cf. Kuhnlein and Turner 1991; Turner 1995; 1997; see Table 1 for references to designated linguistic groups)
Species used as foods or beverages with no reported medicinal, harmful or poisonous properties	91	*Tricholoma populinum* JE Lange (cottonwood mushroom) – eaten by Interior Salish (Turner et al.1985); *Sagittaria latifolia* Willd. (wapato) – tubers cooked and eaten (Secwepemc, Halkomelem); *Corylus cornuta* Marsh. (hazelnut) – nuts eaten; *Claytonia lanceolata* Pursh (spring beauty) – corms eaten
Species with more than one different part used as food or beverage	33	*Pinus contorta* Dougl. ex Loud. (lodgepole pine) – inner bark and seeds eaten; gum chewed (several groups); *Tsuga heterophylla* (Raf.) Sarg. (western hemlock) – inner bark and branch tips eaten and used to collect herring eggs; *Calochortus macrocarpus* Dougl. (mariposa lily) – bulbs and flowerbuds eaten; *Lomatium nudicaule* (Pursh) Coult. & Rose (wild celery) – leaves eaten; seeds used as tea and flavouring (Nlaka'pamux); *Balsamorhiza sagittata* (Pursh) Nutt. (balsamroot) – taproots, young budstalks, and seeds eaten (many groups)
Species used as foods or beverages considered notably health-giving or beneficial when consumed	52	*Equisetum telmateia* Ehrh. (giant horsetail) – young shoots eaten by Saanich; said to be good for the blood; *Pinus contorta* (lodgepole pine) – inner bark considered a good spring tonic, used as laxative and for worms, as a tonic for coughs and stomach medicine for ulcers (several groups); *Lilium columbianum* Hanson (tiger lily) – bulbs considered a good health food, for long life (Secwepemc); *Opuntia* spp. (pricklypear cacti) – food valuable for older men; helps urination (Okanagan)
Species known to be eaten by animals as well as people (may overlap with other categories)	56	*Bryoria fremontii* (Tuck.) Brodo & D. Hawksw. (black tree lichen) – eaten by caribou, deer; *Dryopteris expansa* (C.Presl) Fraser-Jenkins & Jermy (wood fern) – rootstocks eaten by mountain goats as well as people (Turner et al. 1992); *Pinus contorta* (lodgepole pine) – inner bark eaten by grizzlies and black bears;

Categories of food, medicine, poisonous or harmful to eat species (including plants, marine algae, lichens and fungi)	No. of species documented*	Selected examples (cf. Kuhnlein and Turner 1991; Turner 1995; 1997; see Table 1 for references to designated linguistic groups)
		Erythronium grandiflorum Pursh (yellow glacier lily) – bulbs a favouring food of grizzlies; *Zostera marina* L. (eelgrass) – rhizomes eaten by brant geese and ducks; *Corylus cornuta* (hazelnut) – nuts eaten by squirrels and bears; *Trifolium wormskjoldii* Lehm. (springbank clover) – rhizomes eaten by geese and ducks (Turner 1997)
Species with parts known to be eaten by animals but not by people (may overlap with other categories)	41	*Alectoria sarmentosa* (Ach.) Ach. (old man's beard lichen) – eaten by deer (Hesquiaht); *Larix occidentalis* Nutt. (western larch) – buds eaten by blue grouse (Okanagan); *Lysichiton americanus* Hultén & St. John (skunk cabbage) – leaves eaten by deer, bear; *Pseudoroegneria spicata* (Pursh) A. Löve (bluebunch wheat grass) – excellent forage for deer and livestock; *Petasites frigidus* (L.) Fries var. *palmatus* (Ait.) Cronquist (coltsfoot) – elk's food; *Lonicera involucrata* Banks ex Spreng. (black twinberry) – berries eaten by crow's, bears and other animals (many groups) (Turner 1997)
Species used as foods or beverages known to have harmful or toxic properties if consumed or needing special processing or eaten at certain stage only	39	*Pteridium aquilinum* (L.) Kuhn. (bracken fern)– rhizomes never dug in summer; should eat with oil; most people do not eat fiddleheads; rhizomes must be cooked; *Pinus albicaulis* Dougl. ex Hook. (whitebark pine) – seeds eaten, but cause constipation if eaten raw or too many eaten; *Taxus brevifolia* Nutt. (Pacific yew) – outer fleshy part of arils eaten, but may cause sterility if a woman eats too many (Haida); *Lysichiton americanus* (skunk cabbage) – roots eaten only in early spring after prolonged cooking; *Triglochin maritimum* L. (arrowgrass) – very young vegetative plants eaten; older plants toxic; *Heracleum lanatum* Michx.(cow parsnip) – stems and leaves phototoxic – cause spots and mouth sores; young shoots only eaten after peeling (many groups)

(Contd.)

The Food/Medicine/Poison Triangle

Categories of food, medicine, poisonous or harmful to eat species (including plants, marine algae, lichens and fungi)	No. of species documented*	Selected examples (cf. Kuhnlein and Turner 1991; Turner 1995; 1997; see Table 1 for references to designated linguistic groups)
Species used as foods or beverages, also used in medicinal preparation(s), taken internally	55	*Pyropia abbottiae* (Krishnamurthy) S. K. Lindstrom (red laver seaweed) – used as food, eaten for indigestion (Gitga'at) (Turner 2003); *Polypodium glycyrrhiza* D.C. Eaton (licorice fern) – rhizomes eaten as sweetener; appetizer; also medicine for coughs, sore throats, sore gums, digestion, tuberculosis; *Picea sitchensis* (Bong.) Carr. (Sitka spruce) – pitch chewed as gum, also as breath freshener, for colds, coughs, tuberculosis, said to whiten teeth
Species with one or more parts used as food or beverage, other parts used for internal medicine	63	*Equisetum telmateia* (giant horsetail) – young shoots eaten raw; decoction drunk for arthritis (Gitga'at); *Abies lasiocarpa* (Hook.) Nutt. (subalpine fir) – seeds, gum and inner bark edible; pitch, bark, boughs as medicine (Secwepemc and others); *Heracleum lanataum* (cow parsnip) – young shoots eaten; roots used medicinally as purgative and for colds (Nlaka'pamux); *Mahonia aquifolium* (Pursh) Nutt. (Oregon grape) – berries eaten – used as antidote for shellfish poisoning; inner bark used as medicinal tonic (Saanich, Okanagan and others)
Species used for food, other parts said to have toxic or otherwise harmful properties	10	*Pteridium aquilinum* (bracken fern) – rhizomes eaten by some after roasting; fiddleheads eaten after soaking in brine overnight; tops of fiddleheads considered toxic; *Lomatium dissectum* (Nutt.) Mathias & Constance (chocolate tips) – roots and young tops eaten in spring; later used as fish poison and insecticide (Okanagan and others)
Species used for medicine, having some edible qualities (although not necessarily considered "food")	63	*Blechnum spicatum* (L.) Sm. (deer fern) – young leaves and stalks chewed to alleviate hunger; also eaten for stomach and lung troubles (Hesquiaht); *Picea sitchensis* (Sitka spruce) – cambium eaten as laxative
Species used in some way for internal medicine, considered neither food nor poisonous	68	*Picea engelmannii* Parry ex Engelm. (Engelmann spruce) – tea from bark for general medicine (Nlaka'pamux; Okanagan); *Acer glabrum* Torr. (Rocky Mountain maple) – infusion of bark as poison antidote

(Contd.)

Categories of food, medicine, poisonous or harmful to eat species (including plants, marine algae, lichens and fungi)	No. of species documented*	Selected examples (cf. Kuhnlein and Turner 1991; Turner 1995; 1997; see Table 1 for references to designated linguistic groups)
Species used for medicine and also considered to have toxic or harmful properties under some circumstances	80	*Dryopteris filix-mas* (L.) Schott. (male fern)– used as a poison antidote, but also considered poisonous (Gitga'at, Nuxalk); *Juniperus scopulorum* Sarg. (Rocky Mountain juniper)/ *J. maritima* – decoction of branches used as medicine, but also as poison to kill people and animals; *Taxus brevifolia* (Pacific yew) – tea from bark and branches drunk for cancer, ulcers, and tonic, but only in small amounts; *Oplopanax horridus* (J.E. Smith) Miq. (devil's club) – taken in limited doses; said to cause diarrhoea (Lantz et al. 2004); *Artemisia tridentata* Nutt. (big sagebrush) – medicinal tea said to cause sterility (Secwepemc; Nlaka'pamux)
Species considered toxic, not eaten nor used for internal medicine	10	*Streptopus streptopoides* (Ledeb.) A. Nels. & Macbr. (twistedstalk) – berries considered poisonous (Gitga'at; Nuxalk); *Platanthera dilatata* (Pursh) Lindl. Ex Beck(white bog orchid) – poisonous for humans and animals (Okanagan); used as a charm and poison; *Cicuta douglasii* (DC.) Coult. & Rose (water hemlock) – roots and tops extremely poisonous (Haisla, Ulkatcho and many others)
Total number of species inventoried	*375	

Case Examples

To illustrate the complexity and diversity of the relationships, I consider a selection of eight case examples of species that fit in different ways within the food/medicine/poison continuum, according to reports of Indigenous Peoples. These species range across a variety of plant families and are well known among several different cultural groups: *Pteridium aquilinum*; *Juniperus scopulorum* (and its close relative *J. maritima*); *Veratrum viride*; *Lysichiton americanus*; *Achillea millefolium*; *Heracleum maximum*; *Rhododendron neoglandulosum*; *Valeriana sitchensis* (see Figures 1-8). (Note that references for the different groups cited are listed in Table 1, and also that, although the descriptions of the different uses and applications of these plants are written in the past tense, many of these are ongoing to the present time, if not to the same extent as formerly).

***Pteridium aquilinum* (L.) Kuhn; Dennstaedtiaceae (Bracken fern; Hay-scented fern family)** (Figure 1) – Bracken "roots" (rhizomes) were eaten, with caution, by Nlaka'pamux, Saanich, Ditidaht, Hesquiaht (Nuu-Chah-Nulth), Kwakwaka'wakw (Southern Kwakiutl), Nuxalk, and Haida, as well as by First Peoples of Washington (Norton 1979), who used fire to maintain prairies as habitat for bracken and other edible plants such as edible camas (*Camassia* spp.) (Boyd 1999). Bracken is little used today by Indigenous Peoples of western North America. In the past, a number of precautions were associated with the consumption of bracken rhizomes as food, indicating some awareness of its potentially toxic properties. The Nlaka'pamux ate the young shoots, or fiddleheads, which are also commonly eaten by Japanese, Polynesians and Asian peoples. However, the Nlaka'pamux soaked the fiddleheads in brine overnight to get rid of the strong taste, and noted that only the stems were eaten, not the tops. (This use may have been learned relatively recently from Asian immigrants).

The "roots" (rhizomes) were not usually eaten raw but rather were generally prepared by roasting in hot coals of a fire, pit-cooking or steaming them. The cooked "roots" were pounded to remove the black outer "bark" and the central fibres; the whitish

Figure 1: *Pteridium aquilinum* (bracken fern).
Photo: NJ Turner

starchy cortex was the part eaten. Eating bracken "roots" was said by some (e.g. Nlaka'pamux) to give worms. Some people (e.g. Saanich) suggested that the "roots" should be eaten with fish eggs or some kind of oil because they were said to be constipating. The Haida and others further up the coast ate them with oulachen grease. Furthermore, the Saanich maintained that bracken "roots" should never be dug in the summertime, or they would cause a condition called "getting snaked," in which snakes enter one's body and are very difficult to expel. There are several versions of a Saanich story about this condition and its association with bracken "roots" (Turner and Hebda 2012). The Nuxalk also associated bracken "roots" with snakes (Turner 1973). The Gitga'at considered them to be poisonous and apparently did not eat them at all. At least two groups used bracken medicinally: for the Nlaka'pamux, an infusion of the "roots" was drunk for internal injuries and vomiting of blood, as well as for a cold or to improve the appetite. The Hesquiaht ate the young shoots for "troubles with one's insides," such as cancer of the womb.

Bracken fern contains a number of toxic constituents including: a cyanide-producing glycoside (prunasin), which produces a sharp bitter "almond" odor in the shoots when broken; an enzyme, thiaminase, which can deplete the body's thiamine reserves; and at least two carcinogens, quercetin and kaempferol, as well as a radiation mimicking substance (Turner and von Aderkas 2009). It is known to cause poisoning in cattle and other livestock, as well as being potentially harmful to humans.

***Juniperus scopulorum* Sarg. (Rocky Mountain Juniper; Cupressaceae) [and its close relative *J. maritima* R. P. Adams (Coast Juniper)]** (Figure 2) – These junipers are best known as a medicine rather than as a food. The "berries" (berry-like cones) are very strong tasting, and were not eaten in quantity. However some Secwepemc people used them for flavouring tea, and others, such as the Nlaka'pamux and Stl'atl'imx, used to eat a few berries as a hunger suppressant (Sam Mitchell, pers. comm. 1974). The Okanagan, who believed the berries to be poisonous, nevertheless sometimes drank a tea of them in the sweathouse as part of a cleansing ceremonial. Some people observed that birds sometimes eat the berries. The Secwepemc were said to chew the bark as a hunger suppressant.

Figure 2: *Juniperus scopulorum* (Rocky Mountain Juniper). Photo: NJ Turner

The highly aromatic boughs of this juniper were infused as a medicinal tea, for colds, tuberculosis, heart trouble, internal haemorraging, influenza and other ailments (Secwepemc, Nlaka'pamux, Okanagan), as well as for general use as a household disinfectant when there was disease or death in a house, said not only to protect against disease but also to combat evil spirits associated with illness and death (Secwepemc, Nlaka'pamux, Okanagan). One Secwepemc elder warned against making the tea too strong. For the Nlaka'pamux, as an indication of the poisonous properties attributed to this plant, a strong decoction was used to kill ticks on horses and used externally as a wash for seven-year itch, an affliction apparently caused by parasitic organisms, larval forms of flatworms. Juniper boughs were placed on the top of a hot stove as a smudge, "like Lysol", but too much exposure to the smoke is said to irritate the eyes. Juniper tea was drunk by a woman before childbirth, and was considered a good emergency medicine (Okanagan). Nevertheless, it was also regarded as a poison. The Okanagan, for example, soaked arrowheads in a solution of the pounded branches, berries and water to render them poisonous; bullets were also treated with this solution, which was said to kill people quickly in warfare and also to kill deer and other game through causing the blood to coagulate.

Juniper is known for its strong scent, and is named after its odour by the Saanich (referring to the closely related *J. maritima*). Junipers contain a rich complex of aromatic compounds, particularly volatile terpenoids (Powell and Adams 1973). Although juniper berries are a well known flavouring for stews and for the alcoholic beverage gin, they are known to cause kidney irritation, uterine contractions and possible miscarriage in pregnant women if taken in quantity (Tyler 1987), possibly due to the presence of significant concentrations of isocupressic acid, known to cause abortion in late term pregnant cattle (Gardner and James 1999). It is notable that one medicinal application of Rocky Mountain juniper tea is for women before childbirth, evidently to bring on labour (Turner et al. 1980).

***Veratrum viride* Ait. (False Hellebore, or Indian Hellebore; Melianthaceae, formerly in Liliaceae)** (Figure 3) – This plant, named in each of the study languages, is universally known as both a medicine and a poison – the "most poisonous plant you can find" (Turner et al. 1990); "really strong stuff" (Hebda et al. 1996). The only other plant generally considered equally, or more virulent by Indigenous People is water hemlock (*Cicuta* spp.; Apiaceae). Many people throughout the study area have recounted incidents of accidental poisoning, and sometimes death, in humans and animals from *Veratrum*. It has reportedly been used to commit suicide. Even recently, people have been admitted to the hospital due to inadvertent poisoning from improper medicinal use. Some people mentioned that it should never be taken internally, or even used externally if the skin is broken, since it can enter the bloodstream through cuts or abrasions. It is most widely known as an external poultice, wash and soaking solution prepared from the roots and applied to sores, swellings, stiffness, sore joints and arthritis, for which it serves as a local anaesthetic.

For those groups who do take *Veratrum* internally as a medicine, people generally state that great care must be used, and one should never take it without understanding the part used, harvesting procedures, seasonality, concentration and dosage. Secwepemc elder Aimee August, for example, was told not to use it herself, because it was too strong. She noted, "you have to eat something that's greasy when you do use it; don't use it by yourself." She said the old people used to know how to use it, but people today could harm themselves because they don't know the details of its use (Turner et al. n.d.). The Nlaka'pamux took a weak decoction of ashes in very small doses for venereal disease. Some peoples used the powdered roots, or sometimes the powered leaves, as a snuff for colds; it causes violent sneezing and clears the sinuses. The Kwakwaka'wakw reported that women formerly drank an infusion to bring about an abortion – an obviously extremely dangerous practice. Kwakwaka'wakw people sometimes took a weak infusion of the scraped roots to induce vomiting for internal pains. Some people would hold the large fleshy roots in the mouth as a laxative, whereas the secondary roots were not used, as they were considered too poisonous. For the Nuxalk, the inner core of the root is considered most dangerous; it was said that if this part were held in the mouth, it would cause immediate death. The outer root, on the other hand, was boiled and the solution drunk for coughs, gonorrhoea, and stomach pains; it is said to act as an emetic and cause fainting. Shamans and those undergoing ritual training for hunting or other purposes have been known to take *Veratrum* to induce dreams or visions, but obviously tremendous knowledge of dosage and preparation is considered essential. The Kwakwaka'wakw people have described the plant as having a "good hand" and a "bad hand" (Boas 1930).

Figure 3: *Veratrum viride* (false hellebore). Photo: NJ Turner

Veratrum viride was accorded tremendous protective powers; for example, it was used as a fumigant and smudge to ward off illness, and children especially would wear a small pouch containing a piece of the root to protect them against illness during disease epidemics, like chickenpox and measles (Turner and Thompson 2006). The root is also used as a charm to protect against evil and malicious spirits that cause disease and bad luck. A solution of the roots is used to treat scabies and to kill head lice. An Okanagan elder noted that animals do not eat this plant, but that a grizzly bear will eat the leaves so that bullets and arrows will not hurt him due to the plant's anaesthetic qualities (Turner et al. 1980).

Many people have noted that the only known antidote for poisoning by this plant is to drink large amounts of fish oil or grease. Drinking salmon head soup, which is very oily, is sometimes identified as the best treatment and there are stories of it having been used successfully to counteract severe poisoning (Turner et al. 1990).

Veratrum viride and its relatives are known to contain numerous complex alkaloids, including germidine, germitrine, veratridine, veratrosine, and veratramine. It is widely known to cause poisoning in humans and animals, with symptoms ranging from burning sensation in the throat, abdominal pain, vomiting and diarrhea, to blurred vision, hallucinations, general paralysis, convulsions and death. It has been used in the past medically to treat high blood pressure, but the complexity and relative instability of the alkaloids made the drug's use impractical (Turner and von Aderkas 2009).

Lysichiton americanus **Hultén & St. John (Skunk cabbage; Araceae)** (Figure 4) – This plant, like many species in the arum family, contains needle-like raphides of calcium oxalate, which can cause intense burning of the tongue and throat if ingested (Turner and von Aderkas 2009). Nevertheless, it was used in limited circumstances as a famine food by some peoples. For example, according to elder Mary Thomas (pers. comm. 2003), the Secwepemc dug the roots in the spring, just when the flowers came into bloom, dried and powdered them and made them into a kind of bread. Some other people,

Figure 4: *Lysichiton americanus* (skunk cabbage). Photo: NJ Turner

too, ate skunk cabbage "roots" (rhizomes) after prolonged cooking, but even then, they were described as tasting hot, like pepper (Turner 1995). Haida shamans formerly ate the rhizomes because they are so strong, and "burn," and are said to provide good protection against evil. The roots were also used variously for medicine. The Makah of Washington drank a decoction of the root as a blood purifier, and Makah women were said to have chewed the root to bring about an abortion (Turner et al. 1983). The Nuxalk took a tea of skunk cabbage "roots" for stomach troubles (Smith 1928).

Skunk cabbage leaves, while not eaten and considered potentially fatally poisonous by some (Turner 2004), were widely used – without apparent harm – in food preparation: for wrapping food for cooking, as a lining for earth ovens and pits used to make fermented salmon eggs, as a surface on which to dry berries, and for temporary drinking cups and berry picking containers. They were also spread overtop of food being stored in wooden boxes, and, occasionally, were powdered and used as a thickener and preservative for berries. As well as skunk cabbage leaves being used topically by the Secwepemc and others, for treating scabies, burns or broken bones, a solution of the leaves was said to be drunk as a treatment for skin troubles.

Despite the potential toxicity of skunk cabbage for humans, many people have noted that animals – especially bear and deer – at certain times of the year eat skunk cabbage freely, both roots and leaves. According to Haisla plant experts, both black and grizzly bears eat the rhizomes after their winter sleep, to cleanse and strengthen their stomachs (Compton 1993). In one Haida story, Deer [in his role as a human-like character] was described as being able to pull out any skunk cabbage root with his teeth, and in another, Deer was said to care for the skunk cabbage he ate just as if it were growing in a garden" (Turner 2004).

***Achillea millefolium* L. (Yarrow; Asteraceae)** (Figure 5) – Yarrow is one of the most widely known and used medicinal plants in the world (cf. Chandler et al. 1982). It was known and used medicinally by all of the groups in the study, including both external and internal applications. The Ulkatcho Dakelh drank a tea mixture of *Achillea*, *Pyrola asarifolia* Michx. and *Juniperus communis* L. to treat influenza. As a diuretic, they boiled *Achillea* leaves in water for about two hours and drank the decoction. Nlaka'pamux people drank an infusion of *Achillea* flowers for

Figure 5: *Achillea millefolium* (yarrow). Photo: NJ Turner

influenza, colds, and bladder troubles. To treat dysentery, they chewed the leaves raw or made them into a tea, and was drunk. They took an infusion or decoction of the roots, stems or whole plant for stomach cramps, diarrhoea, colds, venereal disease, or generally as a remedy for any undiagnosed sickness. They applied the root to a toothache, and used an infusion of the entire plant to wash the eyes. The Okanagan prescribed drinking an infusion of the roots (steeped in boiling water, as in making tea) for headaches, stomachaches, colds or diarrhoea, and a decoction of the roots (boiled in water) as a physic. They also applied the mashed roots for a toothache. The Saanich and Ditidaht chewed the leaves and swallowed the juice to treat a cold, and the Saanich also used the roots for a toothache. Ditidaht people chewed and swallowed the leaves as a general tonic, to treat colds, coughs, and respiratory ailments, and also to prevent getting a cold. The Hesquiaht chewed the leaves and swallowed the juice for a prolonged cough or to alleviate any kind of internal pain or stomach problems. These are just a few examples from a broad range of medicinal applications for yarrow.

Although yarrow wasn't actually eaten as a food, some people, at least recently, have used it to make a beverage tea. The Haida used the stiff stems as skewers for smoke-drying butter clams; they were said to give a good flavour to the clams, which were stored on the stems, then eaten later from them (Turner 2004). Possibly, the yarrow stems served to help preserve the clams and protect them from insect pests. According to Haisla and Hanaksiala consultants, the plant is eaten by bears (Compton 1993). Because of its unpalatability, however, it is seldom grazed by livestock.

Despite its extensive use, many people warn that yarrow is quite strong and should be taken with caution (Turner et al. 1990). Some remarked on its bitterness and disagreeable taste. One elder noted that if handled it could cause white spots on the skin, all over one's face – something like ringworm; children were warned against touching it (Turner et al. 1983). Yet, a poultice of the chewed leaves was applied to the skin for sores and swellings (Turner and Bell 1973), and the plant was also used as a mosquito repellent, according to Secwepemc elder Mary Thomas (see also Tunon et al. 1994).

Yarrow contains a range of aromatic compounds including sesquiterpene lactones and coumarins, which can cause contact allergic dermatitis, including phototoxicity, in susceptible individuals (Mitchell and Rook 1979; Towers 1979).

Heracleum maximum **Bartr. (syn.** *H. lanatum* **Michx.) (Cow Parsnip; Apiaceae)** (Figure 6) – This is one of many species in the celery or umbel family that strands the food/medicine/poison triangle in various ways. In fact, the family Apiaceae – like Solanaceae and Ericaceae – contains a wide range of species that are edible, as well as some that are deadly poisonous. Many of the edible species, including *Heracleum maximum*, have limitations to their edibility and require special knowledge for safe use. The young shoots – both budstalks and leafstalks – of this species have been eaten for probably thousands of years by First Peoples of the study

Figure 6: *Heracleum maximum* (cow parsnip).
Photo: NJ Turner

region. The plant is wide ranging; it grows in moist prairies, meadows and forest edges from sea level to above the treeline in the mountains. The budstalks and leafstalks are generally designated by different names in the various languages, and are peeled and eaten differently. The budstalks, which are cylindrical and attached at the top to a rounded embryonic inflorescence encased in a sheathing leafbase, are peeled from bottom to top, then the fleshy cylinder eaten. The leafstalks are more difficult to peel, and are generally split along one side, then the edible portion peeled away from the fibrous skin and eaten. Both types were consumed raw or sometimes cooked, dipped in oil or grease, and, more recently, eaten with molasses or sugar. Some people prefer the budstalks, others the leafstalks. Everyone, however, is careful to note that only the young plants in springtime can be used, that the stalks *must* be peeled, and that the peels must never touch the lips, or they will cause discoloration and blistering of the lips and skin around the mouth that will remain for many weeks or months. These peeled shoots are considered to be a good "health food" or spring tonic, and in the past were greatly enjoyed; they were in fact the most extensively used traditional green vegetable of the entire region. As the name "cow parsnip" implies, the plant is known to be eaten by cattle, as well as deer and both black and grizzly bears, for whom it is a major springtime food.

The skin irritation caused by *Heracleum* in the presence of ultraviolet light is due to the presence of phototoxic furanocoumarins (Mitchell and Rook 1979). As well as peeling the shoots, and eating only the young ones, people took a range of other precautions. Nlaka'pamux elders warned, "If you eat too much, it'll burn you" (Turner et al. 1990). Ditidaht plant specialist John Thomas cautioned that if you eat too much it might cause sore, cracked and blistered lips, and warned that people should always wash their lips after eating it or they would get a sore, something like a cold sore, on the lips. Furthermore, he noted that if a pregnant woman eats this food, her child might become "epileptic." He said that plants growing in shade could be harvested later in the season than those in the sun (Turner et al. 1983), presumably because their concentrations of furanocoumarins would be lower. A Haida elder said that the shade growing shoots were good to eat but those growing in sunlight were inedible (Turner 2004). A Manhousaht (Nuu-chah-nulth) elder said that if a pregnant woman ate the bud stalks of cow parsnip, her baby would lose its breath when it cried and would then choke. He said that the leaves and stems should be knocked with a stick before they were harvested, to release the "dust," which would otherwise cause burning and spotting of the skin (Turner et al. 1983). One Gitga'at elder recalled that when she used to peel the stalks for the children to eat, she would get brown stripes across the back of her hand that remained for months afterwards. She said that if the children ate the stalks without peeling them, they would get brown marks around their mouths. Even just touching the leaves could cause this condition, she noted (Turner and Thompson 2006).

The acrid roots were not used as food; in fact, the roots and leaves are considered to be poisonous. However, the roots were often applied medicinally. The Okanagan and others used cow-parsnip roots as a cleansing medicine for the scalp. The Secwepemc drank a solution of the roots as a medicine for bladder infections and for loss of weight, but people warned that you have to drink it with broth. One elder noted, "If you do not get fat, but you get skinny all the time, drink it and drink broth right afterward. It kills all the germs inside you." (Turner et al. n.d.). The Nlaka'pamux used a decoction of the roots as a purgative, for purification rituals for youth reaching puberty, and for hunters and others wishing to gain spiritual powers. Some people also drank a tea from the roots for colds and coughs.

***Rhododendron neoglandulosum* Harmaja (syn. *Ledum glandulosum* Nutt.) (Trapper's Tea, Western Labrador Tea, or "Swamp Tea"; Ericaceae)** (Figure 7) – This shrub of high elevation pond margins and wetlands is generally considered to be toxic, yet it is a major source of beverage and medicinal tea for Interior Salish peoples including the Secwepemc, Nlaka'pamux and Stl'atl'imx. In fact, these leaves and those of wild mint (*Mentha arvensis* L.) are the major traditional types of tea for many people, even today. On the coast and in the north, Labrador tea, or "Hudson's Bay Tea" (*Rhododendron groenlandicum* (Oeder) K.A. Kron & W.S. Judd) is the main beverage tea. The genus *Rhododendron* is well known for the toxicity of the leaves, flowers, pollen and nectar, due to the presents of several toxic diterpenoids called grayanotoxins. Rhododendrons have caused poisoning in humans and animals, with symptoms of initial burning of the mouth, salivation, watering eyes, runny nose, followed several hours later by vomiting, convulsions, headache, muscular weakness, drowsiness, slow and irregular heartbeat, low blood pressure and, potentially paralysis, coma and convulsions. Even honey made from rhododendron nectar is said to be toxic, and chronic poisoning from drinking rhododendron tea is possible, but rhododendron poisoning is rarely known to be fatal for humans (Turner and von Aderkas 2009).

Figure 7: *Rhododendron neoglandulosum* (trapper's tea). Photo: NJ Turner

Trapper's tea (*Rhododendron neoglandulosum*) is a relaxing beverage according to Secwepemc elder Mary Thomas. It is good to drink when one feels cranky or "dopey." It is said to be very soothing. However, she warned that it should not be made too strong, or it will make one light-headed. It is a good drink to take just before bed as a relaxant, and is also taken as a painkiller and relaxant by new mothers just after childbirth. She also said that "when a person has a heart attack, the first thing they make for that person is this tea – or for indigestion. It gives you a good night's sleep" (Mary Thomas, pers. comm. 1997). It is also said to be good for the kidneys.

Rhododendron groenlandicum, too, is known as not only a beverage tea, but as a medicinal tea. The Ulkatcho boil the leaves until the water turns dark, and then use the decoction to treat colds, coughs, sore throats, and lung problems like pneumonia; they also chew the leaves and swallow the juice for the same purpose. The Ditidaht drink the tea as a tonic for lack of appetite or a "run down" feeling, but some warn that drinking too much can make one fat, or give one a "pot-belly." Others take the tea as a blood purifier, or for women after a miscarriage (Turner et al. 1983). The Nuxalk drink the tea as a beverage and specifically for stomach pains. The Haisla and Hanaksiala take it for tuberculosis and to increase the appetite, as well as for a beverage. One of the Haida names for this plant is "medicine for colds," and the Gitga'at drink a tea from the leaves for a general all-purpose medicine. Various preferences are expressed over what age of leaves should be used, and whether the flowers should or should not be included in the tea. Some prefer the tea of the young leaves picked in the spring before the plant blooms; others say leaves picked in the summer are fine, as long as they are still growing upright. One Haida elder said that a beverage tea is made from the leaves alone, whereas a tea from the flowers and flowering twigs is medicinal (Turner 2004). Many people brew Trapper's tea and Labrador tea very strong, simmering it on the back of the stove until it turns deep brown, and adding more water as the tea is consumed or boils down. In recent years, people sweeten the tea with sugar; formerly, some sweetened it with the rhizomes of licorice fern (*Polypodium glycyrrhiza*) (Turner 1995).

Valeriana sitchensis **Bong. (Mountain Valerian; Valerianaceae)** (Figure 8) – The roots of this plant are used by Interior Salish groups and neighbouring peoples for a variety of medicinal preparations. Secwepemc elder Aimee August noted that valerian root it is a multi-purpose remedy, used for a range of ailments, from coughs and fevers, to digestive problems and toothache. It is generally prepared by boiling in water until the solution is a brownish colour; it is said to be bitter tasting. This decoction is considered good to drink simply for good health, and is said to be especially beneficial for the lungs: it "gives you good wind." People formerly gathered large quantities of these roots to dry for winter use, in case of colds, flu or upset stomach. Secwepemc elder Mary Thomas, however, cautioned that one shouldn't take too much of this medicine at once; it should be drunk only in moderation and as a mild solution, or, she warned, "You just feel like you want to sleep, sleep, sleep. So they made you walk around and wake up." The Nlaka'pamux drank a decoction of the roots for colds and tuberculosis, and also used a tea of the leaves and roots together, in a dosage of 3-4 plants per litre of water, to treat diarrhoea, internal pains, stomach problems, ulcers and influenza. Valerian tea was also drunk by hunters for purification, in order to gain special powers, and was considered by the Nlaka'pamux as a valuable horse medicine. Deer are said to browse the leaves of this plant. *Valeriana sitchensis* roots also have a strong, generally unpleasant smell, which is detectable in subalpine meadows when the soil is disturbed, as well as when the roots are dug.

Figure 8: *Valeriana sitchensis* (mountain valerian). Photo: NJ Turner

Another species of this genus, edible valerian, or tobacco root (*V. edulis* Nutt.), has large, pungent smelling roots, which were nonetheless formerly pit-cooked and eaten by indigenous peoples of the western United States. *Valeriana* species have been used for thousands of years as sedatives, to induce relaxation and sleep. In initial studies, *V. edulis*, for example, has been shown to be promising as a safe and effective treatment for sleep difficulties in some children with special needs (Francis and Dempster 2002). European valerian (*Valeriana officinalis* L.) is especially widely recognized as a sedative and sleep-inducing herb, dating back over 2,000 years to the era of Hippocrates, and is still recommended today for restlessness and sleep disorders based on nervous conditions (Blumenthal et al. 2000). A similar sleep inducing effect is recognized for *V. sitchensis*, but as noted, this was seen as potentially harmful for patients being treated with this plant.

Discussion

These eight case examples taken together span the entire food/medicine/poison continuum of indigenous plant use in British Columbia. Together they illustrate the depth and complexity of collective traditional ecological knowledge required for people to live safely and in good health in this environment and region if they are to rely solely on local and regional resources.

Traditional ecological knowledge systems embody empirical information, as well as practice and belief, transmitted culturally, accumulated and adapted over generations (Berkes 1999; Turner et al. 2000). The knowledge reflected here on food, medicines and poisons is well situated within the traditional knowledge complex. The practical information that forms part of this knowledge incorporates many factors key to the proper identification, nutritional value, palatability, medicinal properties and/or toxicity of plants (as well as lichens and fungi). These include the nature and variability of medicinally active or toxic compounds in relation to habitat, growth stage, seasonality, method of harvesting and preparation. For example, being able to recognize the minute differences between bulbs of the edible camas (*Camassia* spp.) and those of the deadly death camas (*Zigadenus venenosus* S. Wats.), even when these plants are not flowering, takes tremendous skill and experience. Knowing the characteristics of the human consumers of these products – age, gender, level of activity, and even mental state – is likewise important. Recognition of synergistic effects of foods and medicines, and of antidotes for poisoning is also critical. All these types of knowledge are informed and mediated by observations of animals, their use of food and medicine and avoidance of toxic species, and by an entire range of incremental experimentation and monitoring, trial and error (Johns 1996).

The different means by which this knowledge is communicated and disseminated over generations and across cultural boundaries are likewise an important component of this knowledge. A single experience of a poisoning of a child by a plant such as water hemlock (*Cicuta douglasii* (DC) Coult. & Rose), for example, can spread rapidly over a wide cultural network and geographical space, and remain in peoples' collective memory for many generations (Turner and Berkes 2006). Knowledge of some types of food use, and poisonous qualities of plants can become embedded and encoded in narratives, songs, dances and ceremonies, and this assists in its retention and transmission. Even the names and descriptions of species in the various indigenous languages provide keys to their properties, whether positive or negative. For example, the Haida name for Saskatoon berry (*Amelanchier alnifolia*), **gaan xaw'laa** (Skidegate dialect), translates as 'sweet berry', whereas the name for twistedstalk (*Streptopus amplexifolius*), **st'aw gaanga**, means 'witch's or owl's berry', reflecting the food quality of the former and the perceived inedibility of the latter (Turner 2004). In the Okanagan-Colville language, death camas is called *yiw'ístn* (literally, 'causing twitching'); eating the bulbs is said to cause twitching, foaming at the mouth, and eventually death (Turner et al. 1980). In every language, the ethnobotanical vocabularies reveal and encode similar types of knowledge relating to taste, medicinal qualities or potential harmful effects.

In relation to the belief component of traditional ecological knowledge, there is a definite emotional and spiritual aspect to these plants and their uses. This is difficult to capture in a descriptive form, but is reflected, for example, in the application of various potentially toxic plants as emetics and purgatives in cleansing and purification ceremonies for shamans, hunters, youth at puberty and others undergoing specialized ritual training. To achieve a state of spiritual awareness and to make one open to envisioning and obtaining the aid of supernatural spirits, such cleansing is undertaken, both externally and internally. In combination with fasting and bathing, taking certain cathartic plants can result in a heightened state of awareness that is said to bring success to healers, hunters and even those who undertake gambling and other forms of contests. This practice is obviously undertaken with deep awareness of the potential negative consequences. There are stories of poisoning – even fatal poisoning – of hunters who attempted to purify themselves with *Veratrum* and used the wrong dosage. Today, elders worry that younger people are not following the appropriate lifestyles or eating the right food so that they might harm themselves if they try to take these products for cleansing and purification.

There are obvious limitations to knowledge of the long term, indirect or compounding effects of certain substances, such as carcinogens. Nevertheless, as can be seen from the case examples, there is a definite congruence of traditional knowledge of the qualities and effects of foods, medicines and toxics relative to scientific knowledge and understanding of these features. A good example of the amazing knowledge and perception applied in distinguishing foods from potentially harmful types, and in processing them for maximum nutritional benefit is seen in the recent study of black tree lichen and its use by Interior Salish peoples (Crawford 2007). This study determined that Secwepemc elders were able to distinguish visually with remarkable accuracy the non-toxic strains of black lichen (*Bryoria fremontii*) from its toxic look-alike forms, such as *Bryoria tortuosa* (G. Merr.) Brodo & D. Hawksw. Furthermore, the study demonstrated that the edible lichen, when pit-cooked together with camas bulbs (*Camassia quamash* (Pursh) Greene), actually sequestered carbohydrates from the bulbs, which were then utilizable as additional food energy instead of being leached out and lost in the cooking process. This knowledge and understanding was acquired and applied by Indigenous Peoples without the aid of chemical analysis or the use of microscopic examination, yet it is detailed enough to have enabled generations of people to obtain necessary nutrients while avoiding poisoning, thus to maintain their health. This is only one illustration of dozens that could be cited to show the depth of such knowledge.

Conclusions

Indigenous Peoples in British Columbia, like populations in many parts of the world, have benefited significantly from modern medical technologies and have enjoyed a host and diversity of new and tasty foods introduced by Europeans over the past two centuries. Nevertheless, the modern diet – particularly that of the last half of the twentieth century to the present time – has not been a healthy one for many Indigenous People (Kuhnlein et al. 2009). Diabetes, unheard of in former generations, is now of epidemic proportions in some communities, and rates of heart disease, high blood pressure and obesity are also much higher, largely due to increased consumption of unhealthy carbohydrates and fats in an increasingly processed and globalized diet (Heffernan 1995; Wong 2003; Kuhnlein et al. 2006). Harvesting and use of local foods, especially plant foods, as well as use of traditional medicines, has declined significantly over the past century. This is not only due to the introduction of new foods and medicines, but to layer after layer of cumulative impacts on traditional food and medicine systems (Parrish et al. 2007; Turner and Turner 2008). This decline has been reflected in a general erosion and loss of knowledge relating to these resources, including the knowledge described in this paper relevant to the food/medicine/poison triangle.

Yet, perhaps not surprisingly, those who hold this knowledge are the very ones who best understand its importance and value and who are determined to keep it alive, to renew and revitalize it. This is not only because it is a part of their cultural heritage and history, but because it is this very knowledge that helps them to maintain their resilience, their adaptability and their connections to their home places. This knowledge constitutes a safety net for future generations. Many feel that their very survival as a people and as a culture depends on the continuation of their ancient and grounded knowledge of food and healing and how to negotiate the relationships between them.

There are new threats to traditional food systems in the form of environmental contaminants like PCBs that can accumulate in the food chain, industrial development, urbanization, and introduction of exotic species that compete with indigenous species, to name just a few. Environmental loss – and cutting off peoples' access to their lands and traditional resources – leads inevitably to knowledge loss. Ultimately, it is peoples' knowledge about these resources, and how to use them and maintain them, that will help to conserve those resources (Etkin 1998, 2006b). This makes such knowledge systems critically important for the future of the planet.

Endnotes

[1] Note that in this paper, I focus on those species that are toxic and/or medicinal when actually consumed, either whole or in the form of infusions or decoctions taken internally. Other substances can cause skin reactions

Acknowledgements

I am indebted to all of the Indigenous elders and cultural specialists whose knowledge about traditional food and medicine systems has contributed to this paper. In particular, I would like to recognize: Clan Chief Adam Dick (Kwaxsistala), Kim Recalma-Clutesi (Ogwalogwa), Dr. Daisy Sewid-Smith (Mayanilth) (Kwakwaka'wakw); Helen Clifton, Chief Ernie Hill Jr. and Mildred Wilson (Gitga'at; Ts'msyen/Coast Tsimshian); Dr. Margaret Siwallace (Nuxalk) and other participants in the Nuxalk Food and Nutrition Program; Dr. Annie York (Nlaka'pamux); Sam Mitchell (Stl'atl'imx); Dr. Mary Thomas and Dr. Aimee August (Secwepemc); John Thomas and Ida Jones (Ditidaht); Elsie Claxton and Violet Williams (Saanich). I also thank Stuart Crawford, Dr. Nina Etkin, Dr. Timothy Johns, Dr. Harriet Kuhnlein, Dr. Lee Oates, and Dr. Andrea Pieroni, for their important contributions. I am grateful to Dr. Ranjay Singh for his patience and skill in editing this volume. This paper was supported in part through a General Research Grant from SSHRC (# 410-2005-1741). I am grateful to Dr. Eric Peterson and Christina Munck of the Tula Foundation for their ongoing and generous support of my research. I would like to dedicate this paper to Dr. Mary Thomas, who passed away in the summer of 2007, at the age of 90. She was an amazing source of knowledge about food, medicine and the environment, and she is missed by so many.

References

Balick MJ, Cox PA (1996) Plants, people, and culture: The science of ethnobotany. Scientific American Library, New York, NY

Berkes F (1999) Sacred Ecology: Traditional ecological knowledge and resource management. Taylor & Francis, Philadelphia, PA

Blumenthal MAG, Brinckmann J (2000) Herbal medicine: expanded commission E monographs. American Botanical Council, Austin, TX and Integrative Medicine Communications, Newton, MA

Boas F (1921) Ethnology of the Kwakiutl. Bureau of American Ethnology 35th Annual Report, Parts 1 and 2, 1913-14. Smithsonian Institution, Washington, DC

Boas, F (1930) The Religion of the Kwakiutl Indians. Parts 1 and 2. Columbia University Contributions to Anthropology Volume X, Columbia University Press, New York, NY.

Boyd R (1999) Indians, Fire and the Land in the Pacific Northwest. Oregon State University Press, Cornvallis

Campbell S, Affolter J and Randlen W (2007) Spatial and temporal distribution of the alkaloid Sanguinarine in *Sanguinaria canadensis* L. (Bloodroot). *Economic Botany,* 61 (3): 223-234

Chandler RF, Hooper, SN and Harvey MJ (1982) Ethnobotany and Phytochemistry of Yarrow, *Achillea millefolium*, Compositae. *Economic Botany,* 36 (3): 203-223

Compton BD (1993) Upper North Wakashan and Southern Tsimshian Ethnobotany: The knowledge and usage of plants. Ph.D. Dissertation, University of British Columbia, Vancouver

Crawford S (2007) Ethnolichenology of *Bryoria fremontii*: Wisdom of elders, population ecology, and nutritional chemistry. Unpublished M.Sc. thesis, School of Environmental Studies and Department of Biology, University of Victoria, British Columbia

Deur D, Turner NJ (2005) "Keeping it Living": Traditions of plant use and cultivation on the Northwest Coast of North America. University of Washington Press, Seattle and UBC Press, Vancouver

Duke JA (1985) CRC Handbook of Medicinal Herbs. CRC Press, Boca Raton, FL

Etkin NL (1998) Indigenous patterns of conserving biodiversity: Pharmacologic implications. *Journal of Ethnopharmacology,* 63:233-245

Etkin NL (2006a) Edible Medicines. An Ethnopharmacology of Food. University of Arizona Press, Tucson

Etkin NL (2006b) Wild plant management in rural Hausaland: Local ecological knowledge ontributes to the conservation of biodiversity. *Proceedings of the IVth International Congress of Ethnobotany* (ICEB 2005, Istanbul, Turkey), Z. Füsun Ertug (ed), Yeditepe University, Istanbul, pp. 359-364

Francis AJ, Dempster RJ (2002) Effect of valerian, *Valeriana edulis*, on sleep difficulties in children with intellectual deficits: randomised trial. *Phytomedicine,* 9(4):273-279

Gardner DR, James LF (1999) Pine needle abortion in cattle: analysis of isocupressic acid in North American gymnosperms. *Phytochemical Analysis,* 10(3):132-136

Hebda RJ, Turner NJ, Birchwater S, Kay M, Elders of Ulkatcho (1996) Ulkatcho food and medicine plants. Ulkatcho Band, Anahim Lake, B.C

Heffernan MC (1995) Diabetes and aboriginal peoples: The Haida Gwaii Diabetes Project in a global perspective. Pp. 261-292 *In:* Stephenson PH, Elliot SJ, Foster LT and Harris J (eds), A Persistent Spirit: Towards understanding aboriginal health in British Columbia, Vol. 31, Canadian Western Geographical Series, Department of Geography, University of Victoria, BC

Johns T (1996) The Origins of Human Diet & Medicine. University of Arizona Press, Tucson

Johns T, Kubo I (1988) A survey of traditional methods employed for the detoxification of plant foods. *Journal of Ethnobiology,* 8(1):81-129

International Society of Ethnobiology (1988) Declaration of Belem. Articulated and ratified by the First International Congress of Ethnobiology, Belem, Brazil [later ratified by the Society of Ethnobiology. *Journal of Ethnobiology,* 8(1):5]

International Society of Ethnobiology (2006) ISE Code of Ethics (with 2008 additions). Online: http://ethnobiology.net/code-of-ethics/ (Accessed 28 September, 2013)

Kuhnlein HV, Erasmus B, Creed-Kanashiro H, Englberger L, Okeke C, Turner NJ, Allen L, Bhattacharjee L (2006) Indigenous Peoples' food systems for health: finding interventions that work. *Public Health Nutrition,* 9(8):1013-1019

Kuhnlein HV, Erasmus B, Spigelski D (2009) Indigenous Peoples' Food Systems: the many dimensions of culture, diversity and environment for nutrition and health. Food and Agriculture Organization of the United Nations, Rome and Centre for Indigenous Peoples' Nutrition and Environment, McGill University, Montreal. (URL: http://www.fao.org/docrep/012/i0370e/i0370e00.htm cited 24 January, 2010)

Kuhnlein HV, Turner NJ (1987) Cow-parsnip (*Heracleum lanatum* Michx.): an indigenous vegetable of native people of northwestern North America. *Journal of Ethnobiology,* 6(2):309-324.

Kuhnlein HV, Turner NJ (1991) Traditional plant foods of Canadian indigenous peoples. Nutrition, Botany and Use. Volume 8. In: Katz S (ed), Food and nutrition in history and Anthropology, Gordon and Breach Science Publishers, Philadelphia, PA

Lantz S, Kristina S, Turner NJ (2004) Devil's Club (*Oplopanax horridus*): An ethnobotanical review. *Herbalgram,* 62:33-48 (Spring 2004)

Martin AD, Zim HS, Nelson AL (1989) American wildlife and plants. A guide to wildlife food habits. Dover Books, New York

Mitchell J, Rook A (1979) Botanical Dermatology. Plants and plant products injurious to the skin. Greengrass, Ltd., Vancouver, BC

Moerman DE (1996) An analysis of the food plants and drug plants of native North America. *Journal of Ethnopharmacology,* 52(1):1-22

Norton HH (1979) Evidence for Bracken fern as a Food for aboriginal peoples of western Washington. *Economic Botany,* 35(4):384-396

Parrish CC, Turner NJ, Solberg S (2007) Resetting the kitchen table: Food security, culture, health and resilience in coastal communities. Nova Science Publishers, New York, NY

Pieroni A (1999) Toxic plants as food plants in the traditional uses of the Eastern Apuan Alps Region, North-West Tuscany, Italy, Pp. 267-272. In: Guerci A (ed.), *Il* Cibo e Il Corpo. Dal Cibo alla Cultura, Dalla Cultura al Cibo. (food and body, from food to culture, from culture to food. Fondazione Cassa di Risparmio de Genova e Imperia. Atti della 2nd Conferenza Internazionale de Antropologia e Storia della Salute e delle Malattie "Antropologia, Alimentazione et Salute." Genova, Italy

Powell RA, Adams RP (1973) Seasonal variation in the volatile terpenoids of *Juniperus scopulorum* (Cupressaceae). *American Journal of Botany,* 60(1):1041-1050

Smith HI (1928) Materia medica of the Bella Coola and neighbouring tribes of British Columbia. National Museum of Canada, Canada Department of Mines, bulletin No. 56 (Annual Report for 1927), pp. 47–68

Towers GN (1979) Contact hypersensitivity and photodermatitis evoked by Compositae. In: Kinghorn AD (ed). *Toxic Plants.* Columbia University Press, New York, NY

Tunon H, Thorsell W, Bohlin L (1994) Mosquito repelling activity of compounds occurring in *Achillea millefolium* L. (Asteraceae). *Economic Botany,* 48(2):111-120

Turner NJ (1973) Ethnobotany of the Bella Coola Indians of British Columbia. *Syesis,* 6:193-220

Turner NJ (1984) Counter-irritant and other medicinal uses of plants in Ranunculaceae by Native peoples in British Columbia and neighbouring areas. *Journal of Ethnopharmacology,* 11:181-201

Turner NJ (1997) "Le fruit de l'ours": Les rapports entre les plantes et les animaux das les langues et les cultures amérindiennes de la Côte-Ouest" ("The Bear's Own Berry": Ethnobotanical Knowledge as a Reflection of Plant/Animal Interrelationships in Northwestern N America). *Recherches amérindiennes au Québec,* 27(3-4):31-48

Turner NJ (1995) Food plants of coastal first peoples. Royal British Columbia Museum, Victoria, B.C. and University of British Columbia Press, Vancouver

Turner NJ (1997) Food plants of interior first peoples. Royal British Columbia Museum, Victoria, B.C. and University of British Columbia Press, Vancouver

Turner NJ (2003) The Ethnobotany of "edible seaweed" (*Porphyra abbottiae* Krishnamurthy and related species; Rhodophyta: Bangiales) and its use by First Nations on the Pacific Coast of Canada. *Canadian Journal of Botany,* 81(2):283-293

Turner NJ (2004) Plants of Haida Gwaii. X̱aadaa Gwaay guud gina k̲'aws (Skidegate), X̱aadaa Gwaayee guu giin k̲'aws (Massett). Sono Nis Press, Winlaw, BC

Turner NJ, Bell MAM (1971) The Ethnobotany of the Coast Salish Indians of Vancouver Island. *Economic Botany,* 25(1):63-104

Turner NJ, Bell MAM (1973) The Ethnobotany of the Southern Kwakiutl Indians of British Columbia. *Economic Botany,* 27(3):257-310

Turner NJ and Fikret B (2006) "Coming to understanding: developing conservation through incremental learning." *Journal of Human Ecology,* 34(4):495-513

Turner NJ, Bouchard R, Kennedy DID (1981) Ethnobotany of the Okanagan-Colville Indians of British Columbia and Washington. British Columbia Provincial Museum Occasional Paper No. 21, Victoria

Turner NJ, Deur D, D Lepofsky (2013) "Plant management systems of British Columbia's first peoples." *In:* Ethnobotany in British Columbia: Plants and people in a changing World, ed. NJ Turner and D Lepofsky. Special issue, *British Columbia Studies,* 179:107-133

Turner N, Efrat BS (1982) Ethnobotany of the Hesquiat Indians of Vancouver Island. British Columbia Provincial Museum, Cultural Recovery Paper No. 2, Victoria

Turner NJ, Gottesfeld LMJ, Kuhnlein HV, Ceska A (1992) "Edible wood fern rootstocks of western North America: solving an ethnobotanical puzzle." *Journal of Ethnobiology,* 12 (1): 1-34

Turner NJ, Hebda RJ (1990) Contemporary use of bark for medicine by two Salishan Native elders of southeast Vancouver Island. *Journal of Ethnopharmacology,* 229: 59-72

Turner NJ, Hebda RJ (2012) Saanich ethnobotany: Culturally important plants of the WSÁNEC' People. Royal BC Museum, Victoria

Turner NJ, Ignace MB, Ignace R (2000) Traditional ecological knowledge and wisdom of Aboriginal peoples in British Columbia. *Ecological Applications,* 10(5):1275-1287

Turner NJ, Loewen DM, Ignace MB (2010) Plants of the Secwepemc People. unpubl. ms. in possession of the editors

Turner NJ, Thomas J, Carlson BF, Ogilvie RT (1983) Ethnobotany of the Nitinaht Indians of Vancouver Island. British Columbia Provincial Museum Occasional Paper No. 24, Victoria

Turner NJ, Thompson JC (2006) Plants of the Gitga'at people. *'Nwana'a lax Yuup*. Gitga'at Nation, Hartley Bay, BC, Coasts Under Stress Research Project (R. Ommer, P.I.), and Cortex Consulting, Victoria, BC

Turner NJ, Thompson LC, Thompson MT, York AZ (1990) Thompson ethnobotany. knowledge and usage of pants by the Thompson Indians of British Columbia. Royal British Columbia Museum, Memoir No. 3, Victoria and University of British Columbia Press, Vancouver

Turner NJ, Turner KL (2007) "Rich in food": Traditional food systems, erosion and renewal in Northwestern North America. *Indian Journal of Traditional Knowledge*, 6(1):57-68

Turner NJ, Turner KL (2008) "Where our women used to get the food": Cumulative effects and loss of ethnobotanical knowledge and practice; case studies from coastal British Columbia. *Botany*, 86(1):103-115

Turner NJ, von Aderkas P (2009) The North American Guide to Common Poisonous Plants and Mushrooms. Timber Press, Portland, OR

Tyler VE (1987) The Honest Herbal. A sensible guide to herbs and related remedies. George F. Stickley Co., Philadelphia, PA

Wong A (2003) First Nations Nutrition and Health Conference Proceedings (Conference June 2003, Squamish Nation, Vancouver, B.C.), Arbokem Inc., Vancouver, BC

Chapter – 2

Integration into the Market Economy and Dietary Change: An Empirical Study of Dietary Transition in the Amazon

Elizabeth Byron and Victoria Reyes-García*

Abstract

Scholars have long been interested in the effects of markets on the diet of native populations because of diet's direct impacts on health. Exposure and assimilation into a market economy result in shifts in modes of subsistence and decision-making about food consumption. But the clear direction of influence is often elusive. Increased market integration predicts expanded access to goods and services that may improve household and individual diet. At the same time, greater involvement in market trade can shift economic priorities away from secure subsistence production toward the lure of material gain and novel items. Here, we analyze the effects of integration into the market economy on household diet using data from a panel study of two villages of Tsimane', an indigenous group of lowland Bolivia. Because the Tsimane' have been undergoing varying degrees of integration into the market economy, they provide an ideal population among which to examine the relation between markets and dietary change. Our findings suggest that greater integration into the market economy is associated with a higher percent of market-procured foods in the household diet, although market-produced foods still do not

---

*victoria.reyes@uab.cat

surpass farm foods among the Tsimane'. We also found that households that are more integrated into the market economy show signs of turning away from local forest foods to incorporate more market foods in the daily diet. Those findings can orient policy-makers in creating programs that would mitigate potential detrimental consequences of integration into the market economy for native cultures and livelihoods.

Keywords: Acculturation, Amazonian indigenous peoples, dietary change, food production, food consumption, Tsimane.

Introduction

Dietary change among indigenous populations undergoing economic and cultural change is a concern to researchers and policy makers because of the growing interest in how economic transitions impact traditional subsistence strategies and nutritional regimes. Changes in the way a population procures food can affect the welfare of individuals and households. While unique ecological and geographic circumstances strongly influence the composition of traditional diets, the changes associated with a shift away from traditional foods toward greater reliance on market foods are strikingly similar across populations (Kuhnlein et al. 2009).

Traditional diets are defined here as diets made up of plant and animal foods harvested from the local, natural environment (Kuhnlein and Receveur 1996). Scholars have argued that changes in human diet have accompanied changes in economic modes of subsistence over time (Cohen 1989). As groups shift from hunting and gathering to small-scale horticulture and onto commodity agriculture there are changes associated to their diet at each stage (Armelagos 1990; Larsen 1995).

Recent shifts away from traditional foods have been attributed to a variety of factors including environmental degradation, land insecurity, market expansion, forced resettlement (Wilkie 1989; Frantkin et al. 1999), ecological constraints (Carneiro 1974; Shell-Duncan and Obiero 2000), land alienation, migration, urbanization, modernization (Ohtsuka 1993; Latham 1997), willing participation (Henrich 1997), and colonization schemes (Ventura Santos and Coimbra 1996).

While scholars agree that changes in human diet accompany changes in economic modes of subsistence, they debate on the effects on human diet that accompany the shift from subsistence to market economy. Some researchers (Holmes 1985; Dennett and Cornell 1988) argue that increased acculturation and integration into the market economy have the potential benefit of greater access to income and thus to purchased foods. For those scholars increased integration into the market economy might have positive effects on indigenous people's diet. Other researchers have argued that the existence of market-traded food sources does not guarantee access to these items, particularly for populations originating in exchange-based economies. Furthermore, the dependency on market-derived foods that may have a lower nutritional value than local or home-produced items can be detrimental to indigenous people's nutrition (Dewey 1985; Follér 1995). Still other researchers acknowledge that the effects of markets in diet are not uniform across a given population, being dependent on a large number of variables (Dennett and Cornell 1988; Leonard and Thomas 1988).

Dietary transition in the Amazon associated with integration into the market economy has not been extensively studied, with some recent exceptions (Kuhnlein et al. 2009; Zycherman 2013). Previous research on the topic focuses on assessing nutritional status and dietary patterns (Johnson and Behrens 1982; Hill et al. 1984; Hurtado and Hill 1990; Dufour 1992), and on the reconstruction of the hunter-gatherer diet (Kaplan et al. 2000; cf. Cordain et al. 2002; for summary of findings) but does not analyze the socioeconomic factors that fuel dietary change. The topic is important because diet is directly linked to the welfare of indigenous people and to their management of the environment. If indigenous people's ability to produce or procure food is compromised, there will be health consequences for individuals in the group. The ecological balance of societies that acquire a large proportion of their food from the immediate environment might therefore be threatened when the population changes its mode of subsistece. Furthermore, changes in traditional resource management are to be expected as societies shift their dependency from the immediate environment to the market for food procurement.

In this chapter, we look at how differences in household measures of cultural and economic change relate to traditional diets. For the empirical analysis we use information from the two communities of Tsimane', a forager-horticuluralist population of lowland Bolivia in the Amazon.

Integration into the Market Economy and Diet: A Review of the Literature

In the literature on markets and the changes in diets of indigenous peoples, researchers have advocated two contrasting perspectives. Some researchers argue that integration into a market economy might improve diets through access to cash income and schooling because income and schooling allow to smooth consumption, e.g., they help to ensure the proper balance in the consumption of basic goods time periods (Holmes 1985; Kennedy 1994; Haddad et al. 2002). Markets can also function as a safety net for fluctuating dietary intake during seasonal shortfalls in resource availability. Increases in household income can augment individual energy intake through improved access to food (Kennedy 1994).

Other researchers maintain that market economies might be detrimental to the diet of indigenous populations because market expansion is often accompanied by dietary alteration (Leonard 1989; Leatherman et al. 1995). This line of research assumes that traditional societies have adapted to their environments in ways that maintain good nutrition (Wirsing 1985; McElroy and Townsend 1996; Kuhnlein et al. 2009).

A possible explanation for the lack of evidence of the impacts of dietary change on the health of indigenous peoples is that integration into the market economy affects in different ways the source, diversity, nutritional content, availability, and social value of food resources. In the rest of the section, we discuss each of these aspects of diet and how they might be affected by transition from traditional food systems to a market-based economy.

Source of dietary resources

In many regions of the world, a shift from local subsistence production to commercial acquisition of food has occurred, in a process termed "the Nutrition Transition" (Kuhnlein et al. 2006). As households increase their participation in the market economy and thereby their access to cash income, they also increase consumption of purchased foods. Changes in economic activities and time allocation affect the sources from which food can be acquired. Competing demands for labor in the market place can result in decreased time allocated to subsistence activities or to the processing of food. For example, time allocated to wage labour pulls the time of forager-horticultural households away from subsistence production of food staples or from hunting and collecting forest resources (Behrens 1992; Putsche 2000; Zycherman 2013).

As home production or wild food procurement declines, the market becomes the alternate source of food. For example, research on Amerindian households intensifying cash cropping activities reveals that as consumption of wild meat declines domesticated meat sources appear as substitutes (Behrens 1986). In a study on the Tsimane' on the influence of meat prices on the consumption of hunted game, Apaza and colleagues (2002) found a strong positive association between the market price of beef and the consumption of game and fish. But market food resources may not be economically accessible to all households. For example, in a recent study among the same population suggest that recent significant decreases in forest resources and an increase in access to market commodities and cash accruing activities have impacted the frequency in which the Tsimane' engage with the market (Zycherman 2013). Additionally, the prices of basic commodities, as wheat, corn and sugar have recently increased in

Bolivia, as the country tries to decrease its national subsidies. As a result the Tsimane' are working more frequently outside of traditional occupations and are acquiring more cash to afford commodities that have become commonplace necessities (Zycherman 2013).

Diversity of foods

Dietary diversity, or the number and variety of different foods consumed in the diet, has long been thought to be a key element to balancing traditional diets (Kuhnlein and Receveur 1996). Consumption of a wide variety of different kinds of foods leads to better nutrition than a monotonous diet (Messer 1989). In general, greater diversity of food is related to wider exposure to micronutrients, which have a positive effect on health (Wirsing 1985). An increasing proportion of dietary intake from non-traditional origins can affect the diversity of foods consumed by households. Integration into the market may lead to a shift away from the use of a wide variety of traditional foods while only a limited number of market foods are available as supplements or substitutes for traditional diet.

When accessing purchased food, households generally exhibit a shift to relatively expensive food items that do not necessarily entail a net increase in diversity or quantity (Kennedy 1994). Furthermore, an increase in diversity from market-foods does not necessarily translate into a better diet. For example, in Peru Leatherman and colleagues (1995) found that while dietary diversity increased with access to income from the market, the overall nutritional value of the diet decreased. Income from the market was used to replace home produced foods of high nutritional value with higher calorie market substitutes lower in nutritional content. The research also showed that the effects of integration to the market on diversity of food may not be equal across socioeconomic levels. Poorer families had lower diversity and fewer high quality market foods than richer families, especially during the lean season (Leonard and Thomas 1988; Leonard 1989). It is worth noting that those changes are not always evident for the population itself. Thus a recent study in the Bolivian Amazon suggest a sharp contrast between people's reported idealized diet as an overflowing cornucopia of fish, wild animal meats, fruits and vegetables, rich in diversity and forest resources, and the typical daily, which in fact is more restricted and repetitive, primarily relying on the staple crops, dry meat purchased in town, seasonal fish (Figure 1), and fruits and vegetable supplements grown in the agricultural fields (Zycherman 2013).

Beyond diversity, the specific types of food in the diet need to be examined to determine whether overall quality of the diet declines with integration into the market economy. Even if a high diversity of market foods is accessible, market foods incorporated into diet may have collectively lower nutritional value than the traditional foods they replace.

Nutritional Content of Diet

Modern tropical foragers illustrate the dietary change accompanying changes from hunting-gathering to agriculture. The

Figure 1: Tsimane' fishing with bow and arrow after enclosing the river (Barbasco). Photo: Authors

diet of foraging populations is typically high in protein, fiber, and carbohydrate of vegetable origin while low in sugar and saturated fats (Dufour 1992; Cordain et al. 2000; Cordain et al. 2002). As foraging groups shift toward the cultivation of food and domestication of animals, qualitative and quantitative changes occur in the macronutrient content of their diet. For example, tropical farming diets include crops high in bulk, but low in nutrients. Many starchy tropical crops such as manioc, sweet potatoes, and plantains are effective sources of energy (calories) because of the high proportion of carbohydrate, but poor sources of protein. In a review of literature on diet and nutrition in the Amazon, Dufour (1994) found the diet of native Amazonians to be high in bulk, with 76% of energy derived from manioc or plantains. Animal sources accounted for the majority of dietary protein (68%). Greater reliance on farm cereals and domesticated animals can have detrimental effects because greater cereal consumption has been shown to result in lower absorption of key nutrients such as iron than in the hunter-gatherer diet (Cohen 1989) and because domesticated animals have a higher fat: protein ratio than wild game (Figure 2). The shift to a diet with lower protein content is particularly significant to the dietary intake of small children who risk filling up in volume before they meet their protein requirements (McElroy and Townsend 1996). Micronutrients are important too: Kuhnlein et al. (2009) have

Figure 2: Tsimane' women preparing fresh meat. Photo: Authors

found that children are vulnerable to vitamin deficiency – e.g. in vitamin C and folates when they shift away from traditional diets.

Similar patterns of change have been documented among acculturating forager-horticulturalists groups. Calorie-rich industrial products available only in the market (i.e., refined sugars, pasta, crackers, carbonated drinks, canned sardines, or candies) have been introduced and incorporated into the diet (Ventura Santos and Coimbra 1996). What remains ambiguous is whether these changes have any significant impact on the nutritional quality of the diet. Understanding whether market substitutes are detrimental or beneficial to consumers depends on examining the nutritional adequacy of substitutes (Messer 1989).

Market income may also be insufficient to adequately substitute traditional foods with nutritionally valuable market foods. If this is the case, then the base of dietary consumption may still hinge on management of farm and forest resources with market foods contributing less to diet. There is some evidence that households are best off when they maintain some home production in addition to accessing market foods (Messer 1989).

Seasonal availability of food resources

While most regions of the Amazon have distinct wet and dry seasons affecting the availability of edible plant and faunal resources (Headland and Bailey 1991), little attention has been devoted to seasonality in Amazonian diets (Hill et al. 1984; Hurtado and Hill 1990; Dufour 1992). Seasonality matters because it exposes populations to cyclical vulnerability rather than to constant food stress. Researchers have examined seasonal reduction in dietary intake as a predictable, annual process (de Garine and Koppert 1990; Hurtado and Hill 1990). For example, we can look at seasonality and the prevalence of vulnerable periods of protein consumption across time. Seasonal hunger is often evident through the consumption of less desirable foods and through the rise of complaints of not eating highly valued foods such as meat (de Garine and Koppert 1990; Pasquet et al. 1993).

In the Amazon region, seasonal availability of staples such as manioc and plantains does not vary much, but other staples such as rice and maize are harvested according to the annual cycle resulting in periods of abundance following the harvest and scarcity in the preharvest season (Vadez et al. 2004). Seasonal availability of fish and game may result in fluctuations in available sources of protein.

Comparisons of seasonal differences in foraging versus farm diets have shown differences in availability of resources. Hill and colleagues (1984) found seasonal variability in content of diet but not in overall caloric intake among the Ache of Paraguay. They also found little seasonal variation in meat consumption, the most important food resource for the Ache. Rather than a lean period, they found a "fat" period, in the beginning of the warm/wet season, when seasonal availability of honey boosted caloric intake.

Agriculture is one proposed means of shoring up seasonal deficits of wild foods, although crops are also subject to loss. In a study in Brazil, Flowers (1983) explored whether agriculture provided a more stable diet than hunting-foraging. She found no evidence that reliance on agriculture is associated with a more stable food supply; rather she observed more seasonal vulnerability in crop-based diets. So the empirical question remains of whether the transition to market-based activities provides a more stable and more complete diet than foraging of wild resources and farm production. In theory, access to markets should provide opportunities for a more predictable and stable source of food capable of withstanding seasonal fluctuations and nutritional vulnerability (Holmes 1985; Dennett and Cornell 1988). In practice, however, differences in socioeconomic status and access to non-local foods at the household-level mediate how seasonal shortage is weathered (Leonard 1989). Furthermore, expanded opportunities to earn income through the market can potentially shore up seasonal dietary shortfalls, but only if cash income is spent on food. Last, any beneficial effect of greater income from market activities may depend on the "regularity, size, and distribution of this income" (Dennett and Cornell 1988, p. 278). Income from sale of agricultural products and market activities is often seasonal and thereby a constraint to year-round purchasing of food resources. Drops in seasonal demand for crafts, agricultural goods, and labor can put households with no access to home food production at risk (Messer 1989). Seasonality in income may render cash income inadequate to replace the home-produced diet with purchased foods (Dewey 1985).

Values and preferences for foods

Research has demonstrated that acculturation leads to the decline in indigenous systems of knowledge and beliefs (Aunger 1996; Kuhnlein and Receveur 1996; Ventura Santos and Coimbra 1996). Awareness of food preferences and values of the dominant culture is enhanced through contact, exchanges, and interactions. The adoption of new values and food preferences can pull members of the acculturating group into the market economy as consumers of market-valued items and away from traditional foods considered primitive by members of the dominant culture (Eder 1988). This can be problematic if high prestige foods have lower nutritional value than traditionally consumed items.

In the Peruvian Amazon, Berlin (1985) noted that dietary practices are modified via competition for resources with other populations and cultural change. Shifts in food taboos or avoidances can redefine what constitutes food resources (Berlin 1985; Aunger 1996). These changes may be evident in younger generations that turn away from traditional knowledge, use, and preparation of foods (Receveur et al. 1997).

Research Methodology

Data for this study forms part of a long-term research project (1999-2010) to measure the effect of a market based-economy on the quality of life of indigenous peoples (Tsimane' Amazonian Panel Study, http://people.brandeis.edu/~rgodoy/). Data for this chapter were collected by two anthropologists, two biologists, and one agronomist who conducted continuous fieldwork during 18 months (May 1999–November 2000) in two Tsimane' villages. Data included baseline information and five repeated measures from the same informants on market integration and dietary intake. The reader can find a more recent ethnographical work on Tsimane' diet in the work of Zycherman (2013).

The study protocol was approved by the University of Florida Institutional Review Board for research involving human subjects. The Tsimane' Grand Council also approved the study and individual consent of participants was obtained before enrollment.

Site selection and sampling

The Tsimane' are a forager-horticulturalist society numbering ~12,000 people and living in the rainforests and savannahs at the foothills of the Andes, mostly in the Department of Beni, Bolivia. Relatively isolated until the mid-twentieth century, they started to engage in more frequent and prolonged contact with Westerners after the arrival of Protestant missionaries in the late 1940s and early 1950s (Daillant 2003). In the past 25 years, the opening of a new road and the boom in the exploitation of precious wood species have provided options for the Tsimane' to enter the market economy (Reyes-García et al. 2012).

The Tsimane' provide an excellent case to study the effects of the market on dietary change because throughout the last century they have been undergoing patterns of dietary alteration that resemble those of other Amazonian groups. The availability of market-derived food resources has influenced Tsimane' traditional diet and shaped the values the Tsimane' place on certain foods (Zycherman 2013). Furthermore, as with other Amazonian groups, integration to market has provided the Tsimane' incentives to sell protein rich foods (i.e., game, rice) and to access supplementary food resources in a region characterized by seasonal fluctuations in availability of wild and cultivated food resources.

For the present study, we selected two villages located along the Maniqui River in the Beni department (Bolivia) but that varied in distance from the closest town, San Borja (population 19,000). The first village, Yaranda, is a three-day canoe trip from San Borja. The second village, San Antonio, is a three-hour walk from the same town. The two villages represent different levels of economic integration and socio-cultural acculturation. We surveyed all the adults in the two villages ($n=108$) every three months over a fifteen-month period for a total of five observations per person. We collected repeated measures because dietary intake and cash income are directly affected by seasonal consumption patterns and resource availability.

Methods

We collected data using quantitative and qualitative methods. Qualitative data collection methods were used to ensure the appropriateness of quantitative methods. Ethnographic information informed the design of formal surveys and helped to interpret results.

Participant observation. The 18 months of research gave us opportunities to participate in the regular activities of the villages. We accompanied the Tsimane' in their festive and working activities such as drinking home-made fermented beer, chatting, going to the field, weaving, or fishing. Participant observation allowed us to see different patterns of food consumption.

Specimen identification. We collected voucher specimens for all wild edible plants. We deposited voucher specimens at the Herbario Nacional de Bolivia, Universidad Mayor de San Andrés, La Paz. We collected specimens of fish and deposited them in the Collección Boliviana de Fauna for identification. A key informant identified specimens in the local vernacular, and taxonomists provided the scientific nomenclature. We have described much of the Tsimane' ethnobiological knowledge in a bilingual book in Tsimane' and Spanish (Nate et al. 2000), thesis and dissertations (Huanca 1999; Apaza 2001; Perez 2001; Reyes-García 2001) and previous articles (Apaza et al. 2002; Reyes-Garcia et al. 2006).

Socio-demographic census. During the first three months, we collected baseline demographic information on all individuals in the sample. Demographic information included names of household members, ages, birth dates, and education (school achievement, literacy, and numeracy).

Dietary information. We conducted twenty-four hour dietary recalls quarterly with the household member primarily responsible for food preparation, most often the female head assisted by family members present at the time of interview (Cassidy 1994). We asked informants to recall all the foods prepared and/or consumed by members of the household in the twenty-four hours prior to the time of the interview. Portion size was aided by the discussion of common containers used by the Tsimane' such as gourds or drinking cups. The specific foodstuffs and modes of preparation were recorded along with the state (e.g., raw, cooked, ripe, or unripe) and portion size to the possible accuracy allowed. We asked whether the items consumed were market, farm, or forest derived foods. Our diet data are taken as a representation of any given day in that quarter (Cassidy 1994).

Integration into the market economy. To assess the level of integration of a household in the market economy, we asked all the individuals in the sample about their income earnings. We measured wage earnings, sale of goods, and the value of goods obtained in barter by asking participants about their earnings from those activities for the two weeks before the day of the interview. The total value of individual income was obtained by adding the different sources of income (labor, sale, and barter). We calculated household income by adding the total income of all adults living in a household.

Hypotheses

The purpose of this work is to estimate whether source, diversity, and seasonality of Tsimane' diet change as households become more articulated with the market economy. Our hypotheses are as follows:

Hypothesis 1: More market-integrated households will have higher proportions of market foods in their diet and rely less on traditional food resources.

Hypothesis 2: More market-integrated households will have greater diversity in their diet as measured by the number of different foods consumed.

Hypothesis 3: More market-integrated households will have greater access to market foods during lean periods as a means of shoring up seasonal fluctuations in availability.

Tsimane' food production and food consumption

Ethnographic description of the Tsimane' can be found in previous work (Reyes-García 2001; Byron 2003; Huanca 2008). So here we focus on 1) describing the Tsimane' traditional mode of food production, 2) examining the relative importance of produced and purchased foods in household consumption, and 3) presenting the nutritional and symbolic value of traditional foods versus market substitutes.

Tsimane' food production

Like most other tropical populations (Shell-Duncan and Obiero 2000), the Tsimane' do not acquire their subsistence exclusively from hunting and gathering wild resources, but rather through a multidimensional strategy of integrating wild, cultivated, and more recently, purchased foods.

Horticulture

Traditional Tsimane' crop production includes cultivation of manioc, maize, plantains, peanuts, cotton, sweet potatoes, pineapple, and peach palm for home consumption (Piland 1991). The Tsimane' grow sweet manioc varieties mainly for consumption as fermented porridge or as processed fine flour. Rice, introduced in the region by Jesuits, is one of the most important crops today for both home consumption and market sale. In a previous study (Vadez et al. 2004), we found that rice was present in 85% of new fields. Another introduced crop, maize, was found in 57% of new fields, and two indigenous crops to Amazonia, plantain and manioc, were found in 33%, and 18% of new fields.

As other Amazonian groups, the Tsimane' practice shifting cultivation. Each year during the dry season (May-August), the Tsimane' open agricultural plots averaging half a hectare (Reyes-García et al. 2008). Once trees in the plot are felled, the field is left to dry out, and later burned over (August to October). After burning, fields are partially cleared, planted, (Figure 3) and left to be rain irrigated. The fields may be weeded once or twice before the harvest season. After one or two seasons of cultivation, a plot is left fallow and often fruit trees are harvested intermittently (Huanca 1999).

Figure 3: Old Tsimane' woman sowing rice with stick.
Photo: Authors

House-gardens

The Tsimane' grow a variety of horticultural crops including onions, squash, watermelons, peanuts, and garlic, as well as medicinal and ornamental plants in small plots close to the residence or in fallows (Piland 1991, Huanca 1999). In a study on Tsimane' agriculture Piland (1991) detailed the species most frequently raised in house-gardens. Those species include fruit trees, especially citrus, mangos, pacay (*Inga* sp.), coquino (*Ardisia* sp.), avocado, and peach palm. Papaya grows rampantly by natural dispersion of seeds and does not require human assisted planting. Tobacco is grown for both recreational and medicinal purposes. Some households also plant sugar cane and cacao trees in their fields. The Tsimane' also plant – or protect when they grow spontaneously – medicinal plants and other useful species such as those that provide dyes, fibers, and fish poisons (Reyes-García et al. 2006).

Gathering of forest plants

As other Amerindians, the Tsimane' have a variety of documented uses of plants gathered from the tropical forests. The Tsimane' use an array of plant species found in the surrounding forests for firewood, fish poisons, housing construction, tool manufacture, craft production, jewelry, medicines, and human consumption (Huanca 1999; Nate et al. 2000; Reyes-García 2001; Reyes-García et al. 2006).

In a previous study (Reyes-García et al. 2006) we documented that the Tsimane' rely on more than 50 wild and semi-domesticated plant species as food (Table 1). From all the wild plant species reported as edible, we observed at least 38 different species being brought home for eating during this research. Gathering of edible plants was particularly important between October and December when most tree species are fruiting.

Plants from the Arecaceae family make up the preferred foods of the Tsimane'. The Tsimane' eat the fruit of *Bactris gasipaes* H.B.K. and *Attalea phalerata* C. Martius ex Sprengel. Unlike other indigenous groups in the area, such as the Tacana (DeWalt et al. 1999), the Tsimane' reportedly do not eat palm hearts. The Clusiaceae family is also well represented in the Tsimane' diet. The Tsimane' eat the fruits of *Rheedia acuminata* (Ruiz and Pavon) Planch & Triana and *R. gardneriana* Miers ex. Planch & Triana. Also important are the Moraceae, from which the Tsimane' eat the fruit of *Pseudolmedia laevis* (Ruiz & Pavon) J.F. Macbr. and Passifloraceae from which they eat the fruit of *Passiflora cossinea* Aubl. and the tuber of *P. triloba* R.&P. ex DC.

Fishing

As most Tsimane' population lives along rivers, fishing is a more important activity for procuring food resources than hunting (Metraux 1948; Pérez 2001). The Tsimane' employ several fishing technologies including metal fish hooks, fishing or mosquito nets, machetes, natural poisons, and weirs. The Tsimane' traditional fishing technology is defined as the use of implements produced with local materials from the natural environment (e.g., poisons and weirs). Fish poisoning techniques provide one of the highest daily yields compared to hunting or hook and line fishing. The Tsimane' employ several techniques for

Table 1: Tsimane' wild edible plants.

Taxonomic family	Scientific name	Tsimane' name	Spanish name
Anacardiaceae	*Anacardium occidentale* L.	Cayon	Cayuj
Anacardiaceae	*Spondias mombin* L.	Moco'	Cedrillo
Annonaceae	*Annona reticulata* L.	Pise'rej	
	Duguetia spixiana Mart	Veya	Chocolatillo
	Xylopia sericea A. St.-Hil.	Tutyi'	Piraquina blanca
Arecaceae	*Acrocomia aculeata*	Catsare'	Totai
	Astrocaryum murumuru C. Martius	Shibo'	Chonta dura
	Attalea phalerata C. Martius ex Sprengel	Mana'i	Motacu
	Bactris gasipaes H.B.K.	Vä'ij	Chonta fina
	Bactris riparia Mart.	Cajna	Chonta chiquita
	Euterpe precatoria C. Martius	Mañere	Asai
	Jessenia bataua (C. Martius) Burret	Jajru'	Majo
	Mauritia flexuosa L. f.	Tyu tyura'	Palma Real
Bombacaceae	*Ochroma pyramidale* (Cav. ex Lam)	Cajñere'	Balsa
Burseraceae	*Protium aracouchini* (Aublet) Marchand	Vi'sison	Uvilla
	Tetragastris altissima (Aublet) Swart	Na'fa	Isigo
Clusiaceae	*Rheedia acuminata* (Ruiz & Pavon) Planch. & Triana	Tsocoi	Ocoro
	Rheedia brasiliensis (Mart.) (Mart.) Planch. & Triana	Corivavaj	Achachairu
	Rheedia gardneriana Miers ex. Planch &Triana	Ibijqui	Achachairu
Dioscoreaceae	*Dioscorea* sp.	Quiñu	
Elaeocarpaceae	*Muntingia calabura* L.	Bojno'ta	Uvillo
	Sloanea fragans Rusby	Copo'tare	Cabeza de mono
Fabaceae	*Hymenaea courbaril* L.	Bejqui'	Paquió
	Inga adenophylla Pittier	I'nishuj vishi'rij	
	Inga crestediona Benth ex Seeman	Cu'na'	Pacay
	Inga sp.	Shabut	Pacay
	Inga sp.	Cojnono	Pacay corto

(*Contd.*)

Taxonomic family	Scientific name	Tsimane' name	Spanish name
Flacourtiaceae	*Lunania parviflora* Spruce ex Benth	Cajnason	Blanquillo
Lauraceae	*Aniba canelilla* (H.B.K.) Mez	Chorecho	Canelon
	Persea caerulea (Ruiz & Pavon) Mez	Cuñuru	Aguacatillo
Meliaceae	*Trichilia rubra* C. DC.	Dabaj dabaj	
Moraceae	*Cecropia concolor* Willd.	Quiruru'	Ambaibo
	Poulsenia armata (Miquel) Standley	Ashaba'	Corocho
	Pourouma cecropiifolia C. Martius	Movai	Ambaibillo
	Pseudolmedia laevis (Ruiz & Pavon) J.F. Macbr.	Ijsi'ta	Nui
	Pseudolmedia macrophilla Trecul.	Pomo	Maximo
	Pseudolmedia sp.	Pimi	
Passifloraceae	*Passiflora cossinea* Aubl.	Joro'co	
	Passiflora sp.	Vadaca	Pachio grande
	Passiflora triloba R.&P. ex DC	Binca'	Pachio Pata de Anta
Rubiaceae	*Alibertia edulis* A. Rich. Ex Dc.	Dyincava'	Tutumillo
	Alibertia pilosa Krause	Shi shibuton	
Sapotaceae	*Chrysophyllum argenteum* Jacq.	Avai	
	Chrysophyllum sericeum A. DC.	Jiji	
	Pouteria macrophylla (Lam) Eyma	Vicoi'	Coquino
	Pouteria torta (Mart.) Radlk.	Cojma	
Solanaceae	*Capsicum annuum* L.	Inoj	Almendrillo
	Cuatresia aff. *forsteriana* A. Huns	Nej jebei	Soliman
Ulmaceae	*Lycianthes asarifolia* (Kunth & Bouche) Bitter	Bo'boty	
	Ampelocera edentula Kuhlm.	Tonoj-tonoj/ soj soty	Cayaya
	Ampelocera ruizii Klotzsch	Vishi Vishi	Palo yodo
	Celtis iguanaea (Jacq) Sarg	Ere' ere'	Blanquillo
	Celtis schippii Standley	Ñove'	
Ulmaceae	*Trema micrantha* (L.)	Dyotoj	Ojoso blanco
Violaceae	*Leonia racemosa* C. Martius	Rojro'	Chumiri
			Huevo de peta

preserving fish including salting and sun-drying or smoking fish wrapped in plantain leaves.

Hunting

Tsimane' men traditionally hunted with hand-made bows and arrows. Today more and more male adults and teenagers use rifles and shotguns to hunt game. While not all Tsimane' men own rifles or shotguns most are competent in their use and borrowing of firearms between relatives is common.

The Tsimane' hunt game in the open savanna, gallery forests, and islands of forest around their settlements. The Tsimane' typically hunt game within a 3-5 km radius from village settlements (Apaza et al. 2002; Luz 2012). Mammals are disproportionately the most common animals hunted. Commonly hunted game include collard peccary (*Tayassu tajacu*), whitelipped peccary (*Tayassu pecari*), agouti (*Dasyprocta* spp.), red brocket deer (*Mazama americana*), howler monkey (*Alouatta caraya*), squirrel monkey (*Saimiri sciureus*), and armadillo (*Dasypus novemcinctus*). In a study of these same two Tsimane' villages, Apaza (2002) found that birds make up only about 15% of game hunted. Meat is preserved through techniques of salting and sun drying raw pieces into *charqui* or by smoking. Hunted game is often shared between extended families.

Animal husbandry

The Tsimane' households raise chickens, pigs, ducks, and occasionally goats or cattle. The Tsimane' do not actively feed their animals nor build specialized structures or corrals; rather, they allow the animals to forage freely about the area surrounding the residential unit and plots. Domesticated animals are not regular part of the Tsimane' diet, although eggs are regularly consumed and whole animals are sometimes eaten during ritual celebrations such as birthdays. Rather, small farm animals function as a stock of wealth. Pigs and chickens are easily converted into cash or bartered for manufactured goods.

Tsimane' food preparation and consumption

Women are primarily responsible for food preparation in the household. Cooking is done directly over exposed logs and fires located in the central part of the residential patio. We observed a general pattern of eating directly from a common cooking pot in the most isolated village of our study, Yaranda, while in San Antonio, the more integrated village, most households divided food into plates or hollowed out gourds. Additionally, our data reveal greater food sharing in Yaranda as evidenced by greater participation of non-household members in the meals recalled as well as meals away from the home.

Table 2 lists the most common food items recalled from the daily household diet in order of prevalence. Despite the introduction of purchased foods and new crops over the last 500 years, we find that the bulk of items in the Tsimane' diet is made up of plantains and fish processed traditionally into a variety of dishes. The Tsimane' diet can be characterized as high in carbohydrates. The main sources of protein are fish and game. Animal lard and vegetable oil contribute to fat intake. This pattern conforms to other findings of Amazonian diets.

Table 2: Most common foods in daily diet of Tsimane' households.

Food item	Scientific name	Source
Plantains (all varieties)	*Musa* sp.	Farm
Fish (all species)		Forest
Sabalo	*Prochiludus nigricans*	
Venton	*Hoplias malabaricus*	
Pacusillo	*Schizodon fasiatum*	
Paleta	*Surubim lima*	
Salt		Market
Rice	*Oryza sativa* L.	Farm
Game meat		Forest
White-lipped peccary	*Tayassu pecari*	
Red brocket deer	*Mazama americana*	
Capuchin monkey	*Cebus apella*	
Collard peccary	*Tayassu tajacu*	
Market meat		Market
Beef		
Chicha (Manioc/Maize/Plantain)		Farm
Onion	*Allium cepa* L.	Market
Animal lard		Market
Sugar		Market
Vegetable oil		Market
Manioc	*Manihot esculenta* Crantz	Farm
White flour/bread		Market
Pasta noodles		Market
Farm animals		Farm
Chicken		

Crops

Plantain has traditionally been the most important staple in the Tsimane' diet. Plantains are available year-round and are easily roasted for quick consumption. They are also the base of a stew made from boiling shavings of unripe plantains in a cooking pot with pieces of fish or meat (Zycherman 2013). After plantains, rice is the next most common food crop in the Tsimane' diet. Unlike plantains or manioc, rice availability is determined by seasonal harvest cycles and exhibits periods of scarcity in the pre-harvest months. Also important are manioc and maize. Manioc is eaten boiled, roasted, shaved into stew, grated and preserved as *chive,* or fermented into beer. Maize was not as important in the diet on the cob or in grain form, but is consumed often as fermented beer (Figure 4).

Figure 4: Tsimane' woman grinding maize to prepare home-made beer.
Photo: Authors

Meats

Meat consumption is nutritionally and culturally important to the Tsimane' (Luz 2012). Table 3 displays the breakdown of sources of meat. At the moment of study, the Tsimane' still depended on the forest for 75% of the different meat foods in their diet, although this trend seem to have recently changed (Zycherman 2013). Sources of meat items recorded in the dietary recalls are fish (53%), game (22%), market (19%), and farm (6%).

Table 3: Type and source of meats in household diet by village.

Meat type	Source	San Antonio	Yaranda	Total
Fish	Forest	25%	80%	53%
Game	Forest	30%	14%	22%
Beef or other purchased	Market	36%	3%	19%
Chickens and pigs	Farm	9%	3%	6%
Total Meat	All	100%	100%	100%

Fish are the most important type of meat despite seasonal fluctuations in availability. Fish resources are exploited more intensively during the dry season (August-October) when low water levels concentrate fish in lagoons and rivers at greater densities. Thirty-seven different fish species appeared in the dietary recalls. Fish accounted for 80% of meat items reported in dietary recalls in Yaranda and for only 25% in San Antonio (Table 3). Residents of Yaranda practice communal fishing techniques such as fish poisoning to a much greater extent than residents downriver in San Antonio.

Hunted game is also important in the Tsimane' diet. Households reported having eaten peccaries, deer, and several species of monkey most often in the dietary recall. Beef and beef products are the most common purchased meats consumed and chickens are the principal livestock. Beef and other purchased meats account for over a third of the total types of meats recorded during dietary recalls in San Antonio, but were less frequent in Yaranda.

Domesticated and wild fruits and vegetables

Domesticated fruits and vegetables are reportedly consumed more than wild plant products. However, the low appearance of fruits and vegetables in our data contrast with our ethnographic observations and is likely due to systematic recall errors. During our time in the villages, we had numerous opportunities to observe first-hand the eating patterns and habits of children and adults. Children and women often graze on wild and domesticated seasonal fruits throughout the day, eating them away from the household during errands or visits to other families. Ethnographic observations suggest that citrus fruits and papaya are regular parts of the individual diet, yet they are not as well-recorded in our data. The absence of data on individual reports of each household member's snacking suggests that there is a degree of underreporting of these foods eaten throughout the course of the day.

The most commonly consumed vegetable reported during 24-hour recalls was onion. The most commonly consumed domesticated fruits reported were grapefruit, orange, and papaya. The most frequent wild fruits reportedly consumed are *Bactris gasipaes* H.B.K, *Pseudolmedia laevis* (Ruiz and Pavon) J.F. Macbr. and *Attalea phalerata* C. Martius ex Sprengel.

Purchased foods (non-meat)

At the moment of the study, the most common non-meat foods in the household diet originating in the market are salt, animal lard, sugar, and vegetable oil. These items are acquired through purchase in town or barter with river traders and are used in the processing of other foods. Salt is relatively cheap, while sugar and oil are more expensive items. Salt is commonly used for the preservation of game and fish resources. Our team observed that Tsimane' in the village closer to the market town, San Antonio, incorporated sugar into their diets and more often considered in a necessity compared to its value in the more distance village of Yaranda. A recent study, however, indicates a sharp increase of purchased food in Tsimane' diet, including dry cow meat, pasta, salt, flour, sugar and oil (Zycherman 2013).

In sum, at the moment of research, the descriptive analysis of the diet data does not support the prediction that the Tsimane' are shifting away from traditional sources of food to market substitutes (but see Zycherman 2013 for updates). The forest was still the most important source of dietary protein for Tsimane' households, with farm production accounting for the largest source of carbohydrate. However, the descriptive account just presented suggests that village differences were already marked.

Nutritional and symbolic value of traditional foods versus market substitutes

Nutritional value

Dietary recall did not yield accurate or specific enough data to determine nutritional content of foods consumed by individuals or households to an acceptable level of certainty. However, a general examination of common foods consumed provides insight into the nutritional content of the Tsimane' diet. Table 4 presents the nutritional content of some of the most common traditional and market foods by 100g edible portions.

The Tsimane' traditional diet consists of starches from cereals and tubers and these are generally low in protein. Plantains and manioc are good sources of carbohydrate but only minimally contribute to intake of protein or iron. The market staples of pasta noodles and white flour are comparable with rice in energy (kcal) contribution and may function as adequate substitutes. However, white flour is a poor source of iron, so bread and fried dough are not nutritional equivalent substitutes for farm starches. A household diet high in market foods would likely include sugar, animal lard or vegetable oil, white flour fried into pancakes or bread, pasta noodles, and beef. This more "market" diet includes more fats, refined sugar, and processed carbohydrates than the traditional diet.

Table 4: Nutritional contribution of common foods in Tsimane' diet, conversions for 100g edible portions.

Food	State	Water (%)	Energy (kcal)	Protein (g)	Fat (g)	Carb. (g)	Iron (mg)
Plantains[a]	ripe	65.6	122	1.0	0.3	32.3	0.8
Plantains[a]	unripe	62.6	132	1.2	0.1	35.3	0.8
Rice[b]	dry	13.1	358	7.1	0.4	78.7	7.8
Manioc[b]	raw	62.2	148	1.1	0.2	35.5	1.2
Maize[b]	raw	71.0	103	2.6	1.0	24.6	1.0
Noodles[b]	dry	13.5	355	11.2	0.9	73.8	7.4
White Flour[a]	unfortified	12.0	364	10.5	1.0	76.1	0.8
Lard (cow)[b]			840	1.7	92.4	0.0	0.3
Vegetable oil[b]	liquid	0.07	881	0.8	99.3	0.0	0.4
Peccary[c]	raw		162				
Sabalo Fish[b]	cooked	66.9	169	21.3	7.0	3.8	4.3
Beef[b]	raw	70.1	135	20.2	2.9	5.8	3.8
Beef[b]	charqui	28.6	256	47.6	3.4	5.7	15.2

a Menchú et al. 1996; b Ministerio de Previsión Social y Salud Pública, Bolivia 1984; c Hill et al. 1984

Among meats, fish and beef are comparable in energy contribution by weight, but fish contribute significantly more grams of desirable fats to the diet than beef. Nutritionally, beef is probably a suitable substitute for fish or game for the Tsimane', however, they most often consume the head, tripe, tongue, and stomach, parts that are not as rich in protein. Price may also result in lower quantities of meat consumed compared to the amount of fish or game a household can acquire. Canned sardines are another popular market food dense in nutrients and fats but expensive by volume compared to beef or forest meats.

Symbolic value

Over the course of 18-months of residence in the two villages, our team compared the observations and impressions of attitudes toward market foods. Market foods have a higher social value and demand in San Antonio than in Yaranda. For example, in San Antonio, the importance of having pasta noodles and beef during village-wide meals to celebrate holidays was unheard of in Yaranda. In San Antonio, cash quotas were gathered and a resident was sent to town to purchase foods for big events, whereas hunted game or a slaughtered pig suffice without question in the more remote Yaranda. The tastes for market items such as pasta, white flour/bread, sugar, and beef were strong in San Antonio, illustrated through mothers' claims that they "needed" sugar for their children. The change in the symbolic status of purchased food is a major topic in the work of Zycherman (2013), who argues that Tsimane' have rethought significant dishes, such as *Jo'na* and *Shogdye*, to maintain importance and incorporate available ingredients.

Market integration and source, diversity, and seasonality of the Tsimane' diet

The discussion now shifts to the testing of the three hypotheses about market integration and diet among the Tsimane'.

Source of foods

The Tsimane' diet is comprised of foods from three sources: forest, farm plots and animals, and the market. To analyze the origin of the Tsimane' household diet, we calculated the percentage of household diet by source. We totaled the number of items from each source and then divided by the total number of distinct items reported in the daily diet. The percentage of market foods compared to locally harvested resources is one indicator that differentiates households' reliance on the market (Receveur et al. 1997). We hypothesized that more market-integrated households will have higher proportions of market foods in their diet and rely less on food resources from the farm or forest.

Table 5 displays summary analysis of the average proportion of food items in the daily household diet by source. Overall, the greatest proportion of items (47%) came from farm production. Market foods represent a slightly larger proportion (29%) of items in the household diet than forest foods (23%). These aggregate values suggest that the market is equally if not more important as a source of foods than the forest.

When we group the data by village (Table 5, section A), a different pattern emerges. The daily diet of households in the more remote village of Yaranda is on average more dependent on foods from the farm (51%) and forest (32%) than on foods from the market. In San Antonio, the farm (44%) and market (40%) account for the bulk of food items in the daily diet with forest foods contributing only marginally (16 %).

To discern whether differences in source of foods are being driven more by village traits or relative location than by household factors, in Table 5 section B we present the same analysis disaggregated by average income levels for households during the study. We ran an analysis of variance (ANOVA) on income grouping for each source of food items. We found that farm foods are still the most important by percent of items in the household diet for all three income groups. We also found a clear pattern of unequal means across all three categories ($P<0.02$ for all three). We ran t-tests to compare the values at the extreme ends, between households in the upper and lower third on average monthly income. The most marked difference is the percentage of market foods between households in the upper and lower average monthly income group ($t = 5.0$; $P<0.001$). Independent of village, poorer households rely less on the

Table 5: Source of Tsimane' diet, average percentage of items in daily household diet by village and income groups.

A. Source of Tsimane' diet, average percentage of items in daily household diet by village

Source	San Antonio	Yaranda	Total	Between village	
				t-stat	P
Forest	16%	32%	23%	8.0	0.001
Farm	44%	51%	47%	2.7	0.01
Market	40%	17%	29%	-9.4	0.001
All sources	100%	100%	100%		

B. Average percent of items in daily household diet by income groups by thirds

Source	Lower	Middle	Upper	ANOVA F-stat	P>F Between upper and lower third	
					t-stat	P
Forest	26%	25%	19%	3.86	-2.56	0.01
Farm	53%	46%	43%	6.26	-3.39	0.00
Market	21%	28%	38%	12.61	5.01	0.00

market for food than richer households. The richest third of the households depend the most on the market. The significant difference by household groups suggests that income is associated with the incorporation of market foods into the household diet.

Dietary diversity

We defined dietary diversity as the number of distinct food items consumed in a single 24-hour period per household. We used a Dietary Diversity Score (DDS) to compare the variety of foods in the diet (Leonard and Thomas 1988). We hypothesized that more market-integrated households will have greater diversity in their diet.

Table 6 presents summary analysis of DDS by income groupings. Analysis of variance indicates that the average DDS values for each third are not equal ($F = 11.67$; $P<0.001$), although overall variation in DDS among Tsimane' households is small. The biggest difference exists between households in the lower and upper thirds on the average monthly income distribution with higher diversity in the upper income household group. The largest variation is found among households in the upper third ($SD = 3.2$). The analysis suggests that more integrated households have, on average, slightly more diverse diets than less integrated households.

Table 6: Dietary Diversity Score (DDS) by income thirds.

Household Grouping	Mean	SD	F-stat	Prob >F
Lower Third	6.1	2.2		
Middle Third	6.9	2.7		
Upper Third	8.3	3.2		
Total Sample	7.1	2.9	11.67	0.00

Seasonal reliance on the market

We hypothesized that more market-integrated households will have greater access to market foods during lean periods as a means of shoring up seasonal fluctuations in availability. Figure 5 shows the average proportion of foods originating in the market included in the household diets for each quarter of observations. We found that in San Antonio, the highest percent of market foods was during the wet/preharvest season (Quarter 2) and the lowest percent of market food was during the dry/cool season (Quarter 4) when hunting is at its peak and rice stores are high. The pattern suggests that the market may be a seasonal alternative to low availability of farm or forest resources for the residents of the village closer to the market town.

Integration into the Market Economy and Dietary Change

There was less seasonal variation in use of market foods in Yaranda. Residents in Yaranda may be adopting a more conservative strategy toward dependence on market items in their dietary portfolio due to their distance and greater isolation from the market town. They may be better-off investing in home production and wild resource acquisition because market access is more variable and less predictable.

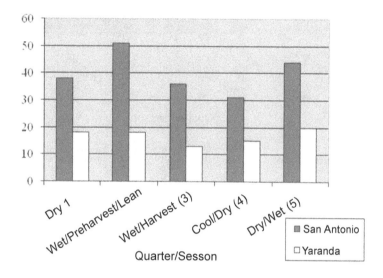

Figure 5: Percent of diet from market foods by village for each quarter of study.

Figure 5 examines seasonal use of market foods by household income groupings. We found a clear pattern of greater reliance on market foods during the wet/preharvest/lean season by households in the upper third of monthly income. Households in the middle third of average monthly income have less variability in use of market foods in the diet. A dip in use of market foods as a percent of total food items occurs during the wet/harvest and cool/dry season among households in the lower third. Households in the lower third increase their use of market foods during the lean period (Quarter 2), but do not reach the levels of households in the upper third. A drop back down to a lower percent occurs with the beginning of the rice harvest. Figure 6 suggests that during the annual lean period households with the highest average monthly income increase their reliance on market foods more than the other two income groups, implying greater access to market foods among better-off households.

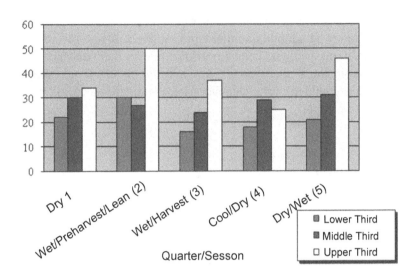

Figure 6: Percent of foods from market by income groupings in thirds.

Discussion and Conclusion

We organize the discussion around three topics that emerge from our analysis. First, our analysis of household diet at the time of research supports the classification of the Tsimane' as forager-horticulturalists, since farm resources were the most important source of foods in the Tsimane' diet. The data imply that Tsimane' households had not become fully dependent on the market as a source of food (but see recent changes reported in Zycherman 2013). Our data suggest that the more market-involved and exposed households are leaning toward shifting their dietary preferences, but even the richest Tsimane' households in the more market integrated village still depend in large part on farming for food. The Tsimane' might have a difficult time transitioning to a completely market-based household diet because sources of cash income are irregular. Thus, for the Tsimane' adequate subsistence depends on a multidimensional strategy of maintaining farm production, continuing to acquire forest resources, including fish and game, and purchasing other necessary items such as salt in the local market towns.

Despite reliance on farm and forest production for food consumption, we also found signs of shifts toward increased consumption of market foods. There may be no notable changes to traditional diet until production patterns are affected by market participation (Guyer 1989). Sale or barter of agricultural goods were demonstrated to be the most important sources of market income for the Tsimane'. These same goods are part of the subsistence production for the household. During our research it was not uncommon for Tsimane' households in San Antonio to sell off nearly all of their rice early in the season and either have to turn to alternate foods or even purchase rice themselves as their own supplies run out. The motivation to initially sell rice was often to access goods and services traded in the market town. Casual observation concludes that this demand is increasing with greater exposure to the market system and culture.

Second, we found that the distribution of secondary reliance on foods from the forest or market is largely dependent on distance and access to those foods. The analysis presented here suggests that greater integration into the market economy is associated with a higher percent of foods from the market in the household diet. In addition, households that are more integrated show signs of turning away from forest foods to incorporate more market foods in the daily diet. These features also pattern use of non-local foods during changes in seasonal availability of local food resources (Leonard 1989). A greater reliance on market foods during the lean/preharvest season noted in San Antonio suggests that market foods become important when availability of farm and forest foods decreases. We found evidence to indicate that households in the upper income third have greater access to market foods during the lean season.

We found village differences in source of meat, one of the most important and socially valued food items. Households in the village in closest proximity to the market town rely more on purchased and farm meats than households in the more distant village. These differences highlight dietary differences associated with proximity, contact, and access to resources of the market economy.

Third, our data do not support the hypothesis that income from the market is used to purchase more nutritious food. For example, we find that flour and bread are status market-substitutes for rice consumption as market access increases, but flour and bread are nutritionally poorer than rice. The analysis presented here is limited in that we are unable to compare quantities of foods eaten between all households in the sample to assess whether the amount of foods consumed is changing. But many of the market foods we casually observed during our residence in the villages can be classified as novelty foods such as sugar, cookies, processed sweets, and alcohol. A growing perceived need for market foods such as sugar and alcohol and greater desire for novel food items were more apparent in San Antonio, particularly among young males who have the greatest exposure to the market economy. Tastes seem to be influenced by contact and exposure to the larger society much more in the village closer to the market town and among the younger generation. Values and preferences are shaped and modified through exposure to and acceptance of the values of members of the dominant society.

In conclusion, indigenous peoples around the world are currently undergoing varying degrees of exposure to and integration into market economies. In many places, these people are not controlling or determining their own relationship with the market economy and are therefore in a disadvantageous position. By identifying patterns of similarity in what is happening as economic integration and social acculturation intensify, findings from this work can help policy-makers to develop programs that would mitigate any detrimental consequences of dietary changes for native cultures and livelihoods.

Acknowledgements

Research was funded by grants from the National Science Foundation (SBR-9731240 and SBR-9904318) and reviewed under the framework of the ERC-FP7 LEK project under grant agreement no261971. We thank T. Huanca, D. Ista, J. Pache, A. Nate, P. Pache, E. Tayo, M. Roca, L. Apaza, E. Pérez, V. Vadez, and Gran Consejo Tsimane' for data collection and logistical support. We thank R. Godoy, W Leonard, and M. Gueze for comments to previous versions. This chapter is based in Byron´s Doctoral Dissertation (Market Integration and Health: The Impact of Markets and Acculturation on the Self-perceived Morbidity, Diet, and Nutritional Status of Tsimane' Amerindians of Lowland Bolivia).

References

Apaza L (2001) Estudio comparativo de la caza y uso de mamíferos en dos comunidades Tsimane'. BA thesis. Universidad Mayor de San Andrés

Apaza L, Wilkie D, Byron E, Huanca T, Leonard W, Pérez E, Reyes-García V, Vadez V, Godoy R (2002) Role of meat prices in household consumption of bushmeat among the Tsimane' Amerindians of Bolivia. *Oryx,* 36(4):382-388

Armelagos G (1990) Health and disease in prehistoric populations in transition. *In:* Swedlund A and Armelagos G (eds) Disease in populations in transition: Anthropological and epidemiological perspectives, Bergin and Garvey, New York

Aunger R (1996) Acculturation and the persistence of indigenous food avoidances in the Ituri forest, Zaire. *Human Organization,* 55(2):206-218

Behrens C (1986) The cultural ecology of dietary change accompanying changing activity patterns among the Shipibo. *Human Ecology,* 14(4):367-396

Behrens C (1992) Labor specialization and the formation of markets for food in a Shipibo subsistence economy. *Human Ecology,* 20(4):435-448

Berlin E (1985) Social implications of dietary patterns in three communities of Amazonian Peru. *In:* Dorothy J and Karl Schwerin (eds) Food energy in tropical ecosystems. Gordon and Breach Science Publishers, New York

Byron E (2003) Market integration and health: The impact of markets on the nutritional status, morbidity, and diet of the Tsimane' Amerindians of Lowland Bolivia. Dissertation. University of Florida

Carneiro RL (1974) The transition from hunting to horticulture in the Amazon. *In:* Murphy R and Steward J. Man in adaptation: The cultural present. Aldeline, Chicago

Cassidy CM (1994) Walk a mile in my shoes: culturally sensitive food-habit research. *American Journal of Clinical Nutrition,* 59:S190-S197

Cohen M (1989) Health and the rise of civilization. Yale University Press, New Haven

Cordain L, Eaton SB, Miller JB, Mann N, Hill K (2002) The paradoxical nature of hunter-gatherer diets: meat-based, yet nonatherogenic. *European Journal of Clinical Nutrition,* 56:S42-S52

Cordain L, Miller JB, Eaton SB, Mann N, Holt S, Speth J (2000) Plant-animal subsistence rations and macronutrient energy estimations in worldwide hunter-gatherer diets. *American Journal of Clinical Nutrition,* 71:682-692

Daillant I (2003) Sens Dessus Dessous. Organization sociale et spatiale des Chimane d'Amazonie boliviane., Recherches Americanes. Societe d'ethnologie, Nanterre

de Garine I, Koppert S (1990) Social Adaptation to season and uncertainty in food supply. In Harrison GA and Waterlow JC (eds) Diet and Disease in Traditional and Developing Societies. Cambridge University Press, Cambridge

Dennett G and Cornell J (1988) Acculturation and health in the highlands of Papua New Guinea. *Current Anthropology,* 29(2):273-298

DeWalt S, Bourdy G, Chavez de Michel L, Quenevo C (1999) Ethnobotany of the Tacana: Quantitative inventories of two permanent plots of Northwestern Bolivia. *Economic Botany,* 53(3):237-260

Dewey KG (1985) Nutritional Consequences of the Transformation from Subsistence to Commercial agriculture in Tabasco, Mexico. *In:* Dorothy J and Karl Schwerin (eds) Food energy in tropical ecosystems. Gordon and Breach Science Publishers, New York.

Dufour D (1992) Nutritional ecology in the tropical rainforests of amazonia. *American Journal of Human Biology,* 4:197-207

Dufour D (1994) Diet and nutritional status of Amazonian Peoples. *In:* Roosevelt A (ed) Amazonian Indians from prehistory to the present. University of Arizona Press, Tucson, AZ

Eder JF (1988) Batak foraging camps today: A window to the history of hunting-gathering economy. *Human Ecology,* 16(1):35-55

Flowers N (1983) Seasonal Factors in Subsistence, Nutrition, and Child Growth in a Central Brazilian Indian Community. *In:* Hames RB and Vickers W (eds) Adaptive responses of Native Amazonians. Academic Press, New York

Follér ML (1995) Future health of indigenous peoples: A human ecology view and the case of the Amazonian Shipibo-Conibo. *Futures,* 27(9/10):1005-1023

Frantkin EM, Roth EA, Nathan MA (1999) When nomads settle: The effects of commoditization, nutritional change, and formal education on Ariaal and Rendille Pastoralists. *Current Anthropology,* 40(5):729-739

Guyer J (1989) From seasonal income to daily diet in a partially commercialized rural economy (Southern Cameroon). *In:* Sahn DE (ed) Seasonal variability in third world agriculture. Johns Hopkins University Press, Baltimore

Haddad L, Alderman H, Appleton S, Song L, Yohannes Y (2002) Reducing child undernutrition: How far does income growth take us? Food consumption and nutrition division discussion paper #137. IFPRI, Washington, DC

Headland TN, Bailey RC (1991) Have hunter-gatherers ever lived in tropical rain forest: Independently of agriculture? *Human Ecology,* 19(2):115-122

Henrich J (1997) Market incorporation, agricultural change and sustainability among the Machiguenga Indians of the Peruvian Amazon. *Human Ecology,* 25(2):319-351

Hill K, Hawkes DD, Hurtado M, Kaplan H (1984) Seasonal variance in the diet of the Ache hunter-gatherers in Eastern Paraguay. *Human Ecology,* 12(2):101-135

Holmes R (1985) Nutritional status and cultural change in Venezuela's Amazon Territory. *In:* Hemming J (ed) Change in the Amazon Basin. University of Manchester Press, Manchester

Huanca T (1999) Tsimane' Indigenous Knowledge. Swidden fallow management and conservation. Dissertation. University of Florida

Huanca T (2008) Tsimane' Oral Tradition, Landscape, and identity in tropical forest. La Paz: Imprenta Wagui

Hurtado A, Hill K (1990) Seasonality in a foraging society: Variation in diet, work effort, Fertility and Sexual division of labor among the Hiwi of Venezuela. *Journal of Anthropological Research,* 46(3):293-346

Johnson A, Behrens C (1982) Nutritional criteria in machiguenga food production decisions: A Linear-Programming analysis. *Human Ecology,* 10(2):167-189

Kaplan H, Hill K, Lancaster J, Hurtado A (2000) A Theory of life history evolution: diet, intelligence, and longevity. *Evolutionary Anthropology,* 9:156-185

Kennedy E (1994) Health and nutrition effects of commercialization of agriculture. *In:* von Braun J and Kennedy E Agricultural commercialization, economic development, and nutrition. Johns Hopkins University Press, Baltimore

Kuhnlein HV, Receveur O (1996) Dietary change and traditional food systems of indigenous peoples. *Annual Reviews of Nutrition,* 16:417-442

Kuhnlein H, Erasmus B, Spigelski D (2009) Indigenous peoples' food systems: The many dimensions of culture and environment for nutrition and health. Rome: FAO Centre for Indigenous peoples' Nutrition and Environment

Larsen CS (1995) Biological changes in human populations with agriculture. *Annual Review of Anthropology,* 24:185-213

Latham MC (1997) Human nutrition in the developing World. FAO, Roma

Leatherman TL, Carey JW, Thomas RB (1995) Socioeconomic change and patterns of growth in the Andes. *American Journal of Physical Anthropology,* 97:307-21

Leonard W (1989) Nutritional determinants of high-altitude growth in nuñoa, Peru. *American Journal of Physical Anthropology,* 80:341-352

Leonard W, Thomas RB (1988) Changing dietary patterns in the Peruvian Andes. *Ecology of Food and Nutrition,* 21:245-263

Luz AC (2012) The role of acculturation in indigenous peoples' hunting patterns and wildlife availability. The case of the Tsimane' in the Bolivian Amazon. Universitat Autònoma de Barcelona

McElroy A, Townsend PK (1996) Medical anthropology in ecological perspective. Westview Press, Boulder, CO

Menchú MT, Méndez H, Barrera MA, Ortega L (1996) Tabla de Composición de Alimentos de Centroamérica, Primera Sección. Guatamala: Oficina Panamericana de la Salud (OPS) and Instituto de Nutrición de Centro América y Panamá (INCAP)

Messer E (1989) Seasonality in food systems: An anthropological perspective on household food security. *In:* Sahn DE (ed) Seasonal Variability in Third World Agriculture. Johns Hopkins University Press, Baltimore

Metraux A (1948) The Yuracare, Mosetene and Chimane. *In:* Steward J (ed) Handbook of South American Indians. Smithsonian Institution, Bureau of American Ethnology, Washington DC

Ministerio de Previsión Social y Salud Pública, Bolivia (1984) Tabla de Composición de Alimentos Bolivianos. La Paz, Bolivia: UNICEF

Nate A, Ista D, Reyes-García V (2000) Plantas Útiles y su Aprovechamiento en la Comunidad Tsimane' de Yaranda. CIDOB-DFID, Santa Cruz

Ohtsuka R (1993) Changing Food and Nutrition of the Gidra in Lowland Papua New Guinea. *In:* Hladik CM, Hladik A, Linares OF, Pagezy H, Semple A, and Hadley M (ed) Tropical Forests, People and Food. Pantheon Publishing Group, London

Pasquet P, Fromet A, Ohtsuka R (1993) Adaptive aspects of food consumption and energy expenditure - background. In Hladik CM, Hladik A, Linares OF, Pagezy H, Semple A, and Hadley M (ed) Tropical forests, people and food. Pantheon Publishing Group, London

Pérez E (2001) Uso de la Ictiofauna entre los Tsimane'. BS. Thesis. Universidad Nacional Mayor de San Andrés

Piland R (1991) Traditional chimane agriculture and its relationship to soils of the Beni Biosphere Reserve, Bolivia. M.A. Thesis. University of Florida

Putsche L (2000) A reassessment of resource depletion, market dependency, and culture change on a Shipibo Reserve in the Peruvian Amazon. *Human Ecology,* 28(1):131

Receveur O, Boulay M, Kuhnlein HV (1997) Decreasing Traditional Food Use Affects Diet Quality for Adult Dene/Métis in 16 Communities of the Canadian Northwest Territories. *Journal of Nutrition,* 127(1):2179-2186

Reyes-García V (2001) Indigenous people, ethnobotanical knowledge, and market economy. A Case study of the Tsimane' Amerindians, Bolivia. Dissertation. University of Florida

Reyes-García V, Huanca T, Vadez V, Leonard W, Wilkie D (2006) Cultural, practical, and economic value of wild plants: A quantitative study in the Bolivian Amazon. *Economic Botany,* 60(1):62-74

Reyes-García V, V Vadez, N Martí, T Huanca, WR Leonard, T McDade, S Tanner (2008) Local ecological knowledge correlates with cultivar diversity. Evidence from a native Amazonian society. *Human Ecology,* 36(4):569-580

Reyes-García V, Ledezma JC, Paneque-Galvez J, Orta-Martínez M, Gueze M, Lobo A, Guinard D, Huanca T, Luz AC, TAPS Bolivia Study Team (2012) Presence and purpose of non-indigenous peoples on indigenous lands. A descriptive account from the Bolivian Lowlands. *Society and Natural Resources,* 25(3): 270-284

Shell-Duncan B, Obungu-Obiero W (2000) Child nutrition in the transition from Nomadic pastoralism to settled lifestyles: Individual, household, and community-level factors. *American Journal of Physical Anthropology,* 113:183-200

Vadez V, Reyes-García V, Apaza L, Byron E, Huanca T, Leonard W, Pérez E, Wilkie D (2004) Does integration to the market threaten agricultural diversity? Panel and cross-sectional evidence from a horticultural-foraging society in the Bolivian Amazon. *Human Ecology,* 32(5):635-646

Ventura Santos R, CEA Coimbra (1996) Socioeconomic differentiation and body morphology in the Surui of Southwestern Amazonia. *Current Anthropology,* 37(5):851-856

Wilkie D (1989) Impact of roadside agriculture on subsistence hunting in the Ituri forest of Northeastern Zaire. *American Journal of Physical Anthropology,* 78:485-494

Wirsing R (1985) The health of traditional societies and the effects of acculturation. *Current Anthropology,* 26(3):303-321

Zycherman A (2013) Shocdye as World: Localizing modernity among the Tsimané Indians of the Bolivian Amazon. Columbia University

Chapter – 3

The Loss of Local Livelihoods and Local Knowledge: Implications for Local Food Systems

Sarah Pilgrim-Morrison* and Jules Pretty

Abstract

For 99% of our time on Earth, humans have been hunter-gatherers, intricately connected to and reliant upon our local lands and waters for food, materials, medicines and shelter. In this time we have evolved a unique and in-depth knowledge base, termed local ecological knowledge or ecoliteracy, which has been essential to sustaining human and ecosystem health over thousands of years. However with livelihood diversification towards non-resource dependent strategies, the emergence of local markets as a consequence of globalisation, and global patterns of economic development, our collective local knowledge is now being lost. With the loss of local livelihoods and knowledge comes a departure from traditional food systems as hunters, fishers, and gatherers and cultivators lose the skills needed to locate, collect, preserve, prepare, consume and manage indigenous foods. Traditional food systems have provided nutritional health and food security for indigenous and marginalised communities for generations, and play a critical role in cultural continuity, social systems and ecosystem biodiversity. The current phenomenon of local knowledge erosion with economic development, and its effect

---
*sepilg@gmail.com; jpretty@essex.ac.uk

on local food systems, is discussed. There has been a global shift from traditional and local foods to industrialised marketed foods, most notable in indigenous and marginalised communities. This shift has consequences for human health (in terms of nutrition and physical health), cultural health (in terms of community identity, ceremonies and social networks) and ecosystem health (in terms of resource management and biodiversity) in both industrialised and developing countries across the world.

Keywords: Culture, diversification, economic development, environment, food system, health, India, Indonesia, livelihood, local knowledge

Introduction

> *"But lo! men have become the tools of their tools. The man who independently plucked the fruits when he was hungry is become a farmer; and he who stood under a tree for shelter, a housekeeper. We now no longer camp as for a night, but have settled down on earth and forgotten heaven"* Thoreau 1847.

For 99% of our time on Earth, humans have been hunters, fishers and harvesters of wild plants, intricately connected to and reliant upon the land and waters for food, materials, medicines, and shelter (Lee and DeVore 1968). In this time we have evolved a unique and in-depth knowledge base, often termed traditional ecological knowledge, local knowledge, or ecoliteracy (Sillitoe 1998; Blench 1999; Olsson and Folke 2001; Davis and Wagner 2003; Pilgrim 2007a; Berkes 2008; Bharucha and Pretty 2010; Pilgrim and Pretty 2010; Pretty 2011). We have also, over time, developed methods of enhancing and improving the quality and productivity of our foods, ultimately domesticating crops and livestock and in some cases entire landscapes. Local knowledge incorporates information on ecosystem dynamics, identification of local species and their uses, and environmental goods and services still invaluable to livelihoods and subsistence strategies in many parts of the world. Unlike modern academic knowledge, local knowledge is rarely written down and instead is acquired through experience and transferred orally between generations, enabling it to remain dynamic and current, despite often being termed 'traditional' (Berkes et al. 1993; Gadgil et al. 2000; Pretty 2007).

As a consequence this knowledge base has been responsible for sustaining the vast majority of the world's biocultural diversity through to the present day (Gadgil et al. 1993; Gilchrist et al. 2005). It has also been recognised as providing innovative strategies for natural resources management, for example at the 1992 UN Conference on Environment and Development (Veitayaki 1997). This chapter describes the shift in resource dependence of low-income households globally from subsistence to income-driven economies as local resources are commodified and develop a monetary value. This, in turn, is creating a departure from local food systems and a consequent loss in associated local knowledge and practice, which is creating repercussions for ecological, social and cultural systems (Pilgrim and Pretty 2010; Pretty 2013).

Livelihood diversification and the effects on local knowledge

Globalisation and marketisation are now creating a general shift in livelihoods away from local resource dependence and towards externally-derived income sources. This can be seen even in developing regions where the majority of the world's biological and cultural diversity is situated. This process is termed livelihood diversification (ODI 2003). In marginalised communities at the early stages of economic development, the current trend towards diversification is well described (Ellis 1999; Béné 2000; Twyman 2000; ODI 2003). Despite the fact that over half of the world's human population still depends directly on natural resources in their daily lives (IUCN 2002), the impacts of livelihood diversification and emerging market pressures on local knowledge and local food systems have received relatively limited attention to date (Kuhnlein et al. 2006; 2009).

A livelihood has been defined as *"the capabilities, assets (stores, resources, claims and access) and activities required for a means of living"* (Chambers and Conway 1992). Therefore livelihood diversification means to expand what people do for a living, for instance by adding new activities, and is usually seen at the early stages of a group's economic development. Livelihood diversification is deemed by many scholars and indigenous rights experts to be a positive occurrence, a means by which marginalised groups and households in the early stages of modernisation can increase their independence, economic security, and life chances (Figure 1) (Ellis 1999; ODI 2003; Pilgrim et al. 2007b). However here we consider the potentially detrimental effects of non-resource dependent diversification on ecological knowledge, local food systems and subsequently human and ecosystem health.

Livelihood diversification has become the primary survival strategy in low-income areas of developing countries, particularly where market pressures are emerging, crop prices are falling, and economies are shifting from subsistence to income-driven (La Rovere et al. 2006). A household's capacity to diversify income is based on five key assets comprising the livelihood strategy framework: natural capital (e.g. land, water, fish stocks and other natural resources), physical capital (e.g. infrastructure, tools), human capital (e.g. skills, knowledge), financial capital (e.g. credit, savings) and social capital (e.g. networks, trust, reciprocity) (Scoones 1998; DFID 1999). With a view to improving their quality of life and/or accumulating assets, low-income households are more frequently adopting economic activities that generate a higher return than their existing activities. Diversification in South Asia, for instance, has led to around 60% of rural households' income being derived from non-farm sources as opposed to traditional farm-based sources (Ellis 1999).

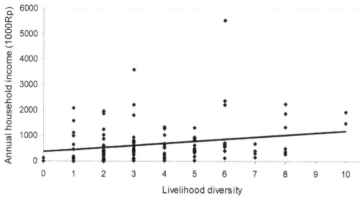

Figure 1: Relationship between mean household income and livelihood diversity (measured by number of income sources) of households in the Kaledupa sub-district of Indonesia (n=144) (Pilgrim et al. 2007b).

Livelihood diversification can occur in one of two forms: positive and negative diversification. Positive diversification takes place when households or individuals choose to exploit new activities as a means of economic advancement and security. Negative diversification, on the other hand, occurs in response to social and ecological stresses, for instance where work is seasonal or crop yields fail (Start 2001; Smith et al. 2005; La Rovere et al. 2006). It is not uncommon for households to combine high risk, high-income activities with low risk, low-income strategies (ODI 2003). By diversifying its livelihood strategies, a household increases its flexibility allowing substitution between different activities at different times of the year, provided that the chosen activities do not rely on the same resources or markets. This makes household income, and subsequent consumption levels, more resilient to externally imposed change and more predictable season to season (Ellis 1999; Start 2001; ODI 2003; FAO 2004; Smith et al. 2005).

Livelihood diversification has been the focus of much empirical and conceptual research for the economic security and development opportunities that it provides, particularly to marginalised communities (Twyman 2000). However the effects of diversification on natural resources and resource knowledge systems have not been widely discussed (Ellis 1999). Since this diversification is now a recognised global trend, it is essential to understand the impact of livelihood diversification on local knowledge bases and traditional food systems, and how this in turn will affect human and ecosystem health in the future.

According to the FAO (2004), livelihood diversification can comprise one of three strategies: (i) expanding the direct utilisation of land or natural resources (e.g. growing new crops or adopting new fishing techniques); (ii) expanding the indirect use of land or natural resources (e.g. trading or processing natural resources); and (iii) adopting externally-derived income sources (e.g. labour or the civil service). This non-natural resource diversification is the most common form of diversification. Today, it is estimated that as much as half of rural households' income in developing countries derives from non-farm activities and/or transfers from urban areas or abroad as a consequence of type (iii), non-natural resource diversification. However despite its provision of increased financial security, relying upon an increased number of externally-derived income sources has also been shown to reduce local resource dependence (see Figure 2) and consequently damage local knowledge bases (see Figure 3) (Pilgrim et al. 2007b).

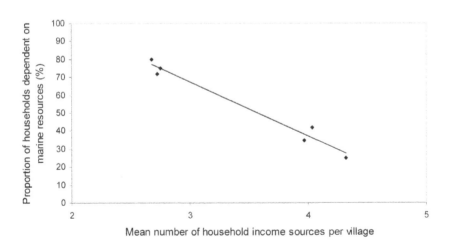

Figure 2: Relationship between the proportion of households dependent on marine resources and the mean number of household income sources across 6 communities in the Kaledupa sub-district of Indonesia (n=144 across 6 villages) (Pilgrim et al. 2007b).

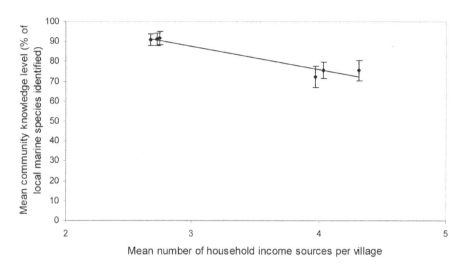

Figure 3: Relationship between mean community knowledge level and the mean number of household income sources across 6 communities in the Kaledupa sub-district of Indonesia (±2SE, n=192 across 6 villages) (Pilgrim et al. 2007b).

Using wild game, fish and plant products for food, like any other natural resource-based livelihoods, centres on a complex relationship between humans and their local ecosystems (Pretty 2007; 2011). Activities like hunting and fishing require an in-depth understanding of environmentally-regulating factors that affect species diversity, abundance, and habitats (Twyman 2000). With over half of the global population still dependant upon local natural resources in their food systems, all forms of local knowledge is still heavily relied upon today (IUCN 2002). However the current trend of non-natural resource diversification in pursuit of economic development is altering peoples' relationships with and connections to the land and local waters. This in turn is leading to reduced oral transfer of knowledge and less time spent experiencing nature causing local knowledge bases to decline in marginalised communities globally (Pyle 2001; Pilgrim et al. 2007a).

Thus wealthier households are those that depend upon a higher number of livelihood activities for income and have, therefore, often lost a portion of their local knowledge as a consequence of a departure from local resource dependence. Since all humans aspire to a better quality of living, the positive correlation between livelihood diversity and income level is likely to lead other low-income households to diversify their externally-derived livelihood strategies in the future. However this will come at a cost to local knowledge systems, since these very households currently hold the greatest knowledge about their local environment and food systems (Pilgrim et al. 2008). These findings are contrary to earlier research that describes livelihood diversification of any form to positively affect human capital by increasing local experience and skills (Ellis 1999; ODI 2003). Instead, Pilgrim et al. (2008) demonstrate the detrimental effect of non-natural resource diversification on specialised local knowledge and the innovative resource management strategies derived from it.

The emergence of local markets and the impact on local knowledge

Livelihood diversification is one symptom of a much larger process that is the shift from a subsistence-driven to an income-driven economy. This is a process affecting developing countries worldwide as well as many rural communities in industrialised countries like Canada, United States and Australia (Samson and Pretty 2006; Pretty 2011). Local co-operatives and international markets are emerging, formal schooling is being introduced and with it come textbooks, uniforms and school fees. Western health clinics are being built that charge for doctor's appointments and prescriptions, and modern gear and equipment such as fishing nets and motorboats are becoming available even in isolated areas (May 2004). Most affected by these changes are likely to be indigenous and marginalised peoples whose livelihoods were previously independent from market prices and consumer groups. Yet with globalisation today, many of these peoples are now trading local products and natural resources of their traditional territories in exchange for money and modern imports of food and other goods from the marketplace.

Previous studies suggest that the emergence of local markets and the commoditisation of natural resources will primarily affect the consumption patterns and food knowledge of wealthy households. This can be explained by the increased purchasing power of these households giving them access to external food systems. However the knowledge of the poorest and most resource dependent people, who lack access to external markets, will remain the least affected (Gadgil et al. 2000; Byg and Balslev 2001; Anishetty 2004; Ladio and Lozada 2004). Knowledge of which wild and domesticated local plants and animals are palatable and nutritious, and how to collect, preserve and prepare them, has long been a survival strategy to those living in harsh environmental conditions frequently affected by insect blights, droughts or adverse weather conditions (Gadgil et al. 1993; Pretty 2002). In these areas, where imported foods are financially inaccessible, wild resources continue to provide food security and often constitute the bulk dietary intake (Anishetty 2004). Hence, wild nutritional resources have often been dubbed the 'hidden harvest' (Gadgil et al. 1993; Nabhan 2002; Pretty 2002). However with globalisation and commodification today, it is unlikely that even the most isolated, resource dependent households will remain unaffected by international market prices in the future.

This concept was explored by Pilgrim et al. (2007c) who looked into the traditional food knowledge of Indonesian tribal communities inhabiting the Kaledupa sub-district, off Southeast Sulawesi. Contrary to previous studies, they found that people of wealthier households held the greatest knowledge about the consumption uses of local species, whereas those of low income households, striving to compete in emerging economic markets, held the most detailed knowledge about the economic uses of local species (Figures 4 & 5). Therefore low-income households in this region of Indonesia are abandoning the cultural and the nutritional values of traditional wild and local crop foods in exchange for their financial value.

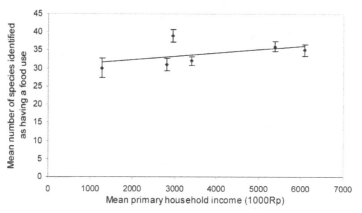

Figure 4.

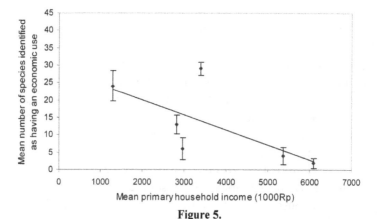

Figure 5.

Figures 4 & 5: Comparing the relationship between mean household income and mean number of food and economic uses for local species across 6 communities in the Kaledupa sub-district of Indonesia ($\pm$2SE, n=192 across 6 villages) (Pilgrim et al. 2007c).

Unlike many low-income households, high-income household members have external income sources and do not have to rely upon local natural resources for income. Therefore they have the option of consuming traditional foods if desired, since their income streams are independent of local natural resource pools, and modern market food staples such as rice and potatoes, are readily accessible to them. However, low income households who want to increase their purchasing power in order to compete in modern economic markets are being forced to sell their local resources as their only available income stream rather than consuming them, and are thus able to eat only the surplus and less palatable species.

This is one example of a more general trend, in which low-income households in developing regions faced with growing economic pressures are being forced to sell more desirable and palatable local species to wealthier households and external markets in exchange for money. In doing so, they are constricting their own dietary intakes to the less palatable, less nutritious local foods combined with small amounts of market-bought rice or other commodities. This shift in primary use from consumption to economic exploitation of local food is indicative of a desire by low-income households to compete in modern markets and purchase imported staples. As a result, local knowledge of traditional foods, such as preservation and preparation methods, is being replaced by knowledge of market values and consumer groups. Therefore, economic development and financial pressures are forcing the local perception of natural resources in developing communities to shift from *hidden harvest* to *hidden revenue*.

In addition to local food systems, this shift in knowledge and dependence is likely to affect local resource management practices (Bharucha and Pretty 2010). Rudd et al. (2003), for instance, attribute the rapid decline of giant clam (*Tridachnidae*) stocks to its high market value in Papua New Guinea. High market prices promoted opportunism and overexploitation amongst local fishers, overriding traditional social norms and subsistence levels of collection practiced when their primary value was in local food systems. Thus a shift towards market-based economies and livelihoods has the capacity to deplete local resource pools through the erosion of local management and conservation practices as well as the knowledge bases that surround them.

The findings of Pilgrim et al. (2007c) study are contrary to some earlier studies that assume traditional food systems persist in low-income households (Gadgil et al. 2000; Byg and Balslev 2001; Anishetty 2004; Ladio and Lozada 2004). Instead, this study demonstrates that the resource dependence of low-income households is, in at least some cases, shifting from subsistence to income-driven as local resources develop a monetary value, creating a departure from local food systems. This is in line with the findings of Jodha's study (1986) that demonstrated the significant contribution of common property resources to household incomes of the rural poor in Rajasthan, India. Therefore knowledge bases and ecosystems are coming under threat where economic development and market pressures are forcing the local view of natural resources to shift from hidden harvest to hidden revenue.

Local knowledge loss and economic development: a global phenomenon?

So far we have shown how local knowledge systems can shift due to emerging economic pressures, firstly through livelihood diversification and, secondly, through commodification of local resources. However these are just two examples of the ways in which economic development and emerging market pressures are altering the local knowledge systems of communities in developing, and even some communities in industrialised, countries across the world. Although developing countries have long been shifting their knowledge bases and adapting to economic development to accommodate pastoralists, agriculturalists, industrialists, and environmentalists (Blench 1999), the research that has been described here demonstrates that even the most isolated indigenous and marginalised communities are now being affected by these pressures.

Current patterns of economic development are therefore eroding local knowledge bases globally (Hamwey 2004). This pattern of loss was quantified at the local and international level in a recent study by Pilgrim et al. (2008) that used ethnobotanical interviews to examine knowledge loss in and between communities in Indonesia, India and the UK. Economic growth can broadly be defined as *"an increase in the production and consumption of goods and services... It is generally gauged by measures of national income such as gross domestic product (GDP) and gross national product (GNP)"* (Czech and Pister, 2005). This study used income as a proxy for wealth and looked to see the effects of small-scale and large-scale differences in economic wealth on local ecological knowledge.

In this study, a strong pattern of local ecological knowledge decline correlating with economic development both within and between countries (Pilgrim et al. 2008). Within country comparisons revealed that people in communities with higher average income levels were able to identify fewer species uses than those of low-income communities (Figure 6). Inter-country comparisons revealed a similar pattern of local knowledge decline with increased Gross Domestic Product (GDP) and Human Development Index (HDI). The researchers found that UK respondents had the lowest knowledge levels coinciding with a high HDI of 0.94 and a mean per capita GDP of US$26,150. Respondents from the India study sites had intermediate knowledge levels, corresponding with intermediate HDI (0.66) and GDP values (US$2892). Respondents from the Indonesia study sites, on the other hand, had the highest local knowledge levels, corresponding with a low HDI score (0.57) and a low GDP (US$2143) (Figure 7). This implies reduced knowledge transfer and substitution for more modern forms of knowledge where economic development is high, threatening local knowledge systems in regions where financial pressures are only recently emerging.

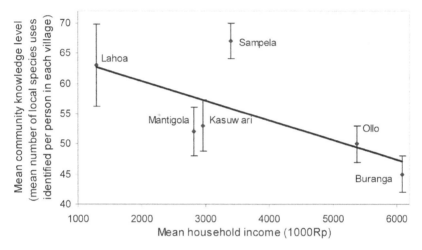

Figure 6: Mean local knowledge level (number of local plant uses known) in relation to mean household income across 6 communities in the Kaledupa sub-district of Indonesia (±2SE, n=192 across 6 villages) (Pilgrim et al. 2008).

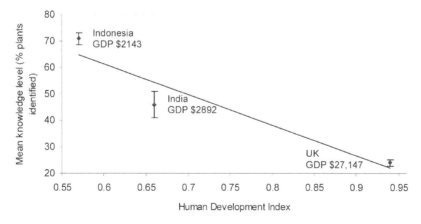

Figure 7: Mean knowledge level (plant name identification) in 3 countries and relationship with Human Development Index and per capita GDP (±2SE, n=1095 across 3 countries) (GDP US dollar estimates derived from purchasing power parity (PPP)) (Pilgrim et al. 2008).

Pilgrim et al. then examined the actual process of knowledge transmission and acquisition to see if transfer was, as suspected, less effective in more wealthy communities. They found that knowledge transfer was far more rapid in Indonesia where knowledge saturation (whereby younger generations have a knowledge level that is equal to or exceeds that of elder generations) was reached by the age of 30 on average, compared with India (where saturation was reached by the age of 50) and the UK (where saturation, if reached at all, was not until the age of 70). This indicates that local knowledge transfer tends to slow as wealth increases. Therefore a pattern of progressive knowledge loss is occurring in wealthier countries, communities and even households today.

As almost half of the world's population is now urbanised (Sustainable Development International 2004) and economic growth continues to be a prominent policy goal in most countries (Bigford et al. 2006), there are concerns for the future of local knowledge bases. These knowledge bases that have been relied upon for generations are now being threatened, as shown here, in the light of economic development and growing market pressures. A shift away from cultural traditions and towards financially-driven livelihoods is causing a growing disconnection from local lands and waters, and local ecological knowledge systems to become dilute and devoid of purpose. Subsequently, local knowledge bases are no longer being transmitted or are being supplanted by modern knowledge, such as the current market value of local resources, even in marginalised grassroots communities (Berkes et al. 2000; Gadgil et al. 2000; MEA 2005; Singh et al. 2006; Pretty 2011). Therefore current patterns of economic development are impacting upon local livelihoods and local knowledge bases globally. However given the importance of local food systems to developing countries and the strong interconnections between local food systems and local knowledge, it is surprising that the implications of widespread knowledge loss for local food systems are rarely discussed (Kuhnlein et al. 2009). In the next section, we consider the consequences of economic development and subsequent knowledge loss for traditional food systems and the wider implications of a shift away from local food systems.

The importance of traditional food systems and what the loss of local knowledge means for human and ecosystem health

Traditional foods have been defined as *"culturally accepted foods available from local natural resources that constitute the food systems of Aboriginal peoples"* (Willows 2005). Traditional food systems comprise of a complex system of collection, harvesting, management, preservation, and preparation techniques as well as the local foods themselves. Food plays a very complex role in our lives, of which nutrition is just one aspect. Local food systems comprise a complex web that links biological, economic, social, and cultural components (Tansey 2004). Food provides an intricate link with the past through cultural continuity, conveys spiritual meaning and reinforces established social systems. For instance collecting and harvesting local foods, preparing those foods and then eating them together strengthens social bonds and networks and instils a sense of personal identity (Pars et al. 2001; Raine 2005). In this way food systems form an interconnected whole; connecting the food, the meal, the local community, the global society, the seed, the soil and the land and water (Mann and Lawrence 1999; Pretty 2011).

A number of reasons have been cited for the general and global shift away from traditional food systems towards industrialised food systems. These include loss of territories, the introduction of new foods, the loss of traditional management practices, and land transformation and degradation. However, perhaps one of the most pressing reasons today for the shift away from traditional foods is knowledge loss related to economic development (Pilgrim et al. 2008). Traditional food systems have existed, and always will exist, as a central part of local knowledge bases. Community knowledge bases contain information on how to collect local foods (e.g. hunting and gathering as well as farming techniques), what to collect (e.g. what species are available and palatable locally and what species provide energy and nutritious content), where to find food, how to store it, how to prepare it, how to cook it and finally how to eat it. Thus local knowledge provides the pathway along which traditional food systems can travel from the land all the way to the plate (Willows 2005; Bharucha and Pretty 2010).

Local knowledge bases are, therefore, essential to the existence, transfer, use and continuation of traditional food systems, and without this knowledge, people will be forced into industrialised food systems. The Greenlandic Inuit hunting narwhal, for instance, need knowledge of where to locate the narwhal and at what time of year (e.g. in spring during the annual migrations through Canada's Admiralty Inlet), what to use to hunt narwhal, how to kill a narwhal in order to retrieve it (by shooting it in the brain or top of the spine at the split second it fills it lungs full of air to ensure that after being shot it can be landed without sinking), how to butcher and store the meat in Arctic conditions (often the blubber is used to store other food items such as sea birds for a season) and finally how to cook all of its different body parts (Kishigami 2000; Nicklen 2007). All of this knowledge requires a deep understanding of the food species itself and the local environment which it inhabits.

Another resource-dependent group is the nomadic boat-dwelling tribe of Indonesia, the *Orang Bajo*. The *Bajo* are so intricately linked with their local marine environment that they know the exact location of fish spawning and nursing grounds, and are able to predict weather patterns and seasonal fluctuations. It has been said that they can locate a school of fish through the vibrations of the water on their paddle and they can navigate their way home by the stars (Adi and Sudarman 1994; Pilgrim and Pretty 2010). This intricate pool of knowledge forms the basis of their traditional food system and if one piece of the jigsaw is lost, such as how to hunt or where to find prey species, then the whole system is at risk of collapse. This exemplifies how local knowledge, traditional food systems, and the land they are based upon are all intricately intertwined, their evolution interconnected and their survival interdependent.

A survey conducted in 2003 by the International Indian Treaty Council (IITC) carried out for the FAO found that cultural foods were considered to be vitally important by 89% of indigenous respondents for their nutritional value, by 80% for food sovereignty and local self-sufficiency, by 87% for the preservation of knowledge, practices and the traditional way of life they represent, by 83% for cultural continuity and ceremonial practices, and by 83% of respondents as an important component for protecting the local land and environment (IITC 2003). However, local knowledge bases crucial to local food systems are today being threatened by economic development and delocalisation. This is creating a shift from traditional food systems to industrialised and globalized food systems, in which the human energy required to obtain food has decreased and yet the energy contained within food has increased (Lieberman 2003). With this shift comes an array of consequences affecting human health and well-being, physically, socially, spiritually, and culturally.

The health benefits of local food systems in comparison with industrialised food systems

Traditional local food systems have sustained human populations for 99% of our time on earth by meeting nutritional requirements even in harsh environmental conditions (Lee and DeVore 1968). In addition to providing the amino acids, calories and other nutrients essential for metabolic functioning, traditional food systems provide a great number of human health benefits that today's industrialised food systems are lacking. For instance, the recent emergence of carbohydrate-dominated diets in many indigenous and marginalised communities across the world has led to metal deficiencies including iron and zinc, dental caries and bone lesions (Waldram et al. 2006).

The rapid shift in diet has also led to an epidemic of chronic and degenerative diseases spreading throughout indigenous and marginalised communities of the world, today known as 'diseases of modernisation'. These include obesity, diabetes and cardiovascular diseases, all a symptom of a shift towards industrialised food systems that contain energy-dense foods high in fats, salts, sugars and oils (e.g. soft drinks and snack foods) in combination with reduced levels of physical activity (Kelm 1998; Willows 2005; Kuhnlein et al. 2006; 2009). Despite this, younger generations in particular are developing a taste for industrialised foods and opting out of local food systems in preference of foods rich in sugars and animal fats (Uauy et al. 2001; Lieberman 2003; Turner and Turner 2008).

Common cancers such as breast, lung, colon and prostate used to be rarely seen in Aboriginal populations dependant on traditional food systems and livelihoods, with the exception of cancer of the kidney in men and cancer of the gallbladder and cervix in women (Waldram et al. 2006). In 1935, cancer accounted for 13% of all deaths in British Columbia, Canada, however it only accounted for around 2% of deaths in First Nations communities. *"Cancer...was cited repeatedly by elders as an affliction that they see as the result of non-Native environmental abuse. Elders attributed the absence of these diseases...as being due to the persistence of an Aboriginal lifestyle based on traditional food sources"* (Kelm 1998). However since the shift towards industrialised food systems more common forms of cancer regularly affect indigenous peoples (Kozlov and Zdor 2003; Waldram et al. 2006). One reason for this is the high consumption of animal fats known to cause breast and colorectal cancer. Vitamin A is known to protect against lung cancer and fibre against colorectal cancer. Both vitamin A and fibre were traditionally ingested in high quantities amongst indigenous groups through the consumption of leafy green vegetables in traditional diets. However, store-bought diets have been found to be high in animal fats but lacking in Vitamin A and fibre (Kelm 1998; Waldram et al. 2006). As a consequence, the proportion of indigenous populations across the world affected by common cancers and other chronic diseases relating to diet is, today, equal to or higher than national populations (Waldram et al. 2006).

In British Columbia in 1935, heart disease and circulatory disease accounted for 11% and 26% of deaths respectively. However in Aboriginal populations at this time, heart disease only accounted for around 5% of deaths and circulatory death for around 4% of deaths (Kelm 1998). The shift from traditional to industrialised food systems today is associated with a rise in cardiovascular diseases amongst indigenous populations, in particular ischaemic heart disease (IHD) or heart attacks (Waldram et al. 2006). The hospital admissions of First Nations peoples suffering from IHD in Ontario, Canada, for instance, increased from less than 80 in 10,000 annually to over 150 in 10,000 from 1981 to 1997 (Shah et al. 2000).

Polyunsaturated fatty acids (PUFAs) protect against IHD but cannot be synthesised within the human body (Kozlov and Zdor 2003), and instead are found in oily fish (Parrish et al. 2007), sunflower oil, and various types of nuts and seeds. The shift away from traditional food systems has reduced PUFA intake as populations have abandoned wild meats and fish in exchange for store-bought foods (Kelm 1998; Waldram et al. 2006). The shift from a high intake of PUFAs to a high intake of saturated fatty acids has also caused a change in the ratio of omega-3 to omega-6 intake (Kozlov and Zdor 2003). The prostaglandins formed from omega-3 are known to protect against cardiovascular diseases and potentially malignant tumours, whereas the prostaglandins formed from omega-6 are associated with increased the risk of IHD and the development of carcinogenic diseases. However with the shift towards market foods, the ratio of omega-6 intake to omega-3 intake has shifted from 4:1 or 2:1 in traditional food systems to 15:1 or 10:1 in modern urban populations, Weil (2000). This is likely to contribute to the increased rates of IHD and common cancers observed in indigenous and tribal populations today.

Risk factors associated with IHD include hypertension, high plasma cholesterol, diabetes and obesity, all symptoms of poor diet and physical inactivity stemming from a shift towards industrialised food systems and lifestyles. Rode and Shephard (1994) revealed an increase in subcutaneous fats coinciding with a decline in physical fitness and muscular strength of residents of Igloolik in Canada since 1990. Reflecting this, obesity has been referred to as the fastest growing epidemic of our time, and is particularly prevalent amongst low-income groups undergoing economic development (Frank and Finegood 2005; Uauy et al. 2001). The shift from lean to obese populations as a result of consuming high levels of refined sugars and fats has led to impaired glucose tolerance and diabetes in many indigenous groups today (Uauy et al. 2001). In fact more than 80% of new diagnoses of non-insulin dependent diabetes mellitus (type 2 diabetes) are associated with obesity (Lieberman 2003).

The last century saw a dramatic increase in the prevalence of the noncommunicable disease, type 2 diabetes, amongst indigenous populations in particular (Thorburn et al. 1987; Lieberman 2003). In 1985, it was estimated that 30 million adults were diagnosed with type 2 diabetes, growing to 135 million adults in 1995 and 151 million in 2000. This figure is expected to double to around 300 million by 2025, of which the majority is expected to be in developing countries (Lieberman 2003). There is now a growing evidence base that shows a genetic susceptibility to diabetes amongst Aboriginal populations, caused by an elevated insulin response to glucose (Thorburn et al. 1987; Lieberman 2003). However research by O'Dea (1984) revealed that Aboriginal diabetes patients can show a marked improvement in health within just 7 weeks upon reversion back to a traditional diet and lifestyle. High levels of exercise, for instance, have been shown to reduce abdominal fat and improve insulin sensitivity by releasing increased oxidative enzymes, glucose transporters and capillarity in muscles (Lieberman 2003).

Despite traditional diets often comprising high fat intake, Aboriginal populations frequently maintained lower levels of blood cholesterol than other populations as a result of consuming high levels of unsaturated fats from wild meats and fish. However store-bought meats contain primarily saturated fats. For instance domesticated livestock contains between 3 and 11 times more saturated fats than whale meat. This, combined with the 50 – 65% reduced energy expenditure of modern lifestyles compared with traditional hunter-gatherer lifestyles, has led to an increase in blood cholesterol of many indigenous groups globally, increasing their risk of IHD (Bogoslovskaya et al. 1997). Other dietary changes that have emerged as a consequence of a shift from local traditional health systems to global industrialised health systems include decreases in protein levels, fibre consumption, vitamin C and D uptake leasing to hypovitaminoses, and phosphorus and calcium intake (Kelm 1998; Lieberman 2003; Willows 2005; Samson and Pretty 2006)

There is an evidence base that suggests the digestive systems of many indigenous groups have not yet evolved the mechanisms to efficiently digest the high levels of carbohydrates that comprise modern industrialised diets, due to a long-term dependence on traditional diets high in meats, fish and wild vegetables (Kozlov and Zdor 2003). In other words many groups are still of the wolf-nutritional type, used to a high intake of proteins and PUFAs which require chemical treatment within the central portion of the stomach in order to protect the stomach lining from the high acidity of the gastric juices. However in order to successfully digest industrialised food systems, indigenous groups need to evolve to the wild boar-nutritional type. This is characteristic of populations used to consuming modern-day diets and enables digestion to occur at the walls of the stomach since gastric acidity is lower (Kozlov and Zdor 2003). Therefore modern diets are creating a physiological and biochemical challenge amongst Aboriginal populations that retain genotypes and phenotypes adapted to low fat/low sucrose diets (Kozlov and Zdor 2003; Lieberman 2003).

In summary, diet, nutrition, and physical health are sensitive to the effects of sociocultural changes reflected in the changing proportions of store-bought foods and foods obtained from the land (Willows, 2005). This shift is occurring at a cost to human health and well-being. Ironically, it seems that that which is essential to life (i.e. food) is today responsible for shortening life in so many regions. Thus the nutritional and health value of land-based foods cannot be matched by market foods. Despite this, a shift towards purchasing store-bought, processed foods has long been thought of as a shift towards the civilisation of marginal groups and, therefore, a positive move. However this push for 'civilisation', exacerbated by the rapid loss of local knowledge bases in indigenous societies today, is occurring at a cost to traditional societies and their ecosystems.

The role of traditional foods in the protection of biodiversity

In hunting, harvesting, gathering, cultivating and collecting traditional foods, human populations have acquired an in-depth knowledge base which, combined with experiential observations, has led to the evolution of intricate management practices (Gadgil et al. 1993; Berkes et al. 2000; MEA 2005). These practices have led to management systems that have enabled human populations to sustain themselves and their local ecosystems for generations, testimony of their effectiveness being the survival of both through to the present day (Olsson and Folke 2001; Gilchrist et al. 2005; Singh et al. 2006). However the value of these specialised practices, self-imposed restrictions and intricate knowledge has gone overlooked for years and it is only recently that their importance to biodiversity conservation in the future has been realised (Veitayaki 1997; Bharucha and Pretty 2010).

Traditional management practices derive from years of living off the land and from local food systems. Examples of such practices include the protection of sacred groves in India, the establishment of *tambu* (that can only be fished for ceremonial purposes several times annually) by inhabitants of Ahus Island, Papau New Guinea (Cinner et al. 2005) and harvesting restrictions placed on particular species or areas at certain times (for instance after the death of an influential community member in the New Island Province of Papua New Guinea (Wright 1985). These are all self-imposed restrictions introduced by communities themselves to regulate levels of wild food collection and local food cultivation as a means of ensuring local foods are available for generations to come. By introducing a variety of management techniques and extracting different resources from different local habitats, it has been suggested that indigenous groups, unconsciously perhaps, manage their ecosystems as an intricate mosaic. This enables them to maximise on the number of habitats found locally, the number of species comprising those habitats and therefore the diversity and quality of resources available locally for extraction (Gadgil et al. 1993).

Local food systems have therefore long formed the basis of human diets and thus, in many parts of the world, management practices have evolved by which to protect the diversity of local foods and maximise their yields and their quality. In doing so, traditional management practices have increased biological diversity in ecosystems across the world. However the shift towards industrialised food systems in marginalised regions expected with the continued loss of local knowledge, is causing these diverse ecosystems to become degraded as traditional management techniques are abandoned and eventually forgotten (Fenta 2004; MEA 2005; Pilgrim et al. 2008). This will cause the loss of biodiversity in developing regions across the world, and with it local food species, pharmaceuticals, and future opportunistic uses (Kaushik 2004; Le Quy 2004; Mhame 2004; Zhang 2004). Therefore, promoting and maintaining local food systems and the knowledge they are based upon is likely to ensure the continuation of local management practices and sustain biodiversity for generations to come.

The cultural values and social networks associated with local food systems

According to Kozlov and Zdor (2003), *"Nutritional traditions are an area where biology and culture are closest."* Food choices are determined by a number of different contextual factors above and beyond availability and nutritional status. They are embedded in and influenced by social systems and status, family and peers, the economic and physical environment, local policy and even spiritual beliefs and cultural meanings (Taylor et al. 2005). For many indigenous and marginalised groups, consuming and celebrating traditional local foods demonstrates a form of cultural expression, personal identity, spiritual well-being, and continuity with the past essential to the holistic health of both humans and culture. Cultural expression improves quality of life and just by eating certain meals together communities can regain their sense of identity and reconnect with their ancestors (Pilgrim and Pretty 2010). Hence local foods are often termed 'cultural foods'. However in many indigenous cultures, food is perceived and accepted as a life-giving substance and, therefore, by its very nature is health-promoting, so the concept that food can be bad for human health and actually shorten life is completely foreign (Taylor et al. 2005).

"Increasingly, we live in a social environment that disconnects us from the source of our food.... Yet, we continue to celebrate life and traditions through sharing food, since food and eating have strong social dimensions" (Raine 2005). In addition to the actual food product having cultural meaning, the activities required to collect, distribute and procure traditional foods play a key role in sustaining the cultural characteristics and social relationships of a community. Therefore the consumption of local or cultural foods is more than just about eating. It is a socially-constructed act that *"is the endpoint of a series of culturally meaningful processes involved in the harvesting, processing, distribution and preparation of these foods"* (Willows 2005). Tending, collecting, processing and sharing local foods between family and community members creates a sense of belonging, strengthens community bonds and family ties, and reinforces relations of trust and reciprocity. Thus the continuation of local food systems provides dietary, biological and cultural diversity.

Local food systems to enhance local food security

Some 800-1000 million people suffer persistent hunger and 1.3 billion live in poverty. Ironically, hunger is most pressing in areas responsible for producing the vast majority of food (i.e. low income rural areas) (Mann and Lawrence, 1999). With substantial population increase still expected by 2050, much of the increased food demand is anticipated to come from developing regions (McNeely and Scherr 2001; Foresight 2012). With the pervasiveness of poverty and population increases expected in the coming years, household income and food prices may well exceed factors such as health, taste, culture and social systems as the key predictors of food choice (Willows 2005). In light of this, local food systems may offer the solution by being low in cost and highly nutritious when compared with store-bought foods. Local foods have traditionally provided food security during times of environmental strain or economic crisis. For instance, many traditional food systems are based around food sharing networks whereby food is shared between family members, disadvantaged community members (such as the widowed or elderly) and/or households that have been unsuccessful in their own attempts to collect food. Some communities even announce the success of a good hunt using a loudspeaker and invite community members to the house of the hunter to come and collect a share (Willows 2005).

However the widespread departure from traditional food systems has reinforced the food insecurity of many Aboriginal groups across the world (Kuhnlein et al. 2009). No longer are these people dependent upon their own skills and knowledge to feed their families, but instead they are reliant on low wage incomes and imports into the region, depleting their self-sufficiency. Canada's National Health Survey of 1998-99, for example, revealed that food insecurity affects 27% of Aboriginal peoples living off reserves, and 24% have a compromised diet (lack of food or lack of variety) as a consequence of low household income and high food prices. To make matters worse, store-bought foods transported to more remote regions suffer lack of variety, lack of fruit and vegetables, due to their high perishability, and poor quality (Willows 2005).

Conclusion and Policy Implications

From this study, we can conclude that the socioeconomic and lifestyle changes stemming from economic development are leading to deterioration of the biophysical environment (e.g. through deforestation, mining, hydroelectric dams) combined with intense knowledge loss, forcing traditional communities away from local food systems to consume processed, store-bought foods often from far distant locations. This is subsequently damaging human health, natural environments, local cultures and food security for the future. For these reasons amongst others, international documents such as the Millennium Ecosystem Assessment and the United Nations Environment Programme are now acknowledging the importance of local knowledge and that action needs to be taken in the future to actively protect community knowledge bases from further degradation in the light of economic development (MEA 2005; UNEP 2006). With growing concerns over food security caused by human population expansion and the impacts of global climate change, local food systems and the knowledge that these are based upon need to be sustainably managed into the future. Policy-makers need to prioritise the protection of local knowledge for use in natural resource and agroecosystem management, particularly where biodiversity is high and financial capacity is limited (Pilgrim and Pretty 2010). Furthermore, local food systems provide a mechanism by which the protection of biological and cultural diversity can be integrated in both policy and practice.

Abbreviations

DFID	Department for International Development
FAO	Food and Agriculture Organisation
GDP	Gross Domestic Product
GNP	Gross Net Product
HDI	Human Development Index
IHD	Ischaemic Heart Disease
IITC	International Indian Treaty Council
IUCN	International Union for Conservation of Nature and Natural Resources
MEA	Millennium Ecosystem Assessment
n	Sample size
ODI	Overseas Development Institute
PUFAs	Polyunsaturated Fatty Acids
PPP	Purchasing Power Parity
PGR	Population Growth Rate
Rp	Rupiah
SE	Standard Error
UK	United Kingdom
UN	United Nations
UNDP	United Nations Development Programme
UNEP	United Nations Environment Programme
US	United States of America

References

Adi CM, Sudarman P (1994) Bajau, Second edition. Yayasan Sejati Publishers, Jakarta.
Anishetty M (2004) Conservation and utilization of plant genetic resources for food and agriculture: Strengthening local capacity for food security. In: Twarog S, Kapoor P (eds). Protecting and promoting traditional knowledge: Systems, national experiences and international dimensions. United Nations, Geneva, pp33-40
Béné C, Mindjimba K, Belal E, Jolley T (2000) Evaluating livelihood strategies and the role of inland fisheries in rural development and poverty alleviation: The case of Yaéré Floodplain in north Cameroon. In: Johnston RS (ed) Proceedings of the 10th Biennial Conference of the International Institute of Fisheries Economics and Trade, Corvallis, Oregon, US, 10-14 July 2000
Berkes F (2008) (first edition 1999) Sacred ecology, Second edition: Traditional ecological knowledge and resource management. Taylor & Francis, Philadelphia, Pennsylvania.
Berkes F, Colding J, Folke, C (2000) Rediscovery of traditional ecological knowledge as adaptive management. *Ecological Applications,* 10:1251-1262
Bigford T, Hyatt K, Dobson T, Poage V, Reynolds L, Czech B, Hughes B, Meldrim J, Angermeier PL, Gray B, Whitehead J, Hushak L, Lupi F (2006) Economic growth and fish conservation. *Fisheries,* 31:404-409
Blench R (1999) Hunter-gatherers, conservation and development: From prejudice to policy reform. Natural Resource Perspectives, Overseas Development Institute
Bharucha Z, Pretty J (2010) The role and importance of wild foods in agricultural systems. *Phil Trans Royal Society of London B,* 365:2913-2926
Bogoslovskaya L, Aleinikov P, Safronov S (1997) Nutritive value of gray whaling products. In: Role of gray whaling in the formation of the modern lifestyle of the indigenous population of Chukotka. *Scientific report of the USSR for the International Whaling Commission Supplement,* 3:53-69
Byg A, Balslev H (2001) Diversity and use of palms in Zahamena, eastern Madagascar. *Biodiversity and Conservation,* 10:951-970
Chambers R, Conway, GR (1992) Sustainable rural livelihoods: Practical concepts for the 21st century. In: Discussion Paper 296, Institute of Development Studies, Brighton
Cinner JE, Marnane MJ, McClanahan TR (2005) Conservation and community benefits from traditional coral reef management at Ahus Island, Papau New Guinea. *Conservation Biology,* 19:1714-1723
Czech B, Pister P (2005) Economic growth, fish conservation, and the American fisheries society. *Fisheries,* 30:38-10
Davis A, Wagner, JR (2003) Who knows? On the importance of identifying "experts" when researching local ecological knowledge. *Human Ecology,* 31:463-489
DFID (1999) Sustainable livelihoods guidance sheets. Department for International Development, London
Ellis F (1999) Rural livelihood diversity in developing countries: Evidence and policy implications. In: Natural Resource Perspectives 40, Overseas Development Institute, London.
FAO (2004) Livelihood diversification and natural resource access. In: Livelihood Support Programme Working Paper, UN FAO, Rome
Foresight (2012) Global Food and Farming Futures. UK Government Office of Science, London
Frank J, Finegood D (2005) Foreword from the Canadian Institutes of Health Research. *Canadian Journal of Public Health 96 Supplement,* 3:S5
Gadgil M, Berkes F, Folke C (1993) Indigenous knowledge for biodiversity conservation. *Ambio,* 22:151-156

Gadgil, M, Seshagiri Rao PR, Utkarsh G, Pramod P, Chhatre A, members of the People's Biodiversity Initiative (2000) New meanings for old knowledge: The people's biodiversity registers program. *Ecological Applications,* 10:1307-1317

Gilchrist G, Mallory M, Merkel F (2005) Can local ecological knowledge contribute to wildlife management? Case studies of migratory knowledge. Ecology and Society 10:20. http://www.ecologyandsociety.org/vol10/iss1/art20/

Hamwey R (2004) Traditional knowledge and the environment: Statement by the United Nations Environment Programme. In: Twarog S, Kapoor P (eds) Protecting and promoting traditional knowledge: Systems, national experiences and international dimensions. United Nations, Geneva, pp345-346

IITC (2003) Results: Questionnaire on indigenous peoples' traditional foods and cultures. Available from http://www.treatycouncil.org/QRE%20RESULTS.pdf

IUCN (2002) Sustainable livelihoods: What makes a livelihood sustainable? In: IUCN news archive 2001-2005. http://www.iucn.org/en/news/archive/2001_2005/mbsustliveli.pdf.

Jodha NS (1986) Common property resources and rural poor in dry regions of India. *Economic and Political Weekly,* 11:1169-1182

Kaushik A (2004) Protecting traditional knowledge, innovations and practices: The Indian experience. In: Twarog S, Kapoor P (eds) Protecting and promoting traditional knowledge: Systems, national experiences and international dimensions. United Nations, Geneva, pp 85-90

Kelm, M (1998) Colonizing bodies: Aboriginal health and healing in British Columbia 1900–50. University of British Columbia Press, Vancouver, BC

Kishigami N (2000) Contemporary Inuit food sharing and Hunter Support Program of Nunavik, Canada. In: Wenzel GW, Hovelsrud-Broda G, Kishigami N. The social economy of sharing: Resource allocation and modern hunter-gatherers. *Senri Ethnological Studies,* 53:171-192, National Museum of Ethnology, Osaka

Kozlov AI, Zdor EV (2003) Whaling products as an element of indigenous diet in Chukotka. In: The Anthropology of East Europe Review: Central Europe, Eastern Europe and Eurasia. Special issue: *Food and Foodways in Post-socialist Eurasia,* 21(1):127-137

Kuhnlein H, Erasmus B, Creed-Kanashiro H, Englberger L, Okeke C, Turner N, Allen L, Bhattacharjee L on behalf of the whole group (2006) Indigenous peoples' food systems for health: Finding interventions that work. *Public Health Nutrition,* 9(8):1013-1019

Kuhnlein H V, Erasmus B, Spigelski D (2009) Indigenous peoples' food systems: The many dimensions of culture, diversity and environment for nutrition and health. Food and Agriculture Organization of the United Nations, Rome, Italy

La Rovere R., Aw-Hassan A, Turkelboom F, Thomas R (2006) Targeting research for poverty reduction in marginal areas of rural Syria. *Development and Change,* 37(3):627-648

Ladio AH, Lozada M (2004) Patterns of use and knowledge of wild edible plants in distinct ecological environments: A case study of a Mapuche community from northwestern Patagonia. *Biodiversity and Conservation,* 13:1153-1173

Le Quy A (2004) The use and commercialisation of genetic resources and traditional knowledge in Vietnam: The case of crop and medicinal plants. In: Twarog S, Kapoor P (eds) Protecting and promoting traditional knowledge: Systems, national experiences and international dimensions. United Nations, Geneva, pp7-14

Lee RB, DeVore I (1968) Man the hunter. Aldine, Chicago, Illinois, p 415

Lieberman LS (2003) Dietary, evolutionary, and modernizing influences on the prevalence of type 2 diabetes. *Annual Review of Nutrition,* 23:345-377

Mann P, Lawrence K (1998) Rebuilding our Food System: the Ethical and Spiritual Challenge. In: Posey DA (ed) Cultural and Spiritual Values of Biodiversity. UNEP/ITP, London

May D (2004) Reef fishing activity in the Kaledupa stakeholder area. Operation Wallacea, Lincolnshire. http://www.opwall.com/2004%20Kaledupa%20reef%20fishing.htm

McNeely JA, Scherr SJ (2001) Common ground, common future: How ecoagriculture can help feed the world and save wild biodiversity. IUCN and Future Harvest, Washington, DC

MEA (2005) Current state and trends, Vol 1: Ecosystems and well-being. In: Millennium Ecosystem Assessment. Island Press, London

Mhame PP (2004) The role of traditional knowledge in the national economy: Traditional medicine in Tanzania. In: Twarog S, Kapoor P (eds) Protecting and promoting traditional knowledge: Systems, national experiences and international dimensions. United Nations, Geneva, pp17-20

Nabhan GP (2002) Coming home to eat: The pleasures and politics of local foods. WW Norton and Company, New York

Nicklen P (2007) Arctic ivory: Hunting the narwhal. *National Geographic,* August 2007:110-129

O'Dea K (1984) Marked improvement in carbohydrate and lipid metabolism in diabetic Australian Aborigines after temporary reversion to traditional lifestyle. *Diabetes,* 33:596-603

ODI (2003) Livelihood diversity and diversification. Understanding rural livelihoods: Key issues. In: Policy Guidance Sheets, Overseas Development Institute, London

Olsson P, Folke C (2001) Local ecological knowledge and institutional dynamics for ecosystem management: A study of Lake Racken watershed, Sweden. *Ecosystems,* 4: 85-104

Parrish CC, Turner H, Solberg S (2007) Resetting the kitchen table: Food security, culture, health and resilience in coastal communities. Nova Science Publishers, New York

Pars T, Osler M, Bjerregaard P (2001) Contemporary use of traditional and imported food among Greenlandic Inuit. *Arctic,* 54(1):22-31

Pilgrim S, Pretty J (2010) Nature and Culture. Earthscan, London

Pilgrim SE, Smith D, Pretty J (2007a) A cross-regional quantitative assessment of the factors affecting ecoliteracy: Policy and practice implications. *Ecological Applications,* 17(6):1742-1751

Pilgrim SE, Cullen L, Smith D, Pretty J (2007b) Hidden harvest or hidden revenue? The effect of economic development pressures on local resource use in a remote region of southeast Sulawesi, Indonesia. *Indian Journal of Traditional Knowledge,* 6:150-159

Pilgrim S E, Cullen L, Smith D J, Pretty J (2008) Ecological knowledge is lost in wealthier communities and countries. *Environmental Science & Technology,* 42(4):1004-09

Pretty J (2002) Agri-Culture: Reconnecting People, Land and Nature. Earthscan, London

Pretty J (2007) The Earth Only Endures. Earthscan, London

Pretty J (2011) Interdisciplinary progress in approaches to address social-ecological and ecocultural systems. *Environmental Conservation,* 38(2):127–139

Pretty J (2013) The consumption of a finite planet: well-being, convergence, divergence, and the nascent green economy. *Environmental & Resource Economics,* 55(4):475-499

Pyle RM (2001) The rise and fall of natural history: How a science grew that eclipsed direct experience. *Orion,* 20:17-23

Raine KD (2005) Determinants of healthy eating in Canada: An overview and synthesis. *Canadian Journal of Public Health 96 Supplement,* 3:S8-S14

Rode A, Shephard RJ (1994) Physiological consequences of acculturation: A 20-year study of fitness in an Inuit community. *European Journal of Applied Physiology and Occupational Physiology,* 69(6):516-24

Rudd MA, Tupper MH, Folmer H, van Kooten GC (2003) Policy analysis for tropical marine reserves: Challenges and directions. *Fish and Fisheries*, 4:65-85

Samson C, Pretty J (2006) Environmental and health benefits of hunting lifestyles and diets for the Innu of Labrador. *Food Policy,* 31(6):528-553

Scoones I (1998) Sustainable rural livelihoods: A framework for analysis. In: Working paper 72, Institute of Development Studies, Brighton

Shah BR, Hux JE, Zinman B (2000) Increasing Rates of Ischemic Heart Disease in the Native Population of Ontario, Canada. *Archives of Internal Medicine,* 160:1862-1866

Sillitoe P (1998) The development of indigenous knowledge: A new applied anthropology. *Current Anthropology*, 39:223-252

Singh RK, Singh D, Sureja AK (2006) Community knowledge and conservation of indigenous biodiversity: Exploration of hidden wisdom of Monpa tribe. *Indian Journal of Traditional Knowledge*, 4:513-518

Smith LED, Khoa N, Lorenzen K (2005) Livelihood functions of inland fisheries: Policy implications in developing countries. *Water Policy*, 7:359-383

Start D (2001) Rural diversification: What hope for the poor? Overseas Development Institute, London

Sustainable Development International (2004) World Urban Forum. http://www.sustdev.org/index.php?option=com_e vents & task = view _ detail&agid = 6 & year =2004 & month = 09 & day=13&Itemid=1

Tansey G (2004) A food system overview. In: Twarog S, Kapoor P (eds) Protecting and promoting traditional knowledge: Systems, national experiences and international dimensions. United Nations, Geneva, pp41-58

Taylor JP, Evers S, McKenna M (2005) Determinants of healthy eating in children and youth. *Canadian Journal of Public Health 96 Supplement*, 3:S20-S26

Thorburn AW, Brand JC, Truswell AS (1987) Slowly digested and absorbed carbohydrate in traditional bushfoods: A protective factor against diabetes? *American Society for Clinical Nutrition*, 45:98-106

Thoreau HD (1847) Walden; or, life in the woods. Ticknor & Fields, Boston

Turner, NJ, Turner KL (2008) Where our women used to get the food: Cumulative effects and loss of ethnobotanical knowledge and practice; case studies from coastal British Columbia. *Botany*, 86(1):103-115

Twyman C (2000) Livelihood opportunity and diversity in Kalahari wildlife management areas, Botswana: Rethinking community resource management. *Journal of Southern African Studies*, 26(4):783-806

Uauy R, Albala C, Kain J (2001) Obesity trends in Latin America: Transiting from under- to overweight. In: Symposium: Obesity in developing countries: Biological and ecological factors. *American Society for Nutritional Sciences Supplement*, 893-899

UNEP (2006) Marine and coastal ecosystems and human well-being: A synthesis report based on the findings of the Millennium Ecosystem Assessment. UNEP, Nairobi, Kenya

Veitayaki J (1997) Traditional marine resource management practices used in the Pacific Islands: An agenda for change. *Ocean and Coastal Management*, 37:123-13

Waldram JB, Herring A, Young, TK (2006) Aboriginal health in Canada, 2nd edn: Historical, cultural, and epidemiological perspectives. University of Toronto Press, London

Weil A (2000) Eating well for optimum health. Alfred A Knopf, New York

Willows ND (2005) Determinants of healthy eating in Aboriginal peoples in Canada: The current state of knowledge and research gaps. *Canadian Journal of Public Health 96 Supplement*, 3:S32-S36

Wright A (1985) Marine resource use in Papua New Guinea: Can traditional concepts and contemporary development be integrated? In: Ruddle K, Johannes R (eds) The traditional knowledge and management of coastal systems in Asia and the Pacific UNESCO, Jakarta Pusat, Indonesia, pp79-100

Zhang X (2004) Traditional medicine: Its importance and protection. In: Twarog S, Kapoor P (eds) Protecting and promoting traditional knowledge: Systems, national experiences and international dimensions. United Nations, Geneva, pp3-6

Chapter – 4

The Seasonal Migration of Thai Berry Pickers in Finland: Non-wood Forest Products for Poverty Alleviation or Source of Imminent Conflict?

*Celeste Lacuna-Richman**

Abstract

With categories including food, building materials, medicinal plants and resins, non-wood forest products are the economically important face of biodiversity. Using non-wood forest products to generate income and alleviate poverty requires different approaches to their special characteristics. Among the features that distinguish non-wood forest products from other goods are the unpredictability of the quantities available, the lack of control of collectors over their growth conditions, and the difficulty of ensuring that the products can be placed on the market efficiently. In developed countries such as Finland, with under-harvesting of non-wood forest products, getting an economic benefit from these involves the seasonal migration of collectors. The latter is illustrated by the case of Thai in a working visit to Finland, to pick the often underutilized wild berry harvest in Lapland. The majority of the workers are from farming backgrounds, but there is a weak positive correlation between their non-wood forest product activity in Thailand and berry picking in Finland. The future of the Thai picking berries in Finland depends partly on how the Finnish work authorities make provisions for future predicaments based on

---

*celeste.richman@uef.fi

present concerns of the berry pickers, such as ensuring their social well-being, and predicting the natural harvest of berries for the coming season. It may also lead to the need for institutionalization of such social safeguards to prevent conflict, abuse, charges of exploitation and the lack of sustainability.

Keywords: Berries, biodiversity, migration, non-wood forest products, income, poverty alleviation, seasonal migrants.

Introduction

The biological diversity in tropical forests is usually much higher than in temperate and boreal forests, with the number of endemic species providing a good indicator of just how valuable these forests are. However, these forests, and by association, the biodiversity these forests contain, are also located in the poorest countries in the world, with large populations still dependent on primary production for their subsistence (Whitmore 1998). The preservation of biodiversity in these countries is then often tied in with other national objectives to justify the opportunity cost, objectives such as the eradication of poverty, the delivery of essential services like drinkable water, and even the preservation of the cultural heritage of indigenous people (Park 1992). The latter objective is particularly controversial, since forests tend to be the last available settlements for indigenous people who are often in the margins of the larger society in terms of political and economic progress. In Brazil (Seeger 1982), the Philippines (Lacuna-Richman 2004) and various other countries, there are national policies wherein the rights of indigenous people are linked closely with the natural resource conservation efforts of the government (UNDP 2004). Non-wood forest products are at the forefront of discussions about the value of biodiversity in such countries, mostly because they provide very solid evidence of biodiversity's importance in economic terms. The justification for conserving forests despite the need for land for other uses inevitably includes non-wood forest products in the argument, because such goods provide an income for the people who live in these forests, particularly when a high deforestation rate compels most of these countries to put a logging ban into place (Park 1992; FAO 1994; FAO 1995a; FAO 1995b).

Discussions on Non-wood forest products (NWFP) often veer into classification of the broad term into more manageable categories. The categories are sometimes based on non-wood forest products' economic importance, and when this is the case, the most basic division is between those products that are used mostly as subsistence goods which have not for various reasons, entered the market, and commodities. Both economic categories for NWFP have been studied and discussed in the literature. In practice, products listed in one are almost always also in the other (FAO 1995b). However, despite the great significance of subsistence NWFP, which in some cases are the lifeline of many people in poverty (Lacuna-Richman 2004), the interest in a NWFP usually increases noticeably only when a market is found for it.

Commoditization of specific non-wood forest products has its dangers, the most obvious of which is probably over-harvesting, but it does allow collectors of the products to earn a money income which may not come from any other source (Belcher and Schreckenberg 2007). There are almost no sizeable groups of people in the world still unaffected by market forces in their efforts to fulfill basic needs (UNDP 2005, p. 20) and because of this, non-wood forest products have to compete in the market if they are meant to be used in mitigating poverty and to preserve forests at the same time.

In rich countries, non-wood forest products play a much more minor role than they do in developing countries (FAO 2007), both in terms of their contribution to the household economy, to the macro level of their proportion of a country's GNP. Yet, the idea of conserving biodiversity for decidedly non-economic reasons may be a larger concern in rich than poor countries (Lacuna-Richman et al. 2004). There is also a connection between NWFP utilization and indigenous people, but it may be weaker than that which exists in tropical countries. In rich countries, the conservation of biodiversity is not only for the indigenous people, but for all the people in the country, and as most environmentalists might argue, for everyone (Saastamoinen et al. 2005), but does this arguably more idealistic view extend to the economic sphere, such as earning from NWFP collection? Should there be two or more international standards for the utilization of NWFP - one standard for rich countries for which biodiversity is an afterthought, albeit a worthy one, after other national goals particularly economic development has been achieved; and another for poor countries wherein biodiversity provides most of the goods for the subsistence of many people? Who sets the standards if this is the case? These questions are faced daily by policy makers in poor countries where different communities share the same forest area, but are still largely rhetorical questions at present in rich countries. However, the accelerated pace of globalization in both goods and labor are making these issues increasingly relevant even in Europe and North America.

In Finland, a country which prides itself on its forests and its forestry industry both, the use of non-wood forest products is strongly rooted in its culture. The seasonal collection of berries such as blueberry (*Vaccinium myrtillys*), lingonberry (*Vaccinium vitas-idaea*) and cloudberry (*Rubus chaememorus*), and mushrooms such as ceps (e.g. *Boletus edulis* and *S. luteus*) and chanterelles

(*Chanterellus cibarius*) is considered an important household activity, as well as a traditional source of stores for the winter. During the recent, more affluent decades though, the economic value of collecting NWFP has become less important for Finns than it was previously. In the past decades, Finland's actual berry harvest has always been less than the biological crop available in its forests every year, despite the incidence of relative poverty in this generally affluent country (Kangas and Aho 2000, Saastamoinen et al. 2005). The income from berry and mushroom picking remain important to some sectors of Finnish society who have limited fixed incomes such as retirees and people living in remote rural areas – but these sectors of society still do not collect enough to take advantage of the natural harvest of NWFP from Finnish forests. Estimates of the percentage actually harvested from the natural harvest are set at 5 – 10% (Saastamoinen et al. 1998), with some other estimates as low as 1%. In the early 2000s, it was even presumed that the future of berry consumption in Finland was questionable due to the lack of interest among young people in harvesting these (Kangas 2001).

Finland does not link its policy regarding non-wood forest products with its indigenous people policies. Like its neighbor Sweden, Finland has an "Everyman's right" policy for collecting non-wood forest products, this right existing not only for Finnish people, but for anyone willing to come and collect the mushrooms and berries. It is not even necessary for people who come to Finland to collect berries and mushrooms to get a visa or a visa waiver to do so, as long as they do not stay longer than three months (*Työministeriö* 2007). Previously, foreign collectors of NWFPs such as berries and mushroom come from neighboring countries such as the Baltic states, Ukraine and Russia. There are additional "tourist pickers" of mushrooms from Italy and other western European countries in certain counties, who usually consume their finds. In 2005, however, the community of Savukoski in Lapland played host to 90 berry pickers from Thailand, who came at the behest of a food processing company, and since then the number of pickers have been increasing more than a hundred percent every year, so that a record number of 430 Thais have gone to Finland in the summer of 2007 to collect berries (Nevalainen 2007), and in the succeeding years till the present, in the thousands, culminating in 3500 people in 2013 (Helsingin Sanomat 2013). Perhaps due to the distance they have traveled and the cost of airfare to Finland, the Thai berry pickers have captured the interest of many Finns initially. The 2005 "working visit" to Finland by the Thai generated great media interest, from local newspapers in Lapland to the Helsingin Sanomat, Finland's foremost newspaper, from television features on public Finnish channels to international television (*Deutche Welle*), and the reports were overwhelmingly positive. The work ethic and industriousness of the Thai were lauded, and their contribution to the Finnish berry processing industry acknowledged by the dealers. In 2007, the media coverage of the Thai who have come to pick berries has spread to almost every Finnish major newspaper and television news program, most of which was positive reiteration of the past years' news, but already some reports discussed the negative social effect of so many determined berry pickers on the traditional summer cottage experience in Finland's western coast. Nevertheless, a number of Finnish berry processing companies acknowledge that after only 3 years, an overwhelming percentage (estimates range from 80-95%) of their berries have been picked by the Thai, and to a lesser extent, other foreign berry pickers from Finnish forests.

Finland has come relatively late to the phenomenon of seasonal migrant workers compared to other developed countries. Significantly, it has happened in the context of non-wood forest products and forestry, rather than agriculture. Although this may not seem like an important distinction at first, the characteristics of non-wood forest products as goods make it very different from farm produce, and greatly increases the economic risk of the Thai and other foreign berry pickers who come to Finland.

Non-wood forest products and the temporary migration of collectors

One approach to using NWFP for poverty alleviation comes from recognizing that in some special cases, the natural resource is under-utilized, but that labor is not readily available (UNDP 2004, p. 11, UNDP 2005 p. 11). For the most part, the forest resource from which NWFP come is not movable capital. Although for centuries and perhaps millennia, plant resources from one area have been successfully introduced and cultivated in new settlements, the conditions for this transfer are more limited at present. The reasons are partly ecological, for example, it is not feasible for certain fruits to grow in conditions with different climate and soils, or at least not without a large input of resources (e.g. greenhouses), but it is also due to bureaucratic factors (Alhojärvi 1998; Vantomme 1998).

Governments and private companies regulate the exchange and propagation of plant materials. Governments do this partly to protect the biodiversity and legal rights of the source countries (e.g. CITES) but also to protect the destination countries from dangers such as diseases and or the uncontrollable spread of plants which may be hardier than local species. Companies which have some rights over plant resources are necessarily protective of these rights, whether they are for the sources of raw materials (e.g. mushrooms in Finland) or for specific processes in completing the finished product (e.g. pharmaceuticals from tropical forests). Whatever the reason, there are situations around the world when the non-wood forest products are not harvested in sufficient quantities by the local population to fulfill either domestic or international demand for the product. Because such situations exist, the temporary migration of collectors to the forest area where the NWFP grows is a rising trend.

The one-time visit or even seasonal migrations of collectors for non-wood forest products have usually occurred between countries with greater income poverty and a country whose minimum wage is substantially higher than that of the collectors' homeland. Two characteristics of this phenomenon are the accessibility of the destination country from the country of origin of the collectors, and some similarity between the ecosystems of both homeland and destination areas. Examples of these include the summer berry collection in Sweden by citizens of neighboring European countries whose GDP is not quite as high or a similar trend in the United States wherein citizens of Central and South American countries harvest both non-wood forest products and agricultural products for a few weeks of each year.

In 2005 however, when the Thai flew to Finland to pick berries in Lapland, the phenomenon of migrant work entered the public realm. Although the local representative of the berry processing company has great appreciation for the almost unbelievable quantities of berries collected per day (e.g. the first day of lingonberry picking resulted in a harvest of over 24,000 kg), little has been said about the effects of this work on the livelihood of the Thai pickers. Discussions on policy are also close on the heels of the berry picking population. Although in 2006, the biological harvest of berries in Finland was considered less than the previous years due to dry weather conditions, there was still a great increase in the number of Thai berry pickers over the year before. In addition, though it seems unlikely that the local berry picking population will increase significantly in the coming years, the increase in berry pickers coming from Thailand yearly has turned their arrival from a novelty to an issue. The question of "Who are the Everyman's Rights for?" is now being asked. If biodiversity is indeed for everyone, theoretically, does this extend to the right to earn an income from an industry based on biodiversity? The answers differ considerably depending on the perspectives of different international property regimes. The Convention on Biological Diversity recognizes the predominance of traditional knowledge and the people who hold this knowledge, while the World Intellectual Property Organization, while also recognizing the above, also encourages that this knowledge be shared. The ideal is not to preserve tradition in a static condition or to block investment, but to acknowledge the holders of the tradition, while expanding possibilities for all humanity (UNDP 2004, p. 93). According to Human Rights principles, this ideal should be held whether people are indigenous to an area or migrants, but putting these principles into operation can be difficult.

From the point of view of the migrant workers, it is crucial that they earn enough to at least cover the airfare and other expenses they incur in going to Finland. The possibility of this happening should the natural harvest of berries be low for a certain year is not assured. Another important issue is the availability of social assistance for those who have taken the risk of earning profitably from berry picking, and failed to earn enough to even cover costs. The responsibility for ensuring that such individuals do not risk more than they can afford to, financially and in other ways, lies solely with them, as there is no other source of help or information. The following is an attempt to explore these aspects of the berry collection by Thais in Finland.

Research Methodology

A sample size of 30 from a population of 90 berry pickers was chosen. Face-to-face interviews were conducted with the respondents, who lived, and measured their day's yield, in the old schoolhouse in Savukoski, Lapland which is currently owned by the berry processing company. The Thai pay a small nominal fee to stay in the schoolhouse, and arrangements for meals, cooked in the Thai manner, are also organized collectively by the Thai and the berry processing company.

The interview schedule was prepared beforehand in English, and the questions asked in Thai. The respondents' answers were also given in Thai, but transcribed and then translated into English. The interview schedule was sent to several experts in interviewing for the content, and to two Thai students for language checking. Interviews were conducted on three evenings, from 4 – 6 September 2005, when the berry collectors returned to their camp after a full day in the forest. The berry picker's yield was recorded both by the camp manager for their records and by the researcher, to verify the berry pickers' estimated amounts of the quantities they picked.

Results

Farming in Thailand and Two Months of Berry Picking in Finland

All ninety of the Thai nationals who came to Finland in the June 2005 did so at the behest of a Finnish berry processing company. The group were almost all male, except for four female personnel – two women who cooked for the whole group, one who picked berries together with her husband, and one who managed the camp and did the record-keeping. The trip to Finland was arranged by the company, but was not a special flight in the sense that the company chartered the airplane. The Thais paid for their two-way fare, albeit at discounted prices, from the airline. Despite the discounted prices, the price of airfare was one of the main reasons given why the pickers have early on decided that they will have to work as long as daylight allows. Table 1 shows a summary of the demographic information of the Thai berry pickers.

Table 1: Thai berry collectors in Savukoski, Finland, 2005.

Characteristics	Categories and percentages			
Age	18–24 yrs. = 3%	25-32 yrs. = 27%	33-40 yrs. = 43%	41-50 yrs. = 23%
Previous trips to Europe for berry collection (mostly in Sweden)	0 = 17%	1–2 = 50%	3–4 = 17%	5+ = 17%
Average wage in Thailand	Range: 4000 – 8000 baht/month (2005 exchange rate: 50 baht = 1 Euro), So 80 – 160 Euros/month			
Expected wage in Finland (after paying for airfare)	Range: 500 – 1000 Euros/month			

The data from the above table were the results of the survey, and therefore came directly from the respondents. The exchange rate came from a foreign exchange conversion website for that month (October 2005).

The berry pickers were mostly farmers in Thailand (93%), with rice as their main crop, although two were also traders, and their camp manager used to work both as an English teacher and in administration at the Bangkok International Airport. All of them have hometowns in the northern provinces of Thailand, 70% from Udonthanee, 23% from Chaiyaphum, 3% from Nakhonratchasima, and 3% from Sukhanthanee. These parts of Thailand still have some forest, characterized by the FAO (2003b) as tropical moist deciduous forest and tropical dry forest.

Seventy-five percent (75%) of the respondents has had previous berry-picking experience, although this figure has a low positive correlation (correlation coefficient = 0.09, r^2 = 0.0085, p= 0.0000) with the 67% who also collect NWFPs in Thailand, which were comprised of mostly mushrooms and bamboo shoots. Those who have had previous berry-picking experience but have not collected NWFPs in Thailand are frequently also the respondents who have gone to Sweden in the preceding years for that goal.

The respondents were asked about the difficulty of collecting specific berries, partly because of the assumption that the majority of them have not picked these berries in their home country, and partly to determine how the perceived difficulty affected the actual amounts of berries they have collected. as the easiest by more than two-thirds (Table 2).

Table 2: Difficulty of collecting selected berries in Finland, using ranking, with 1 as "easy to pick" and 5 as "very difficult to pick".

Berry	Perceived difficulty (by percentage of respondents)				
	1	2	3	4	5
Blueberry(*Vaccinium myrtillys*)	0	10%	53%	27%	10%
Cloudberry(*Rubus chaememorus*)	7%	7%	23%	16%	47%
Lingonberry(*Vaccinium vitas-idaea*)	37%	37%	23%	0	3%

The interviews were conducted before the end of the berry season, and lingonberry was picked the first evening of the interviews. Thus, it may have been premature to report on the relation between perceived difficulty of collecting specific berries and the actual amounts harvested. Nevertheless, the average amount harvested (per picker) of blueberry collected was 1300 kg/wk (100-200kg/day), that of cloudberry, 54kg/day and that of lingonberry, 139kg/day. Impressive as these amounts are, especially considering that these berries come from a different forest environment from that in Thailand, the harvest of each berry species supports the evaluation of the respondents about the ease (or difficulty) of collecting them. Of the three major commercial berries, blueberry was considered moderately difficult to harvest by more than two-thirds of the respondents; cloudberry as the most difficult by half; and lingonberry as the easiest by more than two-thirds of the respondents. The results regarding the difficulty of collecting particular berries will probably be confirmed by Finns who collect berries themselves, in forests they are familiar with. The Thai collected berries using hand-held picks, which go through the leaves and stems of the blueberry and lingonberry plants, plucking the berries from the plant. There were several types of these picks offered to the Thai, and most of them chose the wooden ones with metal tines as the easiest to use.

When asked about whether they would consider returning to Finland to collect berries in the following years should the opportunity arise, the majority of respondents said yes (57%), a few said they were not sure (30%), and a few (3%) said no. The remaining did not give any answer to this question. The majority of respondents considered the work of berry picking itself, and the cold weather of Finland difficult. Although some berry pickers, both in the sample and those who were not, were asked about how the trip to Finland was financed, none volunteered any information about details, saying only that some of them had to borrow money to cover the airfare. Regardless of the difficulties mentioned however, all of the respondents expressed positive attitudes towards the Finnish people they have met, and appreciate the quiet of the Finnish countryside.*

Discussion and Recommendations

The economic viability of a non-wood forest product is a desirable outcome to both the collectors of a product and those who market it. However, it is not an unmixed blessing. There are parts of the world such as Thailand, where the forests produce a limited quantity of non-wood forest products, but with a large number of people willing and able to collect them if they can be sold. In other places such as Finland, the natural yield of NWFP from forests is great, but there is not enough labor to harvest them. For many other commodities, the movement of labor and the match of supply to demand are no longer problematic, but for non-wood forest products there are additional challenges.

The forest resource is not easily manipulated and is not transportable, thus it is the number of collectors which may have to be either limited, in the case of scarce natural yields of NWFP or increased, in situations of under-exploitation. Both limiting the number of collectors and encouraging more people to collect non-wood forest products require that care is taken to avoid unfair practices for market gain. While in developing countries, it is necessary to provide support for income from NWFP if rural people are expected to value forests over other land uses, in affluent countries, it is important to ensure sufficient legal and social provisions for temporary workers who are willing to collect NWFP. These are difficult conditions to ensure, but they can be done with sufficient analysis of the situation and political will.

In the 2005 migration of Thai berry pickers to Finland, the income gained by the respondents' berry picking is approximately five times more than what they would earn farming in Thailand. The correlation coefficient between picking NWFP in Thailand and Finland is so low as to be negligible, and yet all the berry pickers have harvested enormous quantities of berries from "unfamiliar forests". The difference in weather conditions and biodiversity between the forests in Finland, and what the respondents are used to in Thailand, has not proved to be a barrier to their productivity, as proven by the great amount of berries they have harvested daily. Because of these reasons, it seems that there is nothing negative about the temporary migration of NWFP collectors, but there is some displacement which the berry processing company has so far been careful to minimize by providing a common place to stay and food supplies at nominal cost to the berry pickers. However, the cost of airfare and processing fees to travel to and from Finland, are extreme barriers to some for getting any profit from the work, and indeed has led to extreme debt in Thailand for many.

Key concerns for the future of seasonal migration of berry pickers remaining sustainable are three-fold, and all concern building some kind of infrastructure while the phenomenon is not yet an intractable problem. The foremost concern expressed among the berry pickers, is ecological. Yearly, there are estimates in Finland about how productive the forests were in producing the major wild berry species, generated by university research groups, the Finnish Forest Research Institute, and other organizations. This information would help in estimating how much berries there are for the picking, and would be useful if it were made public or even actively publicized in such countries as Thailand, where the berry pickers come from. It would lessen the efforts of berry processing companies to try to recruit for more pickers in a bad year, and increase their recruitment efforts for potentially good years.

The other major concern is economic. There should be enough information available to prospective pickers regarding the prices per kilo of each particular berry they will collect, how much they need to collect to offset their airfare and livelihood costs, and average yield per person of berries collected per year (as opposed to the records of only the most productive berry pickers). Again, this information could be made available in Thailand and other countries where berry pickers come from, courtesy of the companies doing the recruiting, or such non-profit organizations such as *Arktiset Aromit ry* (Arctic Flavours Association).

The third concern is social. Since at present, the laws regarding berry picking are open-ended, including the "Everyman's right" policy for the berry collection itself, and the right to come to Finland for the purpose of picking berries without a visa for three months, the Finnish state is not compelled to assist berry pickers. Despite the income it generates for the Thai, berry picking is still mainly considered a summer recreational activity and not work in Finland. For this reason, there are no guidelines from the Finnish Ministry of Labor (*Työministeriö*) regarding the minimum wage for doing it, the number of hours that berry pickers are supposed to work, and the contract (or lack of it) between the buyers of berries and the pickers. Each berry picker is by all purposes, self-employed, in a "working holiday", responsible for their airfare to and from Finland as any tourist, and therefore is not under the protection of the welfare state of Finland nor the labor laws of Thailand. Nevertheless, the economic risk presented by the 1000 Euro airfare paid by a Thai who means to earn from berry picking is not in any way the same amount as the 1000 Euro airfare paid by a Thai who means only to visit Finland for sightseeing. Incurring such risk will doubtless have large social consequences for the families of the berry pickers, if they are unable to repay the debt. The socio-economic sustainability of seasonal migration of Thais to Finland for berry picking depends to a very large extent on the berry harvest being good enough that whatever debt incurred for travel is paid for, and that the berry pickers earn a certain minimum besides. For future research, this information can be gleaned from further surveys both from berry pickers in Finland and from the villages in Thailand where they come from.

Data from the survey points to a positive aspect of the Thai berry pickers' endeavor, which is their employment in other income-generating activities during the time that they are not in Finland. One reason why seasonal migrant work can be so socially disruptive in other developing countries is the lack of other sources of income once the harvest season ends. Fortunately, the majority of respondents have the option of returning to work on their rice fields once the berry season ends in Finland, and those who were in other lines of work also said that they can resume their regular job in Thailand without too much difficulty. This ideal state of affairs may not continue for later groups of Thai berry pickers though, and should be monitored to some extent, perhaps by the Thai authorities. Although most of the respondents expected to earn in two months of berry picking what they would earn in Thailand for almost a year, berry picking is still a very uncertain source of income because of both natural and economic factors. The peak berry season in Finland of late July to September might also affect other livelihood activities of the Thai berry pickers in their home country. It would be fiscally dangerous for the families of the berry pickers, as well as for the economy of the provinces from which these berry pickers come from, to depend on a two-month job to provide for them for the whole year.

An additional social concern is the effect of the mostly male berry pickers absence from their families for the period that they come to Finland; and consequently, if there is any allowance in the Finnish and Thai social systems for the eventuality that women, or both father and mother of a family decide to go to Finland for the seasonal work. As in many countries with sizeable migrant populations, the social cost to the family of seasonal migrant workers sometimes supersedes the economic benefits that they earn.

Limited ecological and economic information regarding berries and other non-wood forest products are available from Finnish institutions, such as the Finnish Forest Research Institute (METLA), and *Arktiset aromit ry*. Likewise, limited information regarding employment conditions and wages in cases where there are contracts between berry processing companies and pickers are available from the Occupational Safety and Health Administration (www.tyosuojelu.fi). Unfortunately for most foreigners, most if not all of this information is only available in Finnish, and therefore remains inaccessible to those who might need it. *Arktiset aromit ry* provides some travel information to foreign berry pickers who are not affiliated with any company as employed labor. However, so far, the information is only for Russian, Estonian and Polish berry pickers.

There is still very limited information available for potential berry pickers from Thailand, and that which exists is in English for the general public.

If there comes a time when the companies who buy from the Thai pickers can no longer assist them with their needs in Finland because there are too many berry pickers, then providing assistance may become a concern of the state. This is a very likely scenario because the majority of the berry pickers would like to come back, and another large percentage is not averse to it. There is also the undefined number of Thai who would like to come to Finland to pick berries based on the positive experiences of those who have gone before. Whether it is the Finnish state or the Thai government who will provide help and/or limit who can go, would have to be discussed and agreed on. In either case, market forces will probably have the stronger effect than information campaigns on those who would risk the trip, unless some strong policy instruments are used to support the information provided – a situation observable in other countries such as the Philippines, or Indonesia, which have high migrant populations (Mercado 2007). Only wage increases in Thailand at a corresponding rate with the income earned from berry picking in Finland, in tandem with possibly even higher airfares, will keep the number of Thai berry pickers at a level that can be assisted by authorities should the need arise.

It is quite early to judge whether the use of imported labor will become a permanent feature of the NWFP harvest in Finland, or how many people willing to work as berry collectors abroad can have a chance to do so. At present however, the seasonal migration of berry pickers seems to be a progressive approach to the problem of using non-wood forest products for poverty mitigation and utilizing the forests' biodiversity sustainably, that nevertheless has to be gone into with caution. To paraphrase the famous words of writer Max Frisch, to describe the German experience with migrant workers, "We wanted berry pickers, and got people instead" is a situation likely to occur in Finland's berry industry soon. It is contingent on the Finnish forestry and social welfare authorities to make this development benefit both the Finnish berry processing industry and Thai seasonal migrants, when the latter cannot depend on the market alone to correct for deviations from good business practice. At some point in the commodification of a resource, policy provisions not only for the resource, but for the rights of people supplying the market becomes necessary. This is needed not only to prevent conflict, but to ensure the social sustainability that is just as, or more important than, the ecological and economic priorities.

Update: This state of affairs has not continued through the years, however. In 2013, 50 Thai pickers have refused to leave Finland unless they are given their wages, which they have charged the company of withholding in violation of labor laws and human rights. The case is still being contested. There is also increasing hostility towards the pickers in certain parts of Finland (YLE 2013).

Acknowledgements

The author would like to thank the Academy of Finland for the funding of this study under a project led by Prof. Olli Saastamoinen; Municipal Manager Mauri Aarevaara for allowing the research in Savukoski; Jari Huttunen of *Korvatunturin marja oy* for introducing the researcher to the berry pickers; Wrongrong Duangjai, a forestry student in the ERASMUS Mundus Programme, for acting as interpreter; and Prof. Rebecca Richards, a Fullbright Visiting Professor from the University of Montana, for making possible the trip for this research component while conducting her study on berry dealers in Finland.

References

Alhojärvi P (1998) Non-wood forest products in the international cooperation financed by Finland. *In:* Lund, G, Pajari, B and Korhonen, M Sustainable Development of Non-Wood Goods and Benefits from Boreal and Cold Temperate Forests. *EFI Proceedings,* 23:9-18

Arktiset Aromitry (2007) Information for foreign pickers. Available at: http://arktisetaromit.fi/index.php?la=en. Cited 25 Sept 2007

Beaudoin SM (2007) Poverty in world history. Routledge, London

Belcher B, Schreckenberg K (2007) Commercialization of non-timber forest products: a reality check. *Development Policy Review,* 25(3): 355-377

FAO (1995b) Report of the international expert consultation on non-wood forest products. Food and Agriculture Organization of the United Nations, Rome

FAO (2003a) Forests and the forestry sector - Philippines. Available at: www.fao.org/forestry/site/23747/en/phl. Cited 10 Nov 2005

FAO (2007) Food and Agriculture Organization of the United Nations: Forestry

FAO (1995a) Beyond timber: Social, economic and cultural dimensions of non-wood forest products in Asia and the Pacific. Proceedings of a Regional Expert Consultation 28 November – 2 December 1994 FAO/RAP. Food and Agriculture Organization of the United Nations Regional Office for Asia and the Pacific, Bangkok

FAO (1994) Non-wood forest products in Asia. Regional office for Asia and the Pacific (RAPA) Food and Agriculture Organization of the United Nations, Bangkok

FAO (2003b) Forests and the forestry Sector - Thailand. Available at: www.fao.org/forestry/site/23747/en/tha. Cited on November 8, 2005

Hayami Y (2006) Globalization and Rural Poverty: A Perspective from a Social Observatory in the Philippines. United Nations University-World Institute for Development Economics Research (INU-WIDER) Research Paper No. 2006/44

http://www.fao.org/forestry/en/ Cited on 3 October 2007

http://www.jijigaho.or.jp/app/0406/eng/sp10.html. Cited 7 Nov 2005

Helsingin Sanomat (2013) Thaipoimijan tuntipalkka voi jäädä alle euroon. 19 September 2013. http://www.hs.fi/kotimaa/Thaipoimijan+tuntipalkka. Cited 11 Oct 2013

Kangas K (2001) Wild Berry Utilization and markets in Finland. University of Joensuu, Joensuu, Finland

Kangas K, Aho H (2000) The picking of wildberries in Finland in 1997 and 1998. *Scandinavian Journal of Forest Research,* 15:645-650

Lacuna-Richman C (2004) Subsistence strategies of an indigenous minority in the Philippines: Non-wood forest product use by the Tagbanua of Narra, Palawan. *Economic Botany,* 58(2):266–285

Lacuna-Richman C, Turtiainen M, Barszcz A (2004) Non-wood forest products and poverty mitigation: Concepts, overviews and cases. University of Joensuu, Faculty of Forestry Research Notes 166

Mercado J (2007) Migrant Culture. In: The Philippine Daily Inquirer. http://opinion.inquirer.net/inquireropinion/columns/view_article.php?article_id=90466. Cited 27Sept 2007

Nevalainen E (2007) *Poimintakausi lähtökuopissa ennätysvoimin. In:* Karjalainen 27.7.2007, p. 7.

Oliveras-Cunanan B (2005) One Town, One Product pushed. In: The Philippine Daily Inquirer. http://beta.inq7.net/opinion/index.php?index=2&story_id=55156&col=78. Cited 15 Nov 2005

One village one Product. June (2004) One village, one product. Available at: Park, CC (1992) Tropical Rainforests. Routledge, London

Saastamoinen O, Aho H, Kangas K (1998) Collection of berries and mushrooms by Finnish Households in 1997. *In:* Lund G, Pajari B and Korhonen M, Sustainable development of non-wood goods and benefits from Boreal and Cold Temperate Forests. *EFI Proceedings* No. 23:219-226

Saastamoinen O, Lacuna-Richman C, Vaara M (2005) Is the use of forest berries for poverty mitigation a relevant issue in an affluent society such as Finland? *In:* Non-wood forest products and poverty mitigation: concepts, overviews and cases, proceedings of a project workshop in Krakow 2004. Research Notes 166. Joensuu, Finland: Faculty of Forestry, University of Joensuu

Sawada Y, Estudillo JP (2006) Trade migration and poverty reduction in the globalizing economy: The case of the Philippines. United Nations University-World Institute for Development Economics Research (UNU-WIDER) Research paper No. 2006/58

Seeger A (1982) Native Americans and the conservation of flora and fauna in Brazil. In: Hallsworth, EG (1982) Socio-economic effects and constraints in tropical forest management. John Wiley & Sons, Chichester, U.K. pp.177-190

Shetty S (2004) Thai Story: One Village One Product Best Bet for Rural Growth. In: The Financial Express. September 30, 2004. http://www.financialexpress.com/fe_full_story.php?content_id=70124. Cited 15 Nov 2005

Työministeriö (2007) Picking forest berries and mushrooms. Available at: http://www.mol.fi/mol/fi/99_pdf/fi/02_tyosuhteet_ja_lait/marjanpoiminta_en.pdf. Cited 27 Sept 2007

UNDP (2004) The state of human development 2005. Available at: http://hdr.undp.org/reports/global/2005/.Cited 28 Sept 2007

UNDP (2005) The state of human development 2005. Available at: http://hdr.undp.org/reports/global/2005/. Cited 8 Aug 2007

Vantomme P (1998) Other similar international meetings on Non-wood forest products. *In:* Lund, G, Pajari, B and Korhonen, M Sustainable development of Non-wood goods and benefits from boreal and cold temperate forests. *EFI Proceedings* No. 23:19-28

Whitmore TC (1998) An introduction to tropical rainforests. Second edition. Oxford University Press, Oxford, U.K

YLE (2013) Thai berry pickers still unpaid after talks. 23 Sept 2013. http://yle.fi/uutiset/thai_berry_pickers_still_unpaid_after_talks/6844102

Chapter – 5

Sustainable Management of Natural Resources and Biocultural Diversity for Subsistence Livelihoods: A Cross Cultural Study

Ranjay K. Singh*, Anamika Singh, Anshuman Singh and B.S. Dwivedi

Abstract

Traditional communities the world over have been using a variety of indigenous biological resources in farming and life support systems. They are direct stakeholders in the *in situ* conservation and sustainable management of diverse plant species on their farms and in their home gardens. Location specific needs and ecosystem diversity plays a decisive role in maintaining biocultural resources. The present paper, based on six different case studies, describes the practices of traditional and innovative farmers in conservation of indigenous biodiversity. How farmers explore, domesticate and manage wild plant resources to sustain their livelihoods is the main theme of this study. It represents an explorative and qualitative study, carried out with diverse communities living in central India, Indo-Gangetic plains and mountainous ecosystems of Arunachal Pradesh. The study is based on 7 years of collaborative and participatory research with local and native communities. Data was generated using a combination of conventional (personal interviews with open-ended questions and informal interaction) and participatory methods. We found

---

*ranjysingh_jbp@rediffmail.com

that the local farmers and traditional women are the true custodians of landraces and ethnobotanical species, conserving them for their diverse food and nutritional contributions. Besides playing a crucial role in the conservation of bioresources and related cultures, the ecological knowledge of local farmers and women substantially contributes in sustaining vital ecological processes. Irrespective of socio-ecological systems, the elders and women play a leading role in the conservation and sustainable management of plant genetic resources. The number of species being managed for use in food and ethnomedicine is higher in mountainous ecosystems as compared to lower elevations. This could be attributed to the harsh living conditions in the mountainous ecosystems, forcing traditional communities in remote locations to evolve unique and effective ways of conserving and sustainably utilizing their genetic resources.

Keywords: Traditional ecological knowledge, biocultural diversity, farmers, women's wisdom, conservation, livelihoods.

Introduction

There is a growing realization worldwide that safeguarding the planet's biodiversity is fundamental to human food security and environmental sustainability (Pretty 2002; Pretty 2007). A greater degree of diversity in plant and animal genetic resources can enhance sustainable agricultural development through allowing development of integrated production systems through that are both efficient and resilient to climate change (Gladis 2003; Singh et al. 2010). Traditionally, farmers have used a wide variety of indigenous plants and animals and therefore have a direct stake in their conservation (Singh and Dwivedi 2002; Saxena et al. 2002; Singh 2004; Singh and Srivastava 2010a). Traditional knowledge pertains to the culture, history and identity of a particular community living in a certain geographical area with specific ecological characteristics. Subsistence farmers and traditional communities hold a deep interest in the conservation of local agrobiodiversity and in the related natural resources and ecosystems (Singh et al. 2013) to support their livelihood over generations (Singh et al. 2006). The knowledge of the agroedaphic growing conditions and the particular requirements of wild and semi-domesticated plant species assist in their easier selection and management under field conditions (Fingleton 1993). This knowledge enables the conservation of these plant genetic resources by traditional communities and ensures that food and nutritional requirements of farm families and native people are met. Again, such endeavours are of great help in sustainable food production under the changing socio-economic and climatic scenario (Singh et al. 2013).

Out of 10,000 reported edible plant species, only about 200 are used predominantly by humankind. Interestingly, only three plant species – namely rice, maize and wheat – contribute nearly 60 % of the food consumed by humans (Groombridge and Jenkins 2002). Many subsistence farmers, especially those in marginal environments where high-yielding crop varieties and livestock breeds do not flourish, rely on a wide range of crop, ethnobotanical and livestock species (Singh and *Adi* Women 2010). Since the 1900s, farmers have replaced many of the locally-adapted crops and landraces with intensively cultivated genetically uniform, high-yielding varieties that have not only resulted in the erosion of plant genetic diversity but have also caused considerable degradation of the natural resources and ecosystems (Bhagirath et al. 2012). Further, the small scale and integrated food production systems that conserve diverse crop varieties and animal breeds and are unique in maintaining sustainability, have been marginalized (Altieri et al. 2011). Most of the small and marginal farmers, who are innovative in conserving agrobiodiversity, have produced valuable models and insights for sustainable agricultural development (Fan and Chan-Kang C 2005; Altieri and Koohafkan 2008). In spite of small land holdings and scanty resources, these marginal farmers are rich in ecological knowledge and ethical capital instrumental in sustaining a number of plant and animal species and varieties (Brush 2005).

Genetic erosion poses one of the most alarming threats to world food security (Pretty et al. 2010). Biodiversity is the arbiter of the quality of human life, and the risk of species loss undermines the very concept of sustainable development, limits options of the future and robs humanity of a key resource-base for survival (McNeely and Scherr 2001; Groombridge 2002). To confront the erosion of genetic diversity, Thrupp (2000) proposes diversification for sustainable agriculture, the use of participatory approaches and building complementarity between agrobiodiversity and habitat conservation in policy development. However, farmers of developing societies and, in particular, women of traditional cultures have always been at the forefront of domestication, improvement and propagation of wild germplasm according to their needs (Singh and Sureja 2006; Turner and Clifton 2009; Singh and *Adi* Women 2010).

It is important to explore the innovative knowledge of the farmers and women of traditional communities on the management of crop and genetic resources, since this plays a vital role in maintaining the genetic diversity (Singh and *Adi* Women 2010). In the recent past, the roles of knowledgeable farmers and grassroots innovators have been recognized as the backbone for location specific conservation and sustainable adaptations for food security. This recognition has necessitated the exploration of the valuable traditional knowledge maintained by the innovative farmers from different parts of India. Therefore the objective of this article is to: (i) explore farmers and local communities' efforts in conservation of local crop species, and (ii) to understand the dynamics of the conservation process applied by creative farmers and outstanding knowledge holders in local food production.

Research Methodology

Study sites

The present study was carried out with six grassroots biodiversity conservators of India, namely, Mr. Radheshyam Singh, Village Sonpaur, district Azamgarh (Uttar Pradesh), Mr. B. R. Choudary, Village Adhartal, district Jabalpur (Madhya Pradesh), Mrs. Orik Rallen, Village Sibut, district East Siang, Arunachal Pradesh (Ar. P), Mr. Bamang Tanyang, Itanagar region (Ar. P), Late Mrs. Pem Dolma, Dirang Bazar, West Kameng district (Ar.P) and Mrs Adi Modi, Village Napit, district East Siang, Pasighat (Ar.P).

The explorations regarding farmers' creativity in domestication and conservation of indigenous biodiversity took place in two districts: i.e. Azamgarh (UP) and Jabalpur (MP). District Azamgarh, lying between 25^0 38" and 26^0 27' north latitude and 82^0 40' and 83^0 52' east longitude at an elevation of about 60 m above sea level (MSL) falls within the Indo-Gangetic plains (sub-humid climate). The district is one of the most productive agricultural areas of the world and lies at the centre of a rice-wheat based cropping system (Rai 2000). Jabalpur District is located at latitude 23.09°N and longitude 79.15° E at 550 MSL belongs to the central highlands (Malwa and Bundelkhand region), hot sub-humid (dry) and central plateau hills agroecosystems. It covers a geographical area of 10,160 km^2 with a total population of 24, 60,714.

The exploration of farmers' (both men and women both) creativity and knowledge was disclosed while exploring and submitting indigenous agricultural technologies to the Mission Mode Project on Indigenous Knowledge launched by the Indian Council of Agricultural Research (ICAR) in 2002. Both the case study states have a subtropical climate with most of the precipitation falling between July and August. They are both endowed with rich natural resources. Integrated farming systems comprising a variety of crop and animal species dominate the region's agroecosystems. Most of the farmers in these regions still maintain their own gene banks, consisting of local crop varieties and animal breeds. The animals raised include mostly cattle, buffalo, bullocks and goats.

The exploration of women's wisdom on domestication and conservation of biodiversity was made during 2003 to 2008 among the three tribes i.e. *Monpa, Nyshi* and *Adi* of East Siang, Papumpare (Itanagar) and West Kameng districts of Ar.P., respectively, living in fragile mountain ecosystems rich in biodiversity. The state of Ar.P. lies between 26° 28' to 29° 30' N latitude and 91° 30' to 97° 30' E longitude and shares borders with Bhutan in the west, Tibet and China in the north, Myanmar in the east and the Indian states of Nagaland and Assam in the south and southeast. The state covers the highest land area amongst the seven sister states of the North Eastern hill region, extending over about 83,743 km^2 (Rao and Hajra 1980). The topography is characteristically rugged due to lofty, haphazardly arranged ranges and deep valleys criss-crossed by a number of rivers and streams spreading along the southern slopes of the eastern Himalayas to the western slope of the Potkoi hills and around the huge valley of mighty Brahmaputra river.

While the districts of East Siang and Papumpare (Itanagar) have a subtropical climate, West Kameng has sub-temperate and temperate climates. We purposively selected the Dirang circle of the Bomdila subdivision of the West Kameng district. Dirang is predominantly inhabited by the *Monpa* tribe, who are Buddhists. The *Monpa* economy is largely agrarian and rural. Most people practice permanent agriculture and occasionally *jhum* cultivation (slash and burn agriculture). The major crops grown by the *Monpa* are maize, paddy, millets, buckwheat, wheat, barley and soybean.

The Pasighat subdivision (part of our study) of East Sing district is home to the *Adi* tribe famous for *jhum* cultivation system. The *Nyshi* and *Adi* tribes, living in mountainous ecosystems, use a variety of ethnobotanicals as food and medicine (Singh and *Adi* Women 2013). They practice integrated farming systems predominated by forest and animal based livelihoods. *Mithun* (*Bos fontailis*), pigs and ducks are the major domesticated animals. The major crops grown include rainfed and lowland paddy (rice), millets (finger millet, foxtail millet), pineapple, orange and other citrus species (Singh et al. 2013).

An explorative research design and qualitative approach were followed in this study. Districts and grassroots knowledge holders were sampled purposively based on the richness of local biodiversity being conserved. These knowledge holders were interviewed after a long period by door-

door contact with the help of village elders and *Gaon Burha* (customary head of each village in Ar.P.). A number of anthropological and ethnographical tools —including interviews, life histories and direct observations — were adopted to explore relationships between the people and biodiversity conservation. These methods helped us to shape a consistent story of connections between a particular culture and its relationships with biodiversity. An interview schedule with open-ended questions was applied to explore local knowledge and biodiversity conservation. The information about farmers' knowledge and creativity was recorded through personal interviews using a set of questions and informal interactions as well. The exploration of women's wisdom was pursued in participant observations complemented with personal interviews over a period of time. Permission to publish the information explored in study was given and prior informed consent (PIC) of each knowledge holder to convey their knowledge was confirmed.

Results

Case Study 1: Indigenous cotton landrace

Our study revealed that some grassroots innovative farmers have selected and conserved some unique landraces of great agricultural importance. For example, in Azamgarh district (UP), a farmer named Mr. Radheshyam Singh has played a leading role in the conservation of an indigenous perennial cotton variety (*Gossypium* sp.) identified from his own homegarden. He narrated in his experience on conserving this variety as under [translated from *Bhojpuri* language]:

> "Taking into account the *unique phenotype characteristics (Banaspatic bridhi lakshan) in terms of ornamental value of this cotton landrace, I conserved a single plant of this indigenous variety and multiplied it over the years. From the available seeds, multiplication was carried out in the kitchen garden itself. Few best performing plants were again selected based on the plant height, size of leaves and bolls. Observations revealed that a few plants provided quality bolls year round in good quantity at an interval of one and half months. Looking to these peculiar characteristics, in the third year, I decided to multiply the regular boll producing plants. Ultimately, this landrace became a matter of discussion among the villagers. Subsequently, most of the farmers procured its seeds from me and planted them in their kitchen gardens for increasing the aesthetic value".*

This cotton landrace grows to about 8-10 feets and the stem diameter is 4-6 inches. It shows good resistance to the aphids and white flies, and under irrigation, it produces about 350 high quality bolls per plant.

Case Study 2: Indigenous landraces of drumstick and guava

In another example, local perennial varieties of *sahjan (Molinga oleifera* L.) and guava (*Psidium guajava* L.) have been conserved by Mr. B.R. Choudhary (an amateur horticulturist) for over 25 years. He explained in his own words [translated from Hindi] his efforts on the domestication of these two wild types:

> *"Based on continued observations on plant performance and fruit quality of these wild types, I brought them under management in home garden. The ultimate aim was to conserve and multiply them for distribution among the peers. The conserved sahjan variety provides fruits in plenty year round. The taste is quite different from other varieties. The unique attribute of this landrace is that even a single branch simultaneously bears flowers, unripened and ripened fruits throughout the year. The fruits are suitable for making pickle, use as vegetable and in ethnomedicines. A person suffering from influenza, muscle and waist pain (especially women) are given the cooked leaves (ethnic vegetable). A decoction of the tree bark is used for bathing. The guava variety fruit is most suitable for jelly, RTS (ready to serve) and cordial preparation".*

The conservation and distribution of such wild horticultural resources by Mr. Choudhary points to the fact that creative ideas of community members hold tremendous significance in the sustainable management of plant genetic resources. Such initiatives in combination with formal knowledge can provide even better crop ideotypes than the existing ones.

Case Study 3: Domestication and conservation of ethnomedicinally important species

In a series of explorations and learning with grassroots biodiversity conservators, the authors identified Mr. Bamang Taniang, a traditional healer of the *Nyshi* tribe of Papumpare, Itanagar district of Arunachal Pradesh. He is an expert in healing human ailments using ethnomedicines. During his healing work, Mr. Bamang has identified more than 60 indigenous plant species having medicinal and food value. He selects such ethnobotanicals from the biodiversity rich Abotani hills. He has devoted 3 hectares of his land in the 'community forest' to conserving plants through domestication.

He selects the plants on the basis of the nature of a particular species, its population, its ability to adapt to a new habitat, its age and agroedaphic requirements. His initiative has prevented the overexploitation of plants in community forests. In the last 3-4 years, he has expanded his network and founded an association of herbalists of Arunachal Pradesh for the purpose of developing and promoting herbal medicines. His contribution in conserving indigenous plant biodiversity has also been recognized at the national level by the SRISTI (Society for Research and Initiatives for Sustainable Technology and Institutions) and he has been conferred the Biodiversity Conservation Champion award (Figure 1).

Figure 1: Shri Bamang Tanyang. Photo: SRISTI, Ahmedabad

Case Study 4: Conservation of food and ethnomedicinally important species

Mrs. Orik Rallen, an *Adi* woman of Sibut Village, East Siang district of Ar P (Figure 2), is an ethnomedical practitioner and community mobilizer to promote the conservation of indigenous plant biodiversity. She sensitizes the rural women about the importance of domesticating and conserving the forest plant species. Her efforts resulted in the conservation of 23 local crop varieties and 36 local forest species. She has encouraged the *Adi* women in scientific value addition of locally available food resources for enhancing their market value. Her efforts were instrumental in promoting community-based plant conservation.

Recognizing the erosion of indigenous knowledge from the younger community members, she teaches school children about local plant resources, their biocultural and medicinal values and sustainable harvesting methods by taking them to community forests and home gardens.

Figure 2: Mrs Orik Ralen. Photo: SRISTI, Ahmedabad

In search of indigenous plant resources, she often visits the neighbouring areas and other districts. From these visits, she learned from elders about the use of *engin* (local variety of sweet potato) and *singe-engin* (local variety of tapioca) as breakfast foods, as well as their methods of processing; and acquired knowledge on the range of traditional knowledge about use of '*toko*' (*Livistona jekinsiana* Griff) fruits, *onger* (*Zanthoxylum rhetsa*) leaves, *bangko* (*Solanum spirale)* leaves and *dekang* (*Gymnocladus burmanicus*) fruit. She has formed a self help group (SHG) to promote local plant food-based micro-enterprises with the help of R&D Institutions. She has trained women on reducing the content of the toxic cyanogenic glycoside taxiphyllin (which on hydrolysis, releases hydrogen cyanide) in bamboo shoots, consumed frequently by the *Adi* tribe. Excessive consumption of unprocessed bamboo shoots by pregnant women results various deformities in unborn babies[1] (Bhardwaj et al. 2008). She has tirelessly worked to develop value added food products from seeds of *namdung* (*Perilla ocymoides* and *Perilla frutescence*), and stems of *pagi-perang* (a sour-tasting plant found in community forests) to promote livelihoods of *Adi* women. Due to her efforts, about 12 landraces of different crops were deposited to the National Bureau of Plant Genetic Resources, New Delhi (NBPGR) (Table 1). For these contributions, she was conferred the Biodiversity Conservation Champion award by SRISTI, Ahmadabad on 21st December, 2009 at the national food festival.

Case Study 5: Conservation of indigenous agrobiodiversity

Our study was conducted with Mrs. Pem Dolma (Figure 3), a resident of Dirang Bazar, in 2004. She was a unique repository of traditional knowledge and culture. Un-fortunately, she died in 2006. For many years, she was involved in conserving local indigenous varieties of crops and developing cures for human and animal ailments. She propagated a way of living rooted in the culture and tradition of her people, the *Monpa*. Dolma's accomplishments in conserving local, indigenous varieties of crops have been acknowledged and rewarded by SRISTI Samman Award in 2007 at IIM, Ahmedabad. The indigenous varieties (27) of local crop species she conserved were submitted in her and related community's name to the NBPGR for *ex situ* conservation (Table 1).

Figure 3: Mrs Pem Dolma.
Photo: Ranjay K. Singh

Table 1: The local landraces and varieties deposited in National Bureau of Plant Genetic Resources, New Delhi as the folk variety.

IC-Number	Collector number	Botanical name	Common name	Village	District and State	Types of variety
IC-427030	CHF/NIF-1	*Triticum aestivum*	Brad wheat	Dirang Bazar (DB)	West Kameng, Ar P	Folk Variety
IC-427031	CHF/NIF-2	*Hordeum vulgare*	Barley	DB	West Kameng, Ar P	Folk Variety
IC-427032	CHF/NIF-3	*Hordeum vulgare*	Barley	DB	West Kameng, Ar P	Folk Variety
IC-427033	CHF/NIF-4	*Fagopyrum esculentum*	Buckwheat	DB	West Kameng, Ar P	Folk Variety
IC-427034	CHF/NIF-5	*Fagopyrum sp.*	Buckwheat	DB	West Kameng, Ar P	Folk Variety
IC-427035	CHF/NIF-6	*Oryza sativa*	Paddy	DB	West Kameng, Ar P	Folk Variety
IC-427036	CHF/NIF-7	*Elucine corcana*	Finger millets	DB	West Kameng, Ar P	Folk Variety
IC-427037	CHF/NIF-8	*Lab-lab purpureus*	Indian bean	DB	West Kameng/ArP	Folk Variety
IC-427038	CHF/NIF-9	*Lab-lab purpureus*	Indian bean	DB	West Kameng, Ar P	Folk Variety
IC-427039	CHF/NIF-10	*Lab-lab purpureus*	Indian bean	DB	West Kameng, Ar P	Folk Variety
IC-427040	CHF/NIF-11	*Phaseolus vulgare*	Rajma	DB	West Kameng, Ar P	Folk Variety
IC-427041	CHF/NIF-12	*Penicum psitopodim*	Millet	DB	West Kameng, Ar P	Folk Variety
IC-427042	CHF/NIF-13	*Coriandrum sativum*	Coriander	DB	West Kameng, Ar P	Folk Variety
IC-427043	CHF/NIF-14	*Lagenaria siceraria*	Bottle gourd	DB	West Kameng, Ar P	Folk Variety
IC-427044	CHF/NIF-15	*Cucurbita sativus*	Cucumber	DB	West Kameng, Ar P	Folk Variety
IC-427045	CHF/NIF-15A	*Glycine max*	Soybean	DB	West Kameng, Ar P	Folk Variety
IC-427046	CHF/NIF-16	*Pisum sativum*	Field pea	DB	West Kameng, Ar P	Folk Variety
IC-427047	CHF/NIF-17	*Cucurbita moschata*	Pumpkin	DB	West Kameng, Ar P	Folk Variety
IC-427048	CHF/NIF-18	*Cucurbita moschata*	Pumpkin	DB	West Kameng, Ar P	Folk Variety
IC-427049	CHF/NIF-19	*Momordica charantia*	Bitter gourd	DB	West Kameng, Ar P	Folk Variety
IC-427050	CHF/NIF-20	*Brassica spp.*	Mustard	DB	West Kameng, Ar P	Folk Variety
IC-427051	CHF/NIF-21	*Allium sativum*	Garlic	DB	West Kameng, Ar P	Folk Variety
IC-427052	CHF/NIF-22	*Allium sepa*	Onion	DB	West Kameng, Ar P	Folk Variety
IC-427053	CHF/NIF-22A	*Panicum scrobiculatum*	Millet crop	DB	West Kameng, Ar P	Folk Variety
IC-427054	CHF/NIF-23	*Capsicum annuum*	Chillies	DB	West Kameng, Ar P	Folk Variety
IC-427055	CHF/NIF-24	*Oryza sativa*	Paddy	Sibut/ Rasam (SR)	East Siang, ArP	Folk Variety

(Contd.)

IC- Number	Collector number	Botanical name	Common name	Village	District and State	Types of variety
IC-427056	CHF/NIF-25	*Zingiber officinale*	Zinger	SR	East Siang, Ar P	Folk Variety
IC-427057	CHF/NIF-26	*Vigna radiata*	Black gram	SR	East Siang, Ar P	Folk Variety
IC-427058	CHF/NIF-27	*Zea mays*	Maize	SR	East Siang, Ar P	Folk Variety
IC-427059	CHF/NIF-28	*Elusine coracana*	Finger millet	SR	East Siang, Ar P	Folk Variety
IC-427060	CHF/NIF-29	*Coriandrum sativum*	Coriander	SR	East Siang, Ar P	Folk Variety
IC-427061	CHF/NIF-30	*Caianus caian*	Red gram	SR	East Siang, Ar P	Folk Variety
IC-427062	CHF/NIF-31	*Illicium verum*	Star anise	DB	West Kameng, Ar P	Folk Variety
IC-427063	CHF/NIF-32	*Illicium verum*	Star anise	DB	West Kameng, Ar P	Folk Variety
IC-427064	CHF/NIF-33	*Curcuma longa*	Turmeric	SR	East Siang, Ar P	Folk Variety
IC-427065	CHF/NIF-34	*Zingiber officinale*	Zinger	SR	East Siang, Ar P	Folk Variety
IC-427066	CHF/NIF-35	*Allium cepa*	Onion	SR	East Siang, Ar P	Folk Variety
IC-427067	CHF/NIF-36	*Allium* sp.	Onion	SR	East Siang, Ar P	Folk Variety
IC-427068	CHF/NIF-37	*Brassica* spp.	Mustards	SR	East Siang, Ar P	Folk Variety

IC= Indian collection number

Her modes of conservation of local landraces/species and their relative importance for food and livelihood are presented in Table 2. The local varieties conserved by Mrs Dolma are compatible with the socio-economic conditions and food habits of the *Monpa* and are adapted to the local biophysical conditions. She formed an informal network of *Monpa* women with an aim to promote biodiversity conservation. She provided seeds of local crops to needy community members free of cost, with the provision that such seeds be used for ensuring genetic conservation and food security.

Mrs Dolma used to preserve seeds of crops using *timbur* leaves (*Xanthoxylum americanum* Mill.); the fruits are used as a spice, (Figure 4) and dried chilli pods. She perceived that this practice helped to avoid pests during storage. The crop seeds, along with the *timbur* leaves and dry chillies, are kept in an earthen pot and placed near the kitchen where cooking is done on a wood stove. Placement of the earthen pot near kitchen also ensures maintenance of proper moisture levels in seeds throughout the year. To avoid pest infestation, maize cobs are hung over bamboo for storage.

Figure 4: A tree of *timbur*.
Photo: Ranjay K. Singh

Her interest in indigenous crops stemmed from her belief that new varieties and chemical inputs in agriculture destroy the land and environment and also affect the quality of food. From childhood, she had been a part of the *Monpa* practice in of exchanging different maize, barley, wheat and other food crops for seeds of wild plants, dried meat, wool, skin, ghee and wet cheese (chhurpi). She had an in-depth knowledge of diverse local crops and their landraces, domesticated by her ancestors.

Her family possesses five acres of land where they grow a variety of crops including maize, wheat, barley, finger millets and, vegetables like brinjal and potato. She reminisced about the systems of cooperation that her people followed in the past. One such system was *mila* (rural social institution), where neighbours helped each other in sowing seeds and collecting minor forest products. She was saddened to observe the changes and erosion in cultural resources (including knowledge) among younger generation. She realized that the younger community members were moving away from their roots and were not aware of their cultural heritage. She said that:

Table 2: Major local crops conserved by Late Mrs. Pem Dolma.

Local name	Botanical/common name	Mode of conservation	Rank order for importance*
Bean-badi	*Panicum psilopodium* var. *coloratum*	Home garden	II
Bean-chhoti	*Panicum psilopodium* var. *psilopodium, P*	Home garden	II
Bong (with and without awns)	*Hordeum vulgare*	Agricultural field	I
Broomsa peela	*Cucurbita moschata*	Home garden	III
Broomsa saphed	*Cucurbita moschata*	Home garden	III
Bundagmo	*Amaranthus* sp.	Agricultural field and home garden	I
Burangsaga	Zingiberaceae family plant	Home garden	I
Chong	*Allium* sp.	Home garden	I
Foxtail millet	*Setaria italica* L.	Agricultural field	I
Kaibandu	A local bitter gourd like crop	Home garden	III
Lai saa penche	*Brassica* sp.	Home garden	I
Lai saag, leme	*Brassica* sp.	Home garden	I
Lamm- mann bada)	*Allium* sp.	Home garden	I
Lau	*Lagenaria siceraria*	Home garden	II
Lee	*Glycine max* L.	Home garden	I
Lee-bean	*Lablab purpureus*	Agricultural field and home garden	I
Local wheat	*Triticum aestivum* L.	Agricultural field	I
Mandua	*Eleusine coracana*	Agricultural field	I
Mann chhota	*Allium* sp.	Home garden	I
Manthong	*Cucumis sativus* L	Home garden	II
Matar	*Pisum sativum*	Home garden	I
Phaphda meetha	*Polygonum fagopyrum*	Agricultural field and home garden	I
Phaphda teeta	*Polygonum fagopyrum*	Agricultural field and home garden	I
Rajma	*Phaseolus vulgaris*	Agricultural field and home garden	II
Solu	*Capsicum annuum* L (local chilli)	Home garden	I
Taktak	*Basella alba*	Home garden	III
Utush	*Coriandrum sativum* L	Home garden	II

*: Rank was decided by conservator on the scale of most important with score '3' to least important with score '1' for the food, nutritional and medicinal security

"Our culture is getting adulterated. The young people want to eat fast food and are not interested in food made from millets or forest based products. Very few can speak the Monpa language. What will happen to our culture, religion and local traditions?" Clearly, the future of the *Monpa* is at stake!

Our experience with Mrs. Dolma points to the significance of grassroots conservators, particularly the women's role in agrobiodiversity conservation.

Case Study 6: Domestication and Conservation of food and ethnomedicinally important biodiversity

Another woman, famous for biodiversity conservation is Mrs Ade Modi (60 years, Figure 5), of Napit village, East Siang district (Ar.P). She has conserved a number of indigenous plant species used in food, medicine and other livelihood needs. The wife of well-known progressive farmer Mr. Jakut Modi, who himself introduced many horticultural crops like *patchouli* (*Pogostemon cablin*) in East Siang district, Mrs Modi is dedicated to the organic farming of indigenous fruits and vegetables and ethnomedicinal plants. She domesticates forest flora used in food and medicine in her kitchen garden and *jhum* lands (slash and burn agriculture). The contributions of Mrs Modi are presented in the remainder of this results section, under different categories within the overall area of biocultural conservation.

Figure 5: Mrs Ade Modi. Photo: Ranjay K. Singh

Knowledge on biocultural resources

Mrs. Modi observed that practices in '*pahad kheti aur jangal kheti*' (ecosystems where hill farming and forest farming occur) are quite different from the farming practices in her native place. For example, low land paddy cultivation predominates in and around the Napit village, while a great number of local crops and ethnobotanicals (a total of 41 species) are being conserved through various modes (Table 3) in her native place. These local crops and ethnobotanical products are more concentrated in hill ecosystems as compared to the plains as perceived by Mrs Modi.

The tuber called *remet* (similar in appearance with *kachu* - the wild *Colocassia*, or taro), is collected and stored for off-season use. The productivity per plant of this indigenous *Colocassia* is about 6-7 kg of tuber. Similarly, *uli* and *usha* (nutritious wild herbs used as vegetables) have been domesticated and conserved in Mrs Modi's kitchen garden (Figure 6). These bioculturally

Figure 6: Home garden of Mrs Ade Modi. Photo: Ranjay K. Singh

important resources are used as ingredients in different ethnic dishes. These resources are of great help in supporting livelihood security. Mrs Modi knows the methods of propagation through cuttings, selection of plant parts and preservation of seeds for the next cropping. She correlates ecological components with her culture of these plants. For example, her astrological knowledge as part of *Adi* culture, is reflected in her narration:

> "*In our locality, the age of a boy is calculated on the basis of the numbers of jhum, while in and around Pasighat town it is done by looking the movement of sun*".

Mrs Modi explained that *nayang* (*Erigeron canadensis*) is used as a boiled vegetable. The tender leaves of the local shrub *agjok* (*Bauhnia berigata*), used as a living fence, is also used as a green vegetable after boiling. A living fence of *agjok* marks the boundary between lands of two farmers or two communities. Some of the local plants promoted by Mrs Modi are now considered important from a monetary standpoint and are being grown by the other *Adi* women for cash sale in the local market (Pasighat) by women's groups (Figure 7). When the Pasighat market is cut off from the rest of India, the demand for these indigenous vegetables increases considerably due to non-availability of introduced produce. Because of the

Figure 7: A group of *Adi* women formed for conservation of ethnobotanicals and selling it in local markets. Photo: Ranjay K. Singh

Table 3: Major local crops conserved by Mrs. Ade Modi.

Local name	Botanical/common name	Mode of conservation	Rank order for importance*
Akshap	*Mussenda glabra*	Jhum land & home garden	I
Angyat	*Setaria italica* L	Jhum land	I
Bamboo-tenga	*Bambusa arundinacea*	Jhum land & community forest	I
Bangko	*Solanum spirale*	Jhum land & home garden	I
Belang	*Artocarpus heterophyllus* Lam. (wild jackfruit)	Jhum land	III
Engin	*Manihot esculenta* (two varieties)	Jhum land & home garden	II
Gende	*Gynura cripidioides*	Jhum land & home garden	I
Gobar oying	*Amaranthus* sp.	Jhum land & home garden	II
Jabjar	*A herb*	Jhum land & home garden	III
Kachchu	*Colocasia esculenta* L	Community forest & jhum land	II
Kekir	Medicinal ginger	Jhum land & home garden	I
Kirin kero	*A local creeper*	Jhum land & home garden	II
Kopi	*Solanum viarum*	Jhum land & home garden	I
Kppir	*Solanum khasianum*	Jhum land &home garden	I
Marshang	*Spilanthes acmella* (two species)	Jhum land & home garden	I
Mirung	*Eleusine coracacna* L	Jhum land	I
Namdung	*Perilla ocimoides*	Jhum land & home garden	II
Nanung	*Polygonum* sp.	Jhum land & home garden	II
Nayang	*Erigeron canadensis*	Jhum land & home garden	II
Nupuk	*A local herb*	Jhum land & home garden	I
Obur	*Solanum* sp.	Jhum land & home garden	I
Ogik,	*A herb*	Jhum land & home garden	I
Okilibo	*A local herb*	Jhum land & home garden	III
Oko-bodo	*Erigeron canadensis*	Jhum land & home garden	II
Okomama	*A local herb*	Jhum land & home garden	I
Ombe	*Carica papaya* (local variety)	Jhum land & home garden	II

(Contd.)

Local name	Botanical/common name	Mode of conservation	Rank order for importance*
Ongen	*Gynura crepidiodes*	Jhum land & home garden	I
Onger	*Zanthoxylum rhesta*	Jhum land & home garden	I
Ongin	*Clerodendrum colebrookianum*	Jhum land & home garden	I
Oporang	*Litsea polyantha*	Jhum land & home garden	III
Oyik	*Pouzolzia benettiana*	Jhum land & home garden	I
Paksum	*Ocal banana* (red small, black-*kodum*, and white)	Jhum land & home garden	III
Paput	*Gnepalium affine*	Jhum land & home garden	I
Piyak-eup	A local herb	Jhum land & home garden	II
Poi	*Basella rubra*	Jhum land & home garden	II
Rainfed paddy	*Oryza sativa* (6 varieties)	Jhum land	I
Shapa	*Zea mays* Local maize (4 varieties)	Jhum land & home garden	I
Takeng	A variety of local gingers	Jhum land & home garden	I
Tambul	*Areca catechu*		I
Toko-patta	*Livistona jenkinsiana* Griff	Jhum land & home garden	I

*Rank was decided by conservator on the scale of most important with score '3' to least important with score '1' for the food, nutritional and medicinal security.

popularization of these indigenous vegetables, the annual income of local women has now increased up to Rs. 50,0000 to 75,000 annually.

Ethnotaxonomic and ecological knowledge

Mrs Modi had sound knowledge of ethnotaxonomy: the names and classification of plants and fungi. She easily distinguished the local species of mushroom in two broad categories. In the first category, she placed those mushroom species that germinate under normal soil conditions, while mushroom species germinating on wood were placed in the second category of (Figures 8 & 9). While selecting a particular mushroom species for food, one has to be very skillful as consuming the wrong species may be harmful, even deadly. Mushrooms growing in normal soil are collected during May to August. After this period, even some edible species of mushroom may become poisonous (notably, the white-coloured mushrooms locally called *ingaleng*) as perceived by Mrs Modi. The mushroom called *doyak*, with a light yellow cap, found during July-August, is considered one of the best edible types. This species is mostly found under *sirang* (*Castanopsis kurzii*) trees and it may be assumed that soils rich in organic matter due to high quantities of *sirang* leaf litter impart good quality and taste to *doyak*.

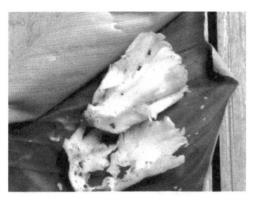

Figure 8: *Innyik tadar* mushroom on *Gorum* tree.
Photo: Ranjay K. Singh

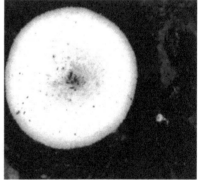

Figure 9: *Lengot tadar* (mushroom) found on *Sisar* tree rotten logs.
Photo: Ranjay K. Singh

The major mushroom species germinating on different woods is locally called *lolem* (light white-coloured caps). The *lolem* having whitish caps most often grows on older *sirang* trees. It is found year round and is eaten after boiling. Another species, called *tapar*, looks like the button mushroom. It is found on *taan* followed by *sirang* and *takuk* trees. This mushroom is most readily available in June and July. Other types, similar to *tapar*, are found more on the *sisar* tree. To generate an income from mushrooms, some of the local women cut the branches of the *sisar* tree and put them for decaying under the shade of trees in a specially made trench. After wood has decayed by about one-third, the mushroom spawn will germinate, and the mushrooms will be ready for sale within 7-10 days.

Indigenous biodiversity and human health

Mrs Modi explained that the local herb called *piyag eup* (Figure 10) is used as a vegetable as well as an ethnomedicine to treat stomach disorders such as indigestion. *Rorum* (Figure 11) is a very effective creeper in curing malaria and jaundice. The tender parts of *rorum* — along with leaves — are boiled and given to the patient for quick recovery from weakness caused by these ailments. In the case of acute malaria, the patient has to bathe in water filtered through a fine cotton cloth collected from the pig shed. Alternatively, in another practice, cloth used to wrap a newborn baby is soaked in water, and this water is sprinkled over the body of a patient suffering from malaria. Mrs Modi mentioned that some people give a lukewarm decoction made from *rorum* to a person suffering from malaria, taken for several weeks as a supplementary treatment.

Figure 10: Local plant *Piyag eup*
Photo: Ranjay K. Singh

Figure 11: *Rorum* (a local creeper)
Photo: Ranjay K. Singh

For cuts and bleeding, an extract of the green leaf of *nomsing-ying* (*Ageratum conyzoides*) is applied. The leaves of this plant are also used to heal blains of human and animals. In the case of severe cuts and bleeding, the epidermal layer of a bamboo stem is crushed, and the powder is applied over the cut, tied in place with a cotton cloth. In the initial days after childbirth, the juice of a local ginger (*Zingiber pardochlamys*) called *kekir* mixed with chicken soup is given to women to help them recover from the birth. Two or three fruits of the seeded banana (*Musa balbisiana*), called *athiya kopak*, are given to patients suffering from dysentery. Some of plant species [(e.g. *oyik* (*Pouzolzia benettiana*), *onger* (*Zanthoxylum rhesta*) and *ogen* (*Solanum nigrum*)] are frequently mentioned in local proverbs, denoting their cultural significance. In the case of *oyik* and *ogen* Mrs Modi narrated the following proverb:

> "*Oyik doboname reyik reyik,*
> *Ogen doboname regen regen.*"

(Namely, those who regularly consume *oyik* become handsome and beautiful, and those who consume *ogen*, maintain good physique and strong muscles.)

The leaves of *ogen*, after being warmed on a wood fire, are slightly pressed on the eyelid to relieve eye pain. The leaves are also useful in treating eye infections. The tender leaves of *ongin* (*Clerodendrum colebrookianum*) are used to treat diarrhoea. The women cook *ongin* with dried meat of *mithun* or dried fish, chilli and bamboo shoots to improve the taste and give it to people who suffer from weakness. To treat diabetes and constipation, *ongin* leaves are boiled with salt and given to the patient. Mrs Modi's knowledge of the different dimensions of even a single plant species, such as *ongin*, points to the need based preparations in different ways to serve different purposes.

Ethnobotanical species and animal rearing

Consumption of meats is highly prevalent among almost all the ethnic communities of Ar.P., including the *Adi*. These tribes primarily depend on three animals: *mithun* (a species of cattle), poultry and pigs. *Mithun* is a semi-wild animal and is not reared, although the *Adi* do provide *mithun* in community forests with common salt. In times of need, *Adi* people catch *mithun* [for sacrificing them] to slaughter on different cultural occasions. Surplus dried *mithun* meat is preserved and used as a contingency food. Different plant products, including tubers, rhizomes and fruits, are fed to the pigs and poultry.

Mrs. Modi reported that poultry reared on maize (called *shapa*) grains grow faster and rapidly develop body weight. Rice grains are the second preferred feed for poultry. The green leaves of onion (*dilap*) and garlic (*Allium sativum*) are given to poultry to promote health and good growth. A number of plant products are collected by Mrs. Modi to feed pigs. The indigenous *Colocasia* varieties called *kochu*, along with fruits of jackfruit (belang, collected from the forest) and unripe papaya (*Carica papaya*) constitute major feed products for pigs. To make these feeds tastier, Mrs Modi adds the pith of wild banana stem (*Musa balbisiana*), and after boiling, gives this to the pigs. She also gives wild growing *shishnu-patta* (*Urtica parviflora*, Figure 12,) to the pigs as feed after boiling its leaves with rice husks and grains. The tender leaves of this species are also used as an ethnic green vegetable. Based on the colour, Mrs. Modi identifies white and black varieties of *kochu*, both of which are considered prime ingredients of pig feed. The

Figure 12: *Shishnu patta* used for pig feed and as vegetable. Photo: Ranjay K. Singh

Adi women form *reglep* (a local informal social institution) for digging and transporting *kochu* tubers from the forest.

Ecoculture and biodiversity conservation

Mrs Modi described how the *Adi* culture is closely intertwined with biodiversity. For example, during the *merum* festival, she stores crop seeds for the next year's sowing in the house (*ekum*). To preserve these seeds, she makes a drier, called *perap*, of bamboo, with 3-4 layers. In the first layer, large-sized fish and meat are preserved, in the second layer, medium to small-sized fish, and in the last, one or two layers of crop seeds are kept. Among these seeds, *lai patta* (*Brassica* sp.), maize, brinjal, *kopi* (*Solanum* sp.), *angyat* (fox millet), *mirung* (finger millet) and chilli are major types.

The festival called *Solung* is celebrated while clearing the forests to establish new *jhum* lands. At this time, the seeds of indigenous crops are sown for the *Rabi* (winter) season. In the following two festivals, called *Aran* and *Etar*, the *Adi* community assembles near the *jhum* lands, where they perform social rituals and worship the *mithun*. This cultural practice is intended to encourage higher crop production in the next year. The *pumeng* festival is celebrated at the time of paddy crop maturity. During this festival, different kinds of ethnic foods are prepared and distributed among the clan members. The *yageng* festival is celebrated at the time of maturity of *angyat* (fox-millet) and *mirung* (finger millet). These two crops are used for making *apong* (traditional alcoholic beverage). The *petpum* festival is exclusively celebrated to encourage higher production of pigs. During the agricultural festival called *pelum*, rice powder is made and distributed among the clan members. Similarly, during other agricultural festivals such as *dorung* (celebrated every five years of *Jhum* cycle) *mithun* is sacrificed and its meat is shared among the clan members, close relatives and friends.

Every *Adi* festival is, in fact, a collective celebration related to nature and culture worship. They provide opportunities for the *Adi* community to learn about traditional ecological knowledge, indigenous biodiversity, knowledge sharing and cultural cohesion. They also ensure collective decision-making on different aspects of biodiversity and natural resources use and management. Unfortunately, this social cohesiveness, collective knowledge systems and eco-literacy are eroding at an alarming rate among the younger generations. This loss is more pronounced in and around the more settled areas where globalization, urbanization and different socio-political factors have adversely affected people's perceptions and knowledge of management and conservation of biodiversity and natural resources.

During discussions on indigenous biodiversity and its conservation, Mrs Modi narrated the worth of ethnoecological and ethnobotanical knowledge in her own words:

> *Si sangisana Omeo*
> *Takamai Nglukai buku Ne Kukai*
> *Manam Lokai Oying Ogoi Dogebonam Em*
> *Takamai Ken Nying Supai Aido*
> *Nugluk Nane Boluke*
> *Manam Lok Oying Oge, Kenbo Naam Sim*

> "*Our ancestors were surviving on the forest and its associated resources like foods, fruits, vegetables, ethnomedicines, etc. They have given us the wisdom of natural resources used in every aspect of life, and hence we should not forget their value. Otherwise, the identity of Adi tribe will be at stake. Sadly, these days our younger generation is losing such invaluable biocultural resources and even they are too poor in ecological ethics, and are blindly adopting the western way of life*".

Local knowledge in risk management

The *Adi* people live in subtropical ecosystems with very unpredictable weather conditions (especially in the Pasighat area). This environmental anomaly compels the local communities to develop location specific ecological knowledge to enable them to manage environmental and food security risks. In this context, Mrs Modi's attempts to increase crop diversification deserve particular merit. She pioneered the practice of relay cropping, in which maize seeds are sown amongst the standing rice crop 20 days before the rice harvest in July. The paddy is harvested at the point of the earhead, and the remaining stalk is cut down to act as mulch for the maize crop and add carbon to the soil. This practice not only maximizes crop productivity in dimensions of time and space but also mitigates crop failures and maintains food security.

In another ecological practice, bamboo and *toko-patta* (*Livistona jenkinsiana* Griff) trees become over mature after 30-40 years, and then they are chopped down and the fields are burned during March-April to maximize the organic ash content in fields (Figure 13). After the first showers of May and June, the seeds of local maize are sown in each hill of the bamboo rhizome with the help of a bamboo stick. Usually, the sowing of maize is preferred at the middle slope of hills, not in the plains. Because the gaps are greater in between two bamboo rhizomes' hills, these spaces are utilized for planting local *Colocasia*, cucumber, ginger, beans and a number of ethnobotanically important species. After harvesting the maize ears, and produce from other local crops, the maize stalks are chopped down and mixed with the soil to improve its quality. This practice helps modify the micro-ecosystems, improves soil fertility and reduces soil erosion. After the maize and cucumber crops are harvested, they are sold in the local market, bringing, on average Rs. 15,000-20,000 annually to each *Adi* woman.

Figure 13: A bamboo garden and *toko-patta* grove after slash and burn. The cleared land will be used for *jhum* cultivation.
Photo: Ranjay K. Singh

Further, Mrs Ade Modi described how in the months of March and April, a group of *Adi* women will visit the community forests (*Morang*) and select patches full of herbs and grasses. These are chopped down and burned after a week. After the rains, a number of ethnobotanicals and local mushrooms germinate in these patches, which are subsequently used as food and medicine. The mushrooms, *lai saag* and local black peppers are intercropped in areca nut plantations for cash sale. In the months of June and July, the *Adi* women collect the dung of elephants and/or cows, and spread it in selected patches of community forests. Local biodiversity enhanced through these place-specific ecological practices helps in diversifying the livelihoods of *Adi* women and to build their resilience.

Discussion

The criteria followed by tribal farmers and outstanding knowledge holders in selecting and conserving local landraces and indigenous plant species are compatible with those formal knowledge systems. These grassroots conservators, who have promoted local biodiversity through informal experimentation and innovative practices, provide insights and lessons for future conservation programmes (Singh et al. 2013c). During the Green Revolution, much emphasis was given to resource-rich farmers through the use of high yielding varieties and application of energy and fertilizer intensive inputs, while resource-poor farmers (poor in wealth but not in creativity and innovations), were mostly by-passed. The new introduced practices and crops resulted in the loss of local and informal ways of promoting and sustaining indigenous biodiversity and sustaining foods and natural resources (Singh et al. 2013c).

Most of the conservation practices or selection of particular species by innovative farmers and knowledge holders we found to be highly location specific. The conservation of these landraces and wild species is governed by a range of sociocultural, ecological, and climatic factors (Singh et al. 2013a). For example, the creative farmers of Indo-Gangetic plains and Central India selected those species for conservation and improvement which were of immediate household use (e.g. drumstick, *Moringa oleifera* L. and guava for year round, use and cotton as an ornamental plant). These species are adapted to sub-humid and subtropical climates. The approach applied by the innovative farmers featured in our study is relevant here. Both the farmers (Mr. Radheshyam Singh and BR Choudhary) first observed phenotypic attributes of plants, kept continuous observations to monitor these attributes, and later propagated and multiplied the species over a period of time.

The approaches and knowledge applied in conserving culturally important landraces and ethnobotanicals by the outstanding knowledge holders from the *Monpa, Adi* and *Nyshi* tribes (Ar.P.) were quite different than those of central India and the Indo-Gangetic plains. Living in fragile ecosystems and highly dependent on nature, these communities have traditionally practiced *in-situ* conservation of local species and landraces (Singh et al. 2013a). The communities' cultures, resource base, intensive interaction with nature and ecological ethics in sustaining bioculturally important resources have made such knowledge holders unique. The efforts of some of these grassroots conservators in biodiversity and natural resource management have been appreciated and recognized at the national level in India

(Singh 2004; Singh and Srivastava 2010b), validating the immense value of their endeavours towards enhancing biocultural diversity.

We found that most of the conservation of indigenous biodiversity in mountainous ecosystems takes place in the *jhum* lands and home gardens. This is a clear indication that agroecosystem diversity plays a considerable role in adaptive practices in conservation. Home gardens are the preferred place where women demonstrate their creativity in plant conservation (Singh 2004; Singh et al. 2013a). For example, the conservation of important food and medicinal plant species by Mrs Rallen, Mrs. Modi and the late Mrs Dolma was mostly undertaken in the home gardens. In a nutshell, these home gardens should not be neglected by those planning and making policies related to biodiversity conservation and climate change adaptations (Roy et al. 2013). The conservation mechanisms applied by these women through their location specific experimentation in home gardens, as well as in the *jhum* lands, has implications for carbon management (Kumar and Nair 2004; Roy et al. 2013), and for sustaining biocultural diversity and ecological services (Pulido et al. 2008).

A comparative analysis among the knowledge holders in our study revealed that the number of species conserved by women was relatively higher than those of men. This indicated a greater role for women in biodiversity conservation, since they not only domesticate and manage many plant resources but also add value to the plants for economic adaptation in changing climatic and development scenarios (Bushamuka et al. 2005). Unfortunately, the voices of such women are often unheard, and their creativities go unnoticed (Singh et al. 2013c). In order to strengthen women's involvement in formal aspects of biodiversity conservation, there is a need to learn from them and promote their participation in decision-making processes relating to environmental conservation. These women could also play a key role in participatory sustainable management of natural resources. However, in the scenario of climate change and environmental degradation caused by various factors, these women will need capacity building support (training, domestication schemes, etc.) to allow an integration of their creativity with formal knowledge. In particular, the younger community members need both cultural and environmental education so that they can build bonds with their elders and cultural experts and participate in learning about and conserving biodiversity. Given the states' fragile ecosystems, those people who rely on a subsistence economic base need to be given options for sustainable livelihoods. These options can never be made by imposing policies and projects from the outside; rather they must be developed at the grassroots level where women's wisdom is widely recognized and respected.

A major concern arises in determining what and how best formal systems can receive insights from the local communities and incorporate their knowledge and practices. It is well recognized that participatory learning and experimentation have a sustainable future and can effectively address existing challenges and those to come. Globally, it has been suggested that small farms and small-holding farmers maintain more resilient and biodynamic agrobiodiversity than large farms and farmers (Altieri and Koohafkan 2008). Incorporating informal local knowledge supported with encouraging policies and scientific research, may indeed lead to better solutions for at least some of the most common problems relating to conservation of indigenous biodiversity and adaptation to environmental change. It is ironic that so much effort and resources are being invested in research and development led by the 'top-to-bottom approach', and minimal attention has been paid to learning from the outstanding local knowledge holders and grassroots conservators. The scientists who develop the variety and package of practices for industrialized food production systems are praised and recognized, but not the farmers and local communities who have given more sustainable solutions to the world (Altieri and Koohafkan 2008). Indeed, this is the real time when lessons for sustainable agriculture, conservation of natural resources and adaptations to variable or changing climate have to be learned from grassroots innovators and outstanding knowledge holders.

Conclusion and Policy Implications

We have concluded in this study that local farmers and communities are able to develop their own crop varieties by domesticating their genetic resources through trial and error and careful and meticulous observation. Such approaches and practices of these farmers and communities that contribute to conservation of local plant biodiversity can be of immense importance in sustaining local crop gene pools, and meeting food and nutrition requirements of local peoples. This knowledge can be of great value for future research and development projects, especially for regions having a subsistence resource base. To limit any probable loss of biodiversity and consequent destruction of natural habitats, farming and land management techniques as adapted by women of *Monpa* and *Adi* tribal communities may be tailored to enhance the conservation of biodiversity (both agrobiodiversity and native species diversity), related cultures and livelihoods. The programmes that facilitate food product development and value addition, particularly for women in more remote mountainous ecosystems, may help strengthening the livelihoods of their communities.

Women were observed to be real custodians of indigenous biodiversity, used for food, ethnomedicines and other purposes. However, the women are often not acknowledged in planning and policies relating to conservation and agricultural development. Such grassroots women, who are actively involved in the domestication and conservation of crop species and ethnobotanicals, may be rewarded and promoted through economic incentives. Any economic benefit, accrued from enhanced local biodiversity, can be equitably shared among such women and other stakeholders.

Policies on farming systems need to be refined, a task for which actions from other stakeholders are needed. For example, developing local markets for promotion of biocultural food products coming from the conserved plant resources can be undertaken. Such products are now becoming more in demand by the emerging urban middle class who themselves are unable to produce such food. This process can further enhance the process of market-led conservation, and women participating in it may receive extra benefits from their home gardens and subsistence farming practices. There is also a need to create incentive mechanisms and promote investments in eco-agriculture, and to ensure intellectual property rights and resource ownership of the local communities. Special attention is needed to improve and sustain the biodiversity-rich ecosystems and save them from the onslaught of industrial development and large-scale, externally driven projects that will destroy both the biodiversity of the area and the knowledge systems that help to create and maintain it.

Endnote

[1] In the past, pregnant women were advised not to consume fresh or unprocessed bamboo shoots, to reduce the probability of any deformity in the fetus. To process and neutralize the cyanogenic glycoside taxiphyllin, the chopped off bamboo shoot is pushed inside the green bamboo and the open mouth portion of the cylinder is made air-tight using *ekkam* leaves (*Phyrinum pubenerve*). The shoots containing the bamboo cylinder are then placed in a water stream under the shade of a tree. After 3-4 months, the fermented and processed bamboo shoots are re-collected and then safely consumed. This practice is no longer followed in the transformed villages.

Acknowledgements

The first author (RKS) is grateful to the outstanding knowledge holders who provided an opportunity to stay with them and learn about the practices related to biodiversity conservation. The financial support obtained from the College of Horticulture and Forestry, Central Agricultural University,

Pasighat Arunachal Pradesh is thankfully acknowledged. The logistic support obtained from Jawaharlal Nehru Krishi Vishwa Vidyalaya, Jabalpur, M.P. and Central Soil Salinity Research Institute, Karnal is also acknowledged. Editorial helps received from Dr Anshuman Singh is thankfully acknowledged.

References

Altieri MA, Koohafkan P (2008) Enduring farms: climate change, smallholders and traditional farming communities. Third World Network, Jalan Macalister, Penang, Malaysia. www.twnside.org.sg. Accessed on 12-12-2013

Altieri MA, Funes-Monzote FR, Petersen P (2011) Agroecologically efficient agricultural systems for smallholder farmers: contributions to food sovereignty. *Agronomy for Sustainable Development*. DOI 10.1007/s13593-011-0065-6

Bhagirath S, Chauhan BS, Mahajan G, Sardana V, Timsina J, Jat ML (2012) Productivity and sustainability of the rice–wheat cropping system in the Indo-Gangetic plains of the Indian subcontinent: problems, opportunities, and strategies. *Advances in Agronomy*, 117: 315-369

Bhardwaj R, Singh RK, Wangchu L, Sureja AK (2008) Bamboo shoots consumption: traditional wisdom and cultural invasion. *In:* Arunachalam A & Arunachalam, K (eds), Biodiversity utilization and conservation. Avishkar Publishers, Jaipur, India, pp. 79-82

Bushamuka VN, de Pee S, Talukder A, Kiess L, Panagides D, Taher A, Bloem M (2005) Impact of a homestead gardening program on household food security and empowerment of women in Bangladesh. *Food and Nutrition Bulletin*, 26(1):17-25

Brush SB (2005) Farmers' right and protection of traditional agricultural knowledge. CAPRi Working paper No. 36, CGIAR Systemwide Program on Collective Action and Property Rights. Secretariat, International Food Policy Research Institute, Washington D.C., U.S.A http://www.capri.cgiar.org/pdf/capriwp36.pdf. Accessed on 16-12-2013

Fan S, Chan-Kang C (2005) Is small beautiful? Farm size, productivity, and poverty in Asian agriculture. *Agricultural Economics*, 32(1):135-146

Fingleton J (1993) Conservation, environmental; protection and custody land tenure. In: Alcorn J (ed), Papua New Guinea conservation needs assessment (Biodiversity Support Programme, Washington DC), pp 31-56

Gladis T (2003) Agrobiodiversity with emphasis on plant genetic resources. *Naturwissenschaften*, 90 (6): 241-50

Groombridge B, Jenkins MD (2002) World atlas of biodiversity: earth's living resources in the 21st century. Berkeley: University of California Press

Janetos S (2005) Participating in the field: research experience in Northern Greece. *Environments*, 33(1):97-114

Kumar BM, Nair PKR (2004) The enigma of tropical homegardens. *Agroforestry Systems*, 61:135–152

Maffi L, Woodley E (2010) Biocultural diversity conservation: a global source book. Earthscan, London. pp. 3-60

McNeely JA, Scherr SJ (2001) Common ground, common future: how ecoagriculture can help Feed the world and save wild biodiversity (IUCN-The World Conservation Union Report No. 5/01. Gland: IUCN and Future Harvest)

Pretty J, Sutherland WJ, Ashby J, Auburn J, Baulcombe D, Bell M, Bentley J, Bickersteth S, Brown K, Burke J, Campbell H, Chen K, Crowley E, Crute I, Dobbelaere D, Edwards-Jones G, Funes-Monzote F, Godfray HCJ, Griffon M, Gypmantisiri P, Haddad L, Halavatau

S, Herren H, Holderness M, Izac AM, Jones M, PKoohafkan P, Lal RL, Lang T, McNeely J, Mueller A, Nisbett N, Noble A, Pingali P, Pinto Y, Rabbinge R, Ravindranath NH, Agnes Rola, Roling N, Sage C, Settle W, Sha JM, Shiming L, Simons T, Smith P, Strzepeck K, Swaine H, Terry E, Tomich TP, Toulmin C, Trigo E, Twomlow S, Vis JK, Wilson J, Sarah P (2010). The top 100 questions of importance to the future of global agriculture. *International Journal of Agricultural Sustainability,* 8(4):219-236

Pretty JN (2002) Agri-culture: reconnecting people, land and nature. Earthscan Publication, London, UK, pp 2-101

Pretty JN (2007) The earth only endures. Earthscan Publication, London, UK, pp 99-139

Pulido MT, Pagaza-Calderón EM, Martínez-Ballesté A, Maldonado-Almanza B, Saynes A, Pacheco RM (2008) Home gardens as an alternative for sustainability: challenges and perspectives in Latin America. In: Paulino U de Albuquerque, Ramos MA (eds) Current topics in ethnobotany. Research Signpost, Trivandrum, Kerala, India

Rai M (2000) A perspective for developing action programme for sustaining productivity of rice-wheat systems in the Indo-Gangetic plains. In: *CYMMIT (*ed*),* Developing an action programme for farm level impact in rice-wheat systems of the Indo-Gangetic plains. Rice-wheat Consortium Paper Series. Rice-wheat Consortium for Indo-Gangetic Plains, New Delhi, India, pp 48-56

Rao RR, Hajra PK (2005) Fern allies and ferns of Kameng District of Arunachal Pradesh. *Indian Forest Record Botany,* 106:327-349

Roy B, Rahman MH, Fardusi J (2013) Status, diversity, and traditional uses of homestead gardens in northern Bangladesh: a means of sustainable biodiversity conservation. ISRN Biodiversity, ID 124103, 11 pages. http://dx.doi.org/10.1155/2013/124103

Saxena S, Chandak V, Ghosh S, Sinha S, Jain N, Gupta AK (2002) Cost of conservation of agro-biodiversity. Working paper No. 2002-05-03, IIM Ahmedabad, India, May, 2-7

Singh RK (2004) Agrobiodiversity: conserving diversity and culture - Pem Dolma. *Honey Bee,* 15 (3):12-13

Singh RK, *Adi* Women (2010) Biocultural knowledge systems of tribes of Eastern Himalayas. NISACIR, CSIR, New Delhi

Singh RK, Igul P (2010) Climate change, REDD and biocultural diversity: Consultations and grassroots initiative with Indigenous People of Arunachal Pradesh. *Current Science,* 99 (4):421-422

Singh RK, Pretty JN, Pilgrim S (2010) Traditional knowledge and biocultural diversity: learning from tribal communities for sustainable development in northeast India. *Journal of Environmental Planning and Management,* 53(4):511-533

Singh RK, Rallen O, Padung E (2013a) Elderly *Adi* women of Arunachal Pradesh: 'Living encyclopedias' and cultural refugia in biodiversity conservation of the eastern Himalaya, India. *Environmental Management,* 52(3):712-735

Singh RK, Shrivastava RC (2010a), Consensus on prior informed consent and conservation of biodiversity based traditional knowledge systems: a step towards protection of intellectual property rights. *Current Science,* 98(5): 1-2

Singh RK, Shrivastava RC (2010b), Grassroots biodiversity conservators of Arunachal Pradesh: National recognition and reward. *Current Science,* 99(2):162

Singh RK, Singh A, Pandey CB (2013b) Agro-biodiversity in rice–wheat-based agroecosystems of eastern Uttar Pradesh, India: implications for conservation and sustainable management. *International Journal of Sustainable Development & World Ecology,* (2013), DOI:10.1080/13504509.2013.869272

Singh RK, Singh D, Sureja AK (2006) Community knowledge and biodiversity conservation by the *Monpa* tribe. *Indian Journal of Traditional Knowledge,* 5(4):513-518

Singh RK, Sureja AK (2006) Community knowledge and sustainable natural resources management: Learning from *Monpa* tribe of Arunachal Pradesh, *TD: The Journal of Transdisciplinary Research in South Africa,* 2 (1):73-102

Singh, RK, Dwivedi BS (2002) Utilization of location specific indigenous paddy varieties for sustainable production: An appraisal of tribal wisdom. *Journal of Asian Agri-History,* 6(3): 261-267

Thrupp LA (2000) Linking agricultural biodiversity and food security: the valuable role of agrobiodiversity for sustainable agriculture'. *Int Affairs,* 76 (2): 265-81

Turner NJ (2005) The earth's blanket: traditional teachings for sustainable living (Culture, place and nature: Studies in anthropology and environment). Douglas and Mckintyre Press, Canada, pp 34-80Turner NJ, Clifton H (2009) "It's so different today": Climate change and indigenous lifeways in British Columbia. *Global Environmental Change,* 19(2):180-190

Chapter – 6

Status and Contribution of Non-cultivated Food Plants Used by Dawro People in Loma District, South Ethiopia

*Kebu Balemie**

Abstract

This study was aimed to document the Dawro people's ethnobotanical knowledge, identify Non-Cultivated Food Plants (NCFPs) and associated challenges to their consumption. Semi-structured interviews, focus group discussions, market survey and field observations were conducted to gather the ethnobotanical data. The study documented a total of 47 non-cultivated food plant species belonging to 44 genera and 34 families. Shrubs (36.7%) and herbs (32. 7%) make up the highest proportion of these plants. The two most frequently used parts were fruits (62%) and leaves (18.4%).Over 70% of the species analyzed have additional uses, primarily for medicine, construction, and firewood. Most (68%) of the recorded species are consumed fresh without further processing and they are mainly fruits. The study found that non-cultivated food plants play a vital role to the Dawro people during food shortages. Of the reported species, *Amaranthus caudatus*, *A. graecizans*, *Corchorus oliterius*, *Erucastrum arabicum*, and *Portulaca quadrifida* were found to have the highest frequency of use report. Five economically important species (i.e. *Aframomum corrorima*, *Curcuma domestica*, *Piper capense*, *Rhamnus*

---
*kebubal@yahoo.com, kebubal@gmail.com

prinoides, and *Zingiber officinale*) were encountered both in wild and under cultivation. The study suggested some promising non-cultivated food plant species for possible domestication.

Keywords: Non-cultivated food plants, ethnobotany, indigenous knowledge, use, Dawro people.

Introduction

The contribution of Non-Cultivated Food Plants (NCFPs) as a local response to food insecurity has been widely documented (Becker 1983; cf. Campbell 1987; IIDE 1995; Zemede 1997; Kebu and Fassil 2006). In this text, non-cultivated, or wild food plants include those undomesticated wild and weedy plants whose leaves, fruits, seeds or roots are consumed seasonally or occasionally by local communities. These plants are part of the cultural diet of local communities due to long-history local use. Documented sources indicate that many wild food plants have important nutritional, economic, ecological, agronomic, food security, and cultural values (Mnzava 1997). In particular, some fruits and leafy vegetables supplement micronutrients and vitamins and thereby combat hidden hunger or malnutrition.

In many African countries, NCFPs make up a portion of people's daily diet during difficult times. For instance, in Botswana, *Amaranthus thunbergii,* fresh or dried, prepared in several traditional ways, is consumed as a relish or used together with sorghum or maize meal to make porridge (Madisa and Tshamekang 1997). In Swaziland, Ogle and Grivetti (1985) reported a wide range of wild foods plants whose use is particularly associated with a shortage of staple foods and noted that these wild species actually contributed a greater share to the annual diet than cultivated species. A similar pattern was observed from Gahana (Abbiww 1996). Fleuret (1979) also discussed wild leafy vegetable relishes as an essential element of the Shamba people's diet in Tanzania. According to the same author, the local people appreciated the taste of traditional leafy vegetables more than that of introduced types. As in many other African countries, the gathering and consumption of NCFPs has been common and central to the culture of many local people in Ethiopia. The level of consumption, however, varies from place to place according to the nature of local customs and beliefs, the availability of grain and other domesticated crops, and climatic conditions. Gathering of wild food plants is common, particularly under difficult conditions as part of survival strategies (Zemede 1995; Zemede 1997; Guinand and Dechassa 2000; Zemde and Mesfin 2001; Kebu and Fassil 2006).

Status of Non-cultivated Food Plants in Ethiopia: A Brief Overview

The food plants of Ethiopia are believed to constitute about 8% of the total number of vascular plant species of the country. Of these, 75% could be categorized as wild, semi-wild or naturalized in different agro-ecological zones or vegetation types (Zemede 1995; Zemede 1997; Zemede and Mesfin 2001). Recent review by Ermias et al. (2011) indicates that a total of 413 wild plants are consumed in various ecological zones of the country. This diversity is valuable for the livelihoods of local people as potential sources of food, nutrition and income. Paradoxically, this rich food plant diversity is in contrast with the persistence of poverty and food insecurity in the country. NCFPs have largely been used as supplementary, emergency or famine food, with elevated intake in times of food scarcity (Zemede 1995; Zemede and Mesfin 2001; Getachew et al. 2005; Kebu and Fassil 2006). In South Ethiopia, where there are many different tribes still living with their indigenous beliefs and traditions, the consumption of wild food plants seems to be one of the important local survival strategies, particularly during food shortages (Guinand and Dechassa 2000; Kebu and Fassil 2006). The Konso people, for example, still have and use a well-developed knowledge concerning which wild food plants can best provide a dietary supplement in periods of food shortage. According to Guinand and Dechassa (2000), the Konso people compensate for damaged or reduced crop harvests by increasing their collection of wild plant foods.

Compared to other African countries with diverse food habits, the consumption of non-cultivated food plants in Ethiopia is however, highly selective and restricted due to traditions, beliefs, attitudes and taboos. As a result, many potentially edible plants have largely been overlooked and this has shrunk the traditional food basket (Guinand and Dechassa 2000; Getachew2001). At the same time, some species, which were at one time part of the cultural diet of local communities are now diminished in numbers or even faced local extinction due to anthropogenic and natural factors including the current climate change. In many cases, loss a species is accompanied with erosion of ethnobotanical knowledge and cultural practices resulting in malnutrition, food insecurity and ecological vulnerability (Getachew 2001; Zemede Asfaw and Mesfin 2001).

Research related to NCFPs in Ethiopia

Previous studies have revealed that Ethiopia's enormous food plant diversity and associated rich cultural practices of the ethnic communities in varied agro-ecological zones of the country are poorly documented (Zemede and Mesfin 2001).To date, most of the ethnobotanical studies in Ethiopia have largely focused on medicinal plants, but a handful of them were specifically conducted on wild food plants. With these limited studies, Amare (1974) reported information on the role of 167 wild edible plant species in the native Ethiopian diet. Zemede (1995, 1997) published a brief note on some cultivated and non-cultivated indigenous food crops of the country. Guinand and Dechassa (2000) categorized and discussed the role of wild food plants in parts of South Ethiopia. Getachew et al. (2005) and Kebu and Fassil (2006) identified and documented detailed uses/contributions of wild edible plants used by studied ethnic groups. Very recently, ethnobotanical studies on edible plants have been conducted in south and central parts of the country (e.g. Tilahun and Mirutse 2010; Debela et al. 2011; Asegid and Tesfaye 2011). Ermias et al. (2011) reviewed wild edible plants of the country and compiled the overall inventory information. Research conducted to date revealed that for many socio-cultural groups of the country, local knowledge, use patterns and distribution of NCFPs species remain largely unstudied. This study was conducted to document ethnobotanical knowledge and identifies the NCFPs and challenges to their consumption in the Loma district of the Dawro zone.

Conservation Status

Considering the diversity of higher plant species in Ethiopia and in light of the growing ecological threats to the habitats where NCFPs grow, the conservation efforts to date are minimal. There is little or no targeted *ex-situ* collection efforts of NCFPs by the national gene bank of Ethiopia. The situation with regards to *in-situ* conservation of NCFPs is not different from *ex-situ* condition. However, some of noticeable conservation of NCFPs might be through the established protected forests. Most of these protected forest areas were intended to protect the broad vegetation types of the country. In addition, there are several area closures and protected areas such as parks designated for conservation of biodiversity, including NCFPs.

Research Methodology

Study area and people

Loma District is located in between 37°16' E and 6°48' N, adjacent to the Omo River Valley, in the Dawro Zone, South Ethiopia (Figure 1). The area is characterized by varied landscapes including valleys, mountains, and steep topography. The altitudes range from 850 meters above sea level (masl) at the Omo River Valley to 2200 masl at Balefulas. Topographically, the area consists of rugged mountains, deep gorges and river valleys. Three agro-climatic zones are found in the areas: highlands (>2000m), mid-altitudes (1500-2000m) and lowlands (<1500m). The flora is mainly mixed Acacia and broad-leaved natural vegetation in the lowland areas. *Syzygium guineense* and *Ehretia cymosa* are common species in mid-altitude areas. Most of the highland areas are devoid of natural vegetation and are exposed to high levels of soil erosion. At hill bottoms and in the valleys, there are patches of natural forest vegetation including bamboo forest.

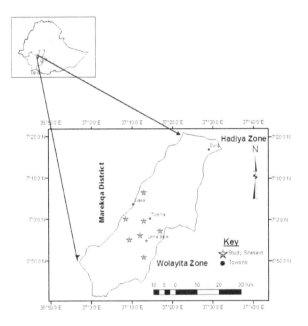

Figure 1: Location of Loma District on Map of Ethiopia.

The human population of the district is about 109,158 (Central Statistics Authority 2007). The Dawro people are the dominant ethnic group and speak Dawro language, which belongs to the Omotic language family. The Dawro people share similar language and culture with neighboring Wolayita and Gamo peoples. Subsistence farming with mixed livestock-crop production is the dominant economic activity. Domesticated species such as *Zea mays*, *Sorghum bicolor*, *Enset ventricosum*, *Ipomoea batatas Solanum tuberosum*, *Colocasia esculenta* and *Manihot esculenta* are common staple (Personal observation).

Methods

Seven localities (Bale Fulasa, Fulas Borzi, GendoWalca, Yelowarbete, Denebe Bola, Eela Becho and Zima Warma) (Figure 1) were sampled in this study. The selection criteria for these localities include natural vegetation cover, agro-ecological zones, drought/famine-prone areas and accessibility. A total of 43 male and female informants of various age groups were interviewed. Information on NCFPs was gathered using semi-structured interviews, market survey, and focus group discussions. The information documented included local names, uses, part used, distribution and perceived threats to NCFPs. Field observations were made to document habitat distribution and herbarium collection on the reported NCFPs. The herbarium specimens were identified and deposited at National Herbarium (NH), Addis Ababa University. Overall use value was calculated for the most frequently reported leafy vegetables as described by Martin (1995).

$$UVis = \frac{\sum Uis}{ni}$$

Where
$UVis$ = Overall value attributed to a particular plant species by one informant
Uis = All of the uses mentioned in each event by informant
ni = Number of events in which an informant gave information on a species.

Results and Discussion

Local knowledge on NCFPs

The rich tradition of using NCFPs still persists among Dawro community, presumably due to their continued interaction with these plants. The local communities were familiar with the uses, parts used, food values and properties, seasonality, phenology and habitat distribution of most recorded NCFPs. In this study, children easily remembered and listed high numbers of edible fruit bearing plants. Likewise, the interviewed women were knowledgeable, particularly about leafy non-cultivated food plants. The variation in knowledge across age groups is perhaps attributed to their daily activities. For example, children spend more time in the bush lands or fields, where they tend cattle, and this might have helped them to gain more experience and easily recall fruit-bearing species whereas women, who are busy with home and garden management, have acquired more experience with leafy greens. Similar findings were reported in Zemede (1995) and Ogle and Grivetti (1985) regarding the skill and knowledge in collection and processing of wild edible plants. On the other hand, local knowledge about certain species varies across ecological zones and with different vegetation types. For example, some species were found to be unique to the lowland vegetation (woodland) type and are known only to inhabitants of this area. Thus, species such as *Dioscorea praehenslis* and *Raphionacme borenensis* are known and used only by low landers. This revealed that variation in ecological zones contributes to variation in occurrence of species and distribution of knowledge on wild food plants (Ogle and Grivetti 1985).

Diversity and uses

A total of 47 non-cultivated food plant species belonging to 44 genera and 34 families were recorded in this study. The life forms of the recorded species were mainly shrubs (n=18, 36.7%), herbs (n=16, 32.7%), and trees (n=14, 28.6%). The edible parts included roots, fruits, leaves, stems, flowers, and seeds. Of these parts, fruits (62%) and leaves (18.4%) make up the highest proportions. Over 70% of these species are also used in other ways, primarily as medicine, in construction, and as firewood (Table 1). Trees and shrubs yield a vast array of fruits, which can provide essential nutrients, especially during food shortages. According to the informants, fruits are used as snacks or supplementary foods by all age groups, especially during herding or field working hours. Some of the widely used fruit bearing trees/shrubs were: *Carissa spinarum, Dovyalis abyssinica, Physalis peruviana, Solanum nigrum, Syzygium guineesne, Ehretia cymosa, Ficus elastica,* and *Rubus apetalus*. Fruits of these species were used by over 50% of the respondents residing at mid-altitude and highland areas. Similarly, the fruits from *Balanites aegyptiaca, Diospyros abyssinica, Sclerocarya birrea, Strychnos innocua,* and *Ximenia americana* were used by over 60% of the respondents in lowland areas.

Leafy vegetables are another type of NCFPs, consumed mainly during food shortage. Most of these leafy edibles are herbaceous and weedy in nature. Often, they accompany staple crops such as sorghum and maize. They are also used in making sauce in the absence of cabbages. Various studies showed that leafy vegetables are a primary source of vitamins A, C and other micronutrients. For example, *Amaranthus* contain more vitamin A precursor, calcium, and iron than green cabbage or *Brassica oleracea* (IPGRI 2003). The present study documented *Amaranthus caudatus, A. graecizans, Corchorus oliterius, Erucastrum arabicum* and *Portulaca quadrifida* as the commonly used (40% report) leafy vegetables (Table 2). A high percentage of reported use can be a measure of how widely a particular species is consumed within the Dawro community. These species can grow quickly and become harvestable even under low rainfall. Through genetic improvement and widespread cultivation they could become promising horticultural vegetable species in Ethiopia. In many parts of Africa, these species are under varying level of cultivation (Guarino 1997) and are favorite food ingredients, especially as condiments and additions to sauces and relishes (Fleuret 1979; Mnzava 1997).

Various authors have discussed seasonality in harvesting and consumption of wild food plants (Campbell 1987; IIDE 1995; Zemede 1997; Kebu and Fassil 2006). In some cultures, gathering for household consumption is a major activity only during difficult times when other staple food crops are in short supply.

Table 1: List of non-cultivted food plant used by Dawro people.

Scientific Names	Vernacular names	Family	Habit	Habitat	Altitude	Uses
Aframomum corrorima (Bram) Jansen	Okashe	Zingiberaceae	Herb	Fallow/farmland	1600-2000m	Sp (F), Me (F)
Amaranthus caudatus L.	Gegebsa	Amaranthaceae	Herb	Fallow/farmland	1250-2900m	Ed (Se), Me (Se)
Amaranthus graecizans L.	Cumadhe	Amaranthaceae	Herb	Farmland/fallow land Fd (Wh)	1250-2900m	Ed (L),
Annona senegalensis Pers.	Monoqo	Annonaceae	Tree	Low land woodland	1250-1500m	Ed (F), Fd (L), Co (S), Fi (S,Br), Ri (Wh)
Balanites aegyptiaca (L.) Del.	Bedena	Balanitaceae	Tree	Low land woodland	1250—1500m	Ed (F,L), Fd (L), So (B), Te (S), Fi (S,Br), Sh (Wh)
Bridelia scleroneura Muell. Arg.	Xema/Uzuzee	Euphorbiaceae	Shrub	Wooded land	1250-2000m	Ed (Se), Po, Fe (S)
Capsicum annuum L.	Mixamixo	Solanaceae	Herb	River bank	1400m	Ed (F)
Carissa spinarum L.	Ladee	Apocynaceae	Shrub	Low/mid-land mixed shrub land	1250-2100m	Ed (F), Fe (S, Br), Me (R)
Cleome gynandra L.	Kembata sansa	Capparidaceae	Herb	Fallow/farmland	1250-1500m	Ed (L)
Cynoglossum amplifolium Hochst.ex DC	Dersa qerchecha	Boraginaceae	Herb	Fallow land	2000m	Me (L)
Corchorus oliterius L.	Joljoluwa	Tiliaceae	Herb	Fallow/farm land	1250-1500m	Ed (L), Fd (L)
Cordia africana Lam.	Meqota	Boraginaceae	Tree	Mixed woodland	1250-2000m	Ed (F), Co(S), Te (S), Fd (L), Fi (Br), Bh (Br)
Crambe hispanica L.	Duriganje	Brassicaceae	Herb	Farmland/ fallow land	1600-2000m	Ed (L), Fd (L)
Curcuma domestica Valeton	Irdiya	Zingiberaceae	Herb	Low land wood land	1250-1500m	Sp (R)
Dioscorea praehenslis Benth.	Sasaa	Dioscoreacaeae	Climber	Low land wood land	1250-1500m	Ed (R)

(Contd.)

Scientific Names	Vernacular names	Family	Habit	Habitat	Altitude	Uses
Diospyros abysinica (Hiern.) F. White	Dul'o	Ebenaceae	Tree	Low/mid-land mixed shrub land	1250-2000m	Ed (F), Co (S), Sh (Wh), Te (S), Ch (S)
Dovyalis abyssinica (A.Rich.) Warburg	Wesiniya/	Flacourtiaceae	Tree	Farm border, fallow land	1600m-2900m	Ed (F), Fe (S), Fi (Br)
Dombeya torrida (G.F. Gamel) P. Bamps.	Unxulle	Sterculiaceae	Tree	Mixed shrub land	1250-2000m	Ed (F), Fi (S, Br), Co (S)
Ehretia cymosa Thonn.	Itriwanje	Boraginaceae	Tree	Mid land wood land	1600-2000m	Ed (F), Co (S)
Erucastrum arabicum Fisch. &Max.	Kefoxugunta	Brassicaceae	Herb	Forest margin	1700-2000m	Ed (L)
Ficus elastica Roxb. ex Hornem.	Eta/Bo'ina	Moraceae	Tree	Farmland/border/ river bank	1400-1700m	Ed (F), CO &Te (S)
Grewia bicolor Juss.	Xewayee	Tiliaceae	Shrub	Low land wood land	1300m	Ed (F), Ro (B), Fe (S, Br)
Hoslundia opposita Vahl	Botana tucadhe	Lamiaceae	Herb	Farmland	1300m	Ed (F)
Lannea barteri (Oliv.) Engl.	Gumade	Anacardiaceae	Tree	Woodland	1300m	Ed (F), Fi (S, Br)
Lantana trifolia L.	Shankshasha	Verbenaceae	Shrub	Farm border	1500m	Ed (Se)
Lepisanthes senegalensis (Poir) Leenh.	Cellee	Bignonaceae	Tree	Low land wood land	1300m	Ed (F), Sh (Wh), Fi (S, Br)
Mussaenda arcuata Poir.	Mirxako	Rubiaceae	Shrub	Low land woodland	1250-1500m	Ed (F), Me (F)
Mimusopus kummel Bruce ex A.DC.	Gurcho	Sapotaceae	Tree	Farm border	1250-1500m	Ed (Se), Te (S)
Onkoba spinosa Forssk.	Hagilowa /Hagilee	Flacourtiaceae	Shrub	Low land woodland	1250-1500m	Ed (F), Fe (Br)
Phoenix reclinata Jaq.	Zemba	Arecaceae	Shrub	Forest margin	2000m	Ed (F), Ba (L), Tb (L)
Physalis peruviana L.	Fugal'ee	Solanaceae	Herb	Farmland/border	1600-2900m	Ed (F)
Piper capense L.f.	Tinja	Piperaceae	Herb	Forest margin/farm border/farmland	2100m	Sp (Se), Me (Se)

(*Contd.*)

Scientific Names	Vernacular names	Family	Habit	Habitat	Altitude	Uses
Portulaca quadrifida L.	Mergude	Portulacaceae	Herb	Farmland/fallow land	1250-1500m	Ed (L), Fd (Ab)
Raphionacme borenensis Venter & M.G.Gilbert	Shomotiree	Asclepiadaceae	Herb	Lowland wood land	1250m	Ed (R)
Rhamnus prinoides L'Herit.	Gisho	Rhamnaceae	Shrub	Mid land wood land	1900m	Flv (L, S)
Rubus apetalus Poir.	Gomoree	Rosaceae	Shrub	Forest margin	1600-2900m	Ed (F), Fe (S, Br), Me (F)
Sclerocarya birrea (A.Rich.) Hochst.	Weslecha/ Tunqalo	Anacardiaceae	Shrub	Low land woodland	1250-1500m	Ed (F), Co (S), Fi (S, Br)
Solanum nigrum L.	Gayeta	Solanaceae	Herb	Farmland/river bank	1250-2000m	Ed (F), Me (L)
Strychnos innocua Del.	Ugugee	Loganiaceae	Shrub	Low land wood land	1250-1500m	Ed (F), Fd (L), Co (S), Te (S) Fl (S,Br)
Syzygium guineense Subsp.afromontanum F. White	Ocha	Myrtaceae	Tree	High land forest	2000m	Ed (F), Me (L, S), Ri (Wh), Co (S)
Syzygium guineense Subsp. macrocarpa (Engl.) F. White	Derensa	Myrtaceae	Shrub	Mixed wood land	1400-2000m Co (S)	Ed (F), Fi (S, Br),
Techlea noblis Del.	Gesa	Rutaceae	Tree	Forest	2000m	Ed (F), Me (Br), Co (S), Te (S), Fi (S,Br)
Uvaria leptocladon Oliv.	Boyna	Annonaceae	Shrub	Low land woodland/fallow land	1300-1500m	Ed (F), Co (S)
Vangueria madagascarensis Gomel.	Deshalomee/ bakka	Rubiaceae	Shrub	Low land wood land	1300-1500m	Ed (F)
Ximenia americana L.	Mulaho	Olacaceae	Shrub	Low land wood land	1300-1500m	Ed (F), Fi (S,Br), Fe (S), Me (F)
Ximenia caffra L.	Aste	Olacaceae	Shrub	Lowland wood land	1300-1500m	Ed (F), Fe (S)
Zingiber officinale Roscoe	Yenjeluwa	Zingiberaceae	Herb	Forest margin	1600-2000m	Sp (Rz), Me (Rz)

NB. (Ed=edible; Me=medicinal; Fuel wood (Fi=firewood, Ch=charcoal); Co=construction (timber, pole, Fe=fencing); Te=technology (farm implements, tool handles, household furniture); Sp=spice; Flv=flavoring; Sh=shade; Ri=ritual; Ba=basket). F=fruit; L=leaves; R=root; Se=seeds; S=stem; B=bark; Br=branches; Wh=whole; Ab=above ground; Rz=rhizome; Fl=flower

The frequency of gathering and quantity of consumption is high during periods of food shortage (Zemede 1995). However, the extent of gathering varies from species to species. Some, such as *Moringa stenopetala* and *Solanum dasyphyllum, which commonly grow near home gardens,* are gathered and used on a year-round basis. Similarly, *Boscia senegalensis* is used year round as dietary staples by Peuhls people of Senegal (Becker 1983).

Table 2: Non-cultivated food plants with high use report frequency.

Edible fruit plants	Local names	% of use report	Use values
Amaranthus graecizans	Gegebsa	100	0.125
Carisa spinarum	Ladee	86	0.038
Diospyros abyssinica	Dul'o	74.4	0.063
Syzygium guineesne	Ocha	74.4	0.069
Erucastrum arabicum	Kefoxugunta	72.1	0.002
Physalis peruviana	Fugal'ee	72.1	0.032
Rubus apetalus	Gomoree	72.1	0.058
Portulaca quadrifida	Mergude	69.8	0.067
Corchorus olitorius	Cumadhe	58.1	0.08
Amaranthes caudatus	Gegebba	46.5	0.17
Balanites aegyptiaca	Bedena	41.9	0.094
Ficus sycomorus	Eta/Bo'ina	41.9	0.033
Solanum nigrum	Gayeta	41.9	0.095
Ximenia americana	Mulaho	41.9	0.087

Apart from food value, the study showed that local people use the reported NCFPs for a variety of other purposes such as for medicine, construction, firewood, forage and other minor uses. Popular medicinal species, each reported by over 70% of informants, include *Amaranthus caudatus, Carissa spinarum,* and *Zingiber officinale*. Similarly, among species used for construction, *Cordia africana, Bridelia scleroneura* and *Sclerocraya birrea* are the ones most preferred for their timber products.

Food Processing and Consumption

The non-cultivated food plants and their parts reported in this study have been consumed in various forms. Processing times for the dishes prepared from the NCFPs, taste and associated side effects after consumption also vary. Species like *Dioscorea praehenslis,* for example, took exceptionally long processing (cooking time) before consumption. The leafy vegetables were assessed as having generally poor taste compared with the fruits, which are usually eaten raw and without further processing, mainly by children, in the field or at home. Leafy edibles and seeds of some species like *Amaranthus caudatus* were used together with grains of staple crops such as maize and sorghum, where it is milled together for usage as porridge. The mixing and consumption of the leafy vegetables or their seeds with other staple grains might improve the taste of these dishes. *A. caudatus* seeds are also used in powder form as a condiment like hot pepper. In the later case, the amaranth seeds are dried, mixed together with crushed and dried garlic containing some salt, and then pounded into powder, which is stored for use in subsequent days or weeks depending on the quantity. This powder is served with *injera*/pancake or unleavened bread. Similar information was reported by other authors (Guinand and Dechassa 2000; Kebu Fassil 2006). Similarly, the leafy and root parts are consumed after further cooking, usually boiling. The processing method for the leafy vegetables involves boiling them in water, then adding flour from grain such as sorghum, maize, and barley, cooking the mixture together, and then mashed it at intervals until it matures for hours. The matured dish, known as *Ibizaa* in the Dawro language, is then served with leavened or unleavened bread. A similar dish, known as *Kurkufa,* is reported in Kebu and Fassil (2006).

Consumption of NCFPs varies with availability of staple crops in stock, as well as among edible parts and species. Some NCFPs, such as weedy vegetables, are mainly collected and used in economically poor households to help alleviate food shortage. Consumption of fruits during normal times depends on their taste and quality. Fruits of *Balanites aegyptiaca, Syzygium guineense,* and *Ximenia mamerican* Figure 2 (a & b) are among edible fruits that are eaten during both normal and hard times.

Fiture 2: (a) *Syzygium guineense.*
Photo: Kebu Balemie

Fiture 2: (b) *Ximenia americana.*
Photo: Kebu Balemie

Local people's perceptions of NCFPs

Local people recognize the contribution of some NCFPs only during hard times. Weedy edibles are given poor image by most of the local communities. They are regarded as "poor peoples' food plants". Generally, there is a declining trend in consumption of these species. Various studies and reports also indicate that most of the nutritious and readily available but cheap food plants that are consumed by rural communities in other countries are discouraged and their consumption is regarded as a source of shame in Ethiopia (Guinand and Dechassa 2000; Getachew 2001; Kebu and Fassil 2006). Campbell (1986) and Tabuti (2001) have reported similar situations. Certainly, various factors have contributed to the present local people's perceptions. Understanding these factors might help in designing strategy for promoting the consumption of NCFPs.

On the other hand, growing global interest in diversification of food crops provide an opportunity to conserve and sustainably utilize these resources. Furthermore, most of these non-cultivated food species can grow on marginal areas even under the changing climate scenario. Promising results have been obtained in Kenya from nutritional analysis of some NCFPs (Kabuye 1997). In this case, the results from improved cultivation technique of some traditional vegetables were taken back to the communities, whose members then showed strong interest in using them as food. Such encouraging results should be promoted through communication of research-based information on the nutritional value and health benefits of these species, together with the preparation and cultivation methods.

Marketable Species

A market survey revealed that some food plants contribute to household income generation, in particular for women (Figure 3). The study showed that about 13 % of the recorded species were commonly sold at the local market. However, the degree of commercialization is different among the different types. Species such as *Aframomum corrorima, Piper capense,* and *Zingiber officinale* have growing market demand. The local market price of *Aframomum corrorima* fruit, for example, was over $2.5/Kg. They have good price at both local and international markets and are one of the country's export products. An equally important species was *Rhamnus prinoides* (leaves, seeds and stems are

Figure 3: *Moringa stenopetala* at Gato market place (Derashe District, South Ethiopia).
Photo: Kebu Balemie and Fassil Kebebew (2006)

pounded together to be used for flavoring *Tella*, common local beer in Ethiopia) and turmeric *Curcuma longa* (used for coloring sauces) was also among the commonly sold species locally and nationally. In fact, these species are horticultural spices in most parts of the country including the Dawro zone. They might be a source of potential genetic material for improvement of cultivated spices against changing climate and for resistance to associated pests or disease outbreaks. Apart from these, the Dawro people earn an income from sale of some popular leafy vegetables such as *Solanum dasyphyllum*. Children also sold fruits of *Syzygium guineense* subsp. *macrocarpa* at local market. In sum, the study revealed that income derived from the sale of the NCFPs makes a significant contribution to those poor households who lack any alternatives to sale in periods of hardship.

Habitat Distribution and Threats to NCFPs

The habitat distribution of the reported non-cultivated food plants were found diverse. Most of the fruit-bearing trees and shrubs were distributed within different forest patches along forest margin and riverbanks, and in woodlands. The leafy vegetables, on the other hand, grow mainly on fallowed lands, farm borders, farmlands and back yards. Some of these leafy vegetables have broad ranges of altitudinal distribution (Table 1). However, from our observation and informants' responses, the abundance of most NCFPs has greatly decreased in recent years. In addition, the availability of many fruit-bearing trees and shrubs is declining. Forests and woodlands, which were the main source of NCFPs, were increasingly degraded. The interviewees cited clearing of the natural vegetation for agriculture, overgrazing and overharvesting, commercialization of selected species, and uncontrolled fires as principal threats to wild plant biodiversity in the study area. Some species, such as *Cordia african*, *Bridelia scleroneura*, and *Sclerocraya birrea*, are threatened because of harvesting pressure for their good quality timber or use for poles. Furthermore, over collection of plants such as roots, seeds and fruits would have negative effects on the regeneration potential of these plants. Most estimates indicate that climate change is likely to affect the availability of NCFPs, particularly in drought-prone area. Effects include changes in growing seasons and growth cycles (phenology), and impacts on species distribution and diversity.

Challenges to the Consumption of NCFPs

Despite their major role as supplementary foods, NCFPs have received little/ or no attention by researchers, extension workers, policy makers, NGOs, or institutions in Ethiopia (Zemede 1995). Efforts made towards their conservation, promotion, value addition and domestication are lacking (Getachew 2001). As a result, the utilization and conservation of these food plants are challenged by several factors that vary from place to place and from region to region. Interview responses indicated that deliberate search for NCFPs from wild/ or from agricultural fields for home consumption has declined in recent years even during food shortage. According to informants, availability of relief foods, decline or loss of edible plants from wild/ or field, changing taste preferences, labor intensive harvesting and processing time requirement, side effects after consumption, people's perceptions and attitudes (viewed as poor people's food, backward food), fear of shame and other social stigma attached to the consumption of some NCFPs category (like famine NCFPs) were among the reported challenges. Seasonal availability of some preferred species and lack of awareness of the high nutritional value of NCFPS could be additional factors limiting their utilization.

Conclusion and Recommendation

The Dawro people are vested with a comprehensive knowledge of NCFPs in their environment. Their ethnobotanical knowledge has been contributed to their survival during food shortages. Nevertheless, these potential food plant resources, which could significantly help in combating nutritional, health and food insecurity, are not fully or adequately utilized. The present study found that the gathering and consumption of NCFPs is declining in the study areas. Several factors have contributed to the declining trend in consumption of NCFPs. Some of the plant species in question are threatened and have become locally extinct due to conversion of vast areas of woodlands and forests into cropland, over-harvesting, climate change, bush fires and overgrazing. Their loss in turn might have threaten ecological services, indigenous knowledge and the future food and health security of local people who have depended on these resources in the past. A better knowledge of the nutritional potential of these species and understanding the constraints attached to their consumption would favor their utilization.

Hence, farmers, extension workers and researchers should strive towards identification, development, conservation and sustainable use of the NCFPs. Species with high nutritive values can be recommended for a domestication program as garden crop or agroforestry species, which is important to diversify agricultural production and to reduce poverty. In addition, sensitization about nutritional importance, health benefits or other values would lead to both increased consumption and conservation of the resources. Further investigation into the nutritional, health, agronomic, value addition and processing of promising NCFPs is suggested for possible domestication and diversification of people's diets. In addition, the study suggests urgent collection of germplasm of threatened NCFPs for *ex situ* conservation.

Acknowledgements

I would like to thank Dawro informants who have shared their empirical knowledge about non-cultivated food plants in Loma district. The Institute of Biodiversity Conservation (IBC) is acknowledged for over all support during field data collection.

References

Abbiw D K (1996) Non-cultivated vegetables in Ghana. *In:* Proceedings of the IPGRI international workshop on genetic resources of non-cultivated vegetables in Africa: Conservation and Use: 29–31 August 1995. Nairobi: Institute of Plant Genetic and Crop Plant Research/International Plant Genetic Resources Institute

Amare G (1974) The role of wild plants in the native diet in Ethiopia. *Agro ecosystems,* 1: 45–56

Assegid A, Tesfaye A (2011) Wild edible trees and shrubs in the semi-arid lowlands of Southern Ethiopia, *Journal of Science and Development,* 1(1):5-19

Becker B (1983) The contribution of wild plants to human nutrition in the Ferlo, Northern Senegal. *Agroforestry Systems,* 1:257–267

Campbell B M (1986) The importance of wild fruits for peasant households in Zimbabwe. *Food and Nutrition,* 12(1):38-44

Campbell B M (1987) The use of wild fruits in Zimbabwe. *Economic Botany,* 41(30): 375-385

Central Statistical Authority (2007) Summury and Statistical Report of the 2007 Population and Housing Census. Dec. 2008, Addis Ababa, Ethiopia

Debela H, Njoka JT, Zemede A, Nyangito MM (2011) Wild edible fruits of importance for human nutrition in semi-arid parts of east Shewa Zone, Ethiopia: Associated indigenous knowledge and implications to food security. *Pakistan Journal of Nutrition,* 10(1):40-50

Ermias L, Zemede A, Ensermu K, Patrick V D (2011) Wild edible plants in Ethiopia: a review on their potential to combat food insecurity. *Afrika Focus,* 24(2):71-121

Fleuret A (1979) The role of wild foliage in the diet: A case study from Lushoto, Tanzania. *Ecology of Food and Nutrition,* 8(2):87-93

Addis G, Urga K, Dikasso D (2005) Ethnobotanical study of edible wild plants in some selected districts of ethiopia. *Human Ecology,* 33(1):83-118

Getachew O (2001) Food source diversification: Potential to Ameliorate the Chronic Food Insecurity in Ethiopia. *In:* Kenyatta C and Henderson A (eds.) Proceedings of the Potential of Indigenous Wild Foods Workshop, 22–26 January 2001 held in Diana, Kenya

Guarino L (1997) Proceedings of the IPGRI international workshop on genetic resources of non-cultivated vegetables in Africa: Conservation and Use: 29–31 August 1995. Nairobi: Institute of Plant Genetic and Crop Plant Research/International Plant Genetic Resources Institute

Guinand Y, Dechassa L (2000) Wild food Plants in Southern Ethiopia: Reflections on the role of 'famine foods' at the time of drought, UNEUE Survey, January 2000, Addis Ababa

Internationals Institute for Environment and Development (IIED) (1995) The value of wild resources in agricultural systems. The hidden harvest, London

International Plant Genetic Resource Institute (IPGRI) (2003) Rediscovering a forgotten treasure. [Internet] IPGRI Public Awareness. Rome, Italy

Kabuye CHS (1997) Potential wild food plants of Kenya. In: Kinyua A M, Kofi-Tsekpo W M, Dangana L B (eds.) Proceeding of the Conservation and utilization of indigenous medicinal plants and wild relatives of food crops. Nairobi, UNESCO, pp. 107–112

Kebu B, Fassil K (2006) Ethnobotanical study of wild edible plants in Derashe and Kucha Districts, South Ethiopia. *Ethnobiomed Journal* http://www.ethnobiomed.com/content/2/1/53

Madisa M E, Tshamekang M E (1997) Conservation and utilization of indigenous vegetables in Botswana. In: Guarino L (ed.) Proceedings of the IPGRI international workshop on genetic resources of non-cultivated vegetables in Africa: Conservation and Use: 29–31 August 1995

Martin GJ (1995) Ethnobotany: A method manual World Wide Fund for Nature. chapman and Hall, London

Mnzava NA (1997) Conservation and utilization of indigenous vegetables in Botswana. In: Proceedings of the IPGRI international workshop on genetic resources of non-cultivated vegetables in Africa: Conservation and Use: 29–31 August 1995. Nairobi: Institute of plant genetic and crop plant research/International Plant Genetic Resources Institute.

Ogle BM, Grivetti LE (1985) Legacy of the chameleon edible wild plants in the Kingdom of Swaziland, South Africa. A cultural, ecological, nutritional study. Parts II-IV, species availability and dietary use, analysis by ecological zone. *Ecology of Food and Nutrition*, 17:1–30

Tabuti J (2001) Traditional Food Plants of Bulamogi County, Kamuli District (Uganda): Preliminary Findings. *In:* Kenyatta C and Henderson A (eds.) Proceedings of the potential of indigenous Wild Foods Workshop, 22–26 January 2001, South Sudan

Tilahun T, Mirutse G (2010) Ethnobotanical study of wild edible plants of Kara and Kwego semi-pastoralist people in Lower Omo River Valley, Debub Omo Zone, SNNPR, Ethiopia. *Journal of Ethnobiology and Ethnomedicine,* 6:23

Zemede A (1995) Current Status of Eco-geographical Survey of Non-cultivated Food Plants. *In:* Hirut K, Dawit T, Fassil K (eds.) Proceeding of the workshop on planning and priority setting in ecogeographic survey and ethnobotanical research in relation to genetic resources in Ethiopia, February 1995, Addis Ababa

Zemede A (1997) Indigenous African food crops and useful plants: Survey of indigenous food crops, their preparations and home gardens. Nairobi: The United Nation University Institute for Natural Resources in Africa

Zemede A, Mesfin T (2001) Prospects for sustainable use and development of Wild Food Plants in Ethiopia. *Economic Botany,* 55:47–62

Chapter – 7

Biocultural Resources and Traditional Food Systems of *Nyishi* Tribe of Arunachal Pradesh (India): An Empirical Learning on the Role of Mythology and Folklore in Conservation

Hui Tag*, P. Kalita, Ranjay, K. Singh and A.K. Das

Abstract

Cultural knowledge on food and beverages is as old as human civilization. However, the traditional knowledge related to food habits and ecosystem conservation has been found to be unique in each cultural and ethnic group across the world. From time immemorial, the Nyishi tribe of Arunachal Pradesh has developed an unique set of traditional knowledge that allows the better utilization of food bioresources of their natural surrounding to sustain their livelihood. Historically, such knowledge was transmited through customary laws, folklore, and oral literature. Literature studies have revealed the lack of scientific investigation on ethnobiology of traditional food bioresources used by the Nyishi tribe of Arunachal Pradesh. Hence, in the present investigation, an attempt has been made to unveil the hidden age-old cultural food and beverage knowledge practised among the Nyishi of Papum Pare and Lower Subansiri district of Arunachal Pradesh. A total of 63 plant and 41 animal species have been reported during field survey in 4 circles covering a total of 12 different localities which includes rural, semi-urbans and urban type ecosystem in the 2 selected districts. In rural settings, most of the foods are boiled, fermented,

*huitag2008rgu@gmail.com

and roasted, and fried foods are nearly insignificant. Each of the plants and animal species used has a deep rooted cultural history and most of them were found to be very rare and threatened in their natural habitat. Folklore, myths and legends play a significant role in the conservation of the traditional food bioresources in *in-situ* condition. People's perceptions and methods of harvesting bioresources from natural ecosystem are simple, holistic and ecofriendly. Impact of alien culture and modern education method seems to play a significant role in attitudinal change in traditional belief and food habits among the younger generation of the *Nyishis* especially in semi-urban and pure urban localities.

Keywords: Nyishi tribe, biocultural resources, traditional knowledge, food security, conservation, Arunachal Pradesh.

Introduction

Humankind has learnt to use plants as food and nutraceutical agents from time immemorial. Prehistoric men and women used plants and animal products as food, cloths, and medicinal sources to sustain their livelihood. They also used them for alleviation of various epidemic diseases and inherited chronic ailments. Such age-old traditional food and ethnomedical knowledge, acquired through generations of close observation and experimentation in nature, has later evolved as powerful knowledge bank for the future generations of the human races. Such knowledge has mostly been retained through customary laws, folklore, and oral literature. The States of Northeast India is a home to about 8 million tribal people clubbed within 110 ethnic communities with diverse ethnocultural heritage. About 1953 ethnomedicinal plants have been reported from the NE region. About 1400 species, out of the 1953 reported are used as both food and ethnomedicinal sources (Dutta and Dutta 2005; Albert and Kuldip 2006). In particular, the state of Arunachal Pradesh (83,743 sq.km), with a total forest coverage of 82%, is rich in flora and faunal diversity of both ethnocultural and ecological significance. The state has also been characterized as one of the top 12th Global Biodiversity Hotspots (Myers et al. 2000). This high diversity is due to the presence of all climatic zones ranging from tropical to snow-clad alpine mountains (Kaul and Haridasan 1987; FSI 1999). Climatically, the region is mainly characterized by the occurrence of heavy rainfall, moderate temperature, and high relative humidity during summer, which favours luxuriant growth of biodiversity in its diverse ecosystems ranging from rivers, stream, ponds and marshy land to dense terrestrial rainforest and snow-clade alpine meadow. Almost 50% of higher plants and 60% of orchid species of India are found in the forest of Arunachal Pradesh (Eastern Himalaya) (Mao and Hyniewta 2000; Hegde et al. 2002).

The cultural and biological diversity of Arunachal Pradesh is well evident through its 26 major tribes and 110 subtribes with age-old indigenous knowledge system living in harmonious relation with nature from time immemorial (Murtem 2000; Tag and Das 2005). Being mountain forest dweller, the *Nyishi* of Arunachal Pradesh have developed food and beverage processing techniques by employing plants and animal ingredient in order to support the physiological and biological needs and to ensure overall food security in rural localities. However, proper documentation of traditional food knowledge system of *Nyishi* of Arunachal Pradesh in scientific literature is still not available.

The *Nyshis* are settled in 5 districts covering a vast geographical area of 24,176 sq.km which accounts for 29.76% of total land area of Arunachal Pradesh (83,743 sq.km). Anthropologically, the *Nyishi* tribe comprises of four sub-groups or races. They are known as Dupum, Dudum, Dol and Aniya Hari. They are the largest tribe in the state in terms of demographic status and mostly found in Kurung Kame; Papum Pare, East Kameng, Lower Subansiri and Upper Subansiri districts of Arunachal Pradesh. They belong to Mongoloid racial stock and speak Tibeto-Burman dialects with rich heritage of folklore and oral literature. They are believed to have migrated from Provinces of the South China and North Eastern Tibet through the courses of River *Siniek* (Subansiri), *Pare, Panior, Kurung* and *Kamle* (*Kamla*) which are tributaries of Mighty River *Sha* (Brahmaputra). This tribe is also popularly known as *Abo-Tani* tribe because they traced their genealogy through mythical and legendry forefather *Abo-Tani*, whom they consider progenitor of the present day *Tanii* races settling in central Arunachal Pradesh. The Nyshis are animist and perform various rituals to propitiate Almighty *Dony-Polo* Gods, other demi-Gods and Goddesses through animal sacrifices. They have been relaying on forest resources, agricultural products, and animal husbandry as primary sources of livelihood from time immemorial. Both men and women equally take part in daily activities such as festivals and rituals, agriculture, fishing, and animal husbandry while men mostly go for hunting and fishing.

The present paper is a first hand report based on ethnobiological study conducted in *Nyishi* dominated Lower Subansiri and Papum Pare Districts of Arunachal Pradesh during 2003-2006. Study discusses some aspects of gaps in traditional food processing knowledge witnessed among the old and the young generation *Nyishis* residing in rural and urban sites. Apart from traditional food knowledge, the gap in ethnoecological and ethnotaxonomical knowledge related to conservation and utilization of food plants and animal species from natural environments, associated taboos, myths and legends behind each of the plant and animal species used in theoretical, and practical, social and economic dimension are discussed in current context.

Study Sites

The selected two districts (Papum Pare and Lower Subansiri) fall within a geographical tract of $26^0 55'$ and $28^0 21'$ N latitude and $92^0 40'$ E and $94^0 21'$ E longitude with an area of 6335 sq.km, which accounts for 7.56% of total land area of Arunachal Pradesh. The district headquarter of Lower Subansiri (Ziro) is located at an altitude of about 1564m, Raga (1000m) and Talley Valley wildlife sanctuary (1800m) while the districts headquarter of Papum Pare (Yupia) is located at an altitude of 530m from the mean sea level (MSL). The combined total population of the two district recorded during the 2001 census was 1,77,729 persons out of which 92,609 were male and 85,120 were female members, with an average sex ratio of 942 female per 1000 male and an average literacy rate of 57.95% (2001 Census). The area receives heavy rainfall from both Northeast and Southwest monsoon (1000-1400mm) annually with a relative humidity of 80-90% during the summer months (May – September) (Anonymous 2001). Forest vegetation is mostly dominated by semi-evergreen and tropical evergreen subtropical broad leafed forest evergreen temperate broad leafed forest subtropical and temperate coniferous, alpine forest dominated by different species of *Rhododendron, Abies* and *Juniperus* coupled with tropical and subtropical deciduous and secondary bamboo barrack exceptionally rich in different species of flora and fauna (Champion and Seth 1968).

Brief Review on Traditional Food Research

In recent decades, there has been a growing consciousness among the researchers about the loss of traditional knowledge and biodiversity based food systems used by the tribal communities and forest dwellers. Turner (2001) and Turner and Turner (2004) had discussed the process of indigenous biodiversity based traditional food systems. Current consensus opinon emphasized that the traditional food knowledge system prevalent among the tribal communities requires quantitative ethnobiological investigation for update and use because food knowledge developed by ethnic group is indigenous and specific to given geographical area which is much diverse as well as fascinating (Turner 2001; Singh et al. 2007). Singh (2004) and Turner (2005) further pinpointed some of the vital factors which contribute to the loss of traditional knowledge related to tribal foods and beverages preparation in the 21st Century. The factors includes rapid erosion of cultural languages and disintegration of the learning chain between the old and the young generation. Interestingly, most of the plants and animal species used in ethnofood systems are found to have both medicinal and cultural significance (Bhaskar et al. 2007). Hence, documentation and bioprospecting initiative at national and regional level is urgently needed at current situation to conserve the vital food bioresources based on indigenous knowledge system for the sustainable use.

In the state of Arunachal Pradesh, documentation effort on ethnology, ethnobotanical and ethnozoological lines was only started during late 70s. Panigrahi and Naik (1961), Jain (1980) and Pal (1984, 1990) pioneered preliminary ethnobotanical research in the Subansiri District of Arunachal region of NE India. During late 80 and 90, Das (1986), Kholi (1996), Rawat and Chowdhury (1996), Haridasan (2000), Tag and Das (2004), Tag et al. (2005), Pallabi et al. (2005), Singh et al. (2007), Angami et al. (2007) and Bhaskar et al. (2007) have reported some traditional food plants used by the Monpa and *Abotani* tribes of Arunachal Pradesh and *Tai Ahom* of Assam. The *Tanii* tribes comprises of Adis, Apatani, Tagin, Hill Miri, Nyishis and Missing which are mostly settled in Lower Dibang Valley, Siang, Kurung Kame, Subansiri, Kameng and Papum Pare district of Arunachal Pradesh, and the Missing are mostly found scattered in the border area between Assam and Arunachal Pradesh and in the flood plain of Brahmautra Valley of Assam. However, earlier researchers have not discussed the food and beverage processing techniques practiced among the *Nyishis* of the Lower Subansiri and Papum Pare districts, women's role in the conservation of traditional food knowledge and their close perception on natural ecosystem and other associated parameters linked with traditional food system such as rituals, social and cultural taboos. Furthermore, at present, Western science has prove n the fact that the nutritional information used by tribals has been derived from observation, experience and clinical trial rather than from molecular and biochemical research with which modern food development is typically associated (Milburn 2004). Against these backdrop, the present ethnobiological investigation attempts to unveil some of the hidden traditional food and beverage knowledge of the Nyishis. Hypothesis and rationale were drawn at the end how each of the participants interviewed perceive their traditional food knowledge system based on practical observation and experimentation, mythology, customary laws, folklore and oral literature.

Research Methodology

In the present investigation, the ethnobiological research methodologies developed by Rao and Jain (1977), Martin (1994), Phillips and Gentry (1996) and Reyes-Garcia (2003) were followed. Before extensive field survey, a rapport was established with local administration and *Gora Aab* (village headman) of Lower Subansiri and Papum Pare districts of Arunachal Pradesh for the geographical, ethnographic, climatic and ethnobiological information pertaining to the area under study. The report is based on three years (2003-2006) of continuous participant observation. Information was gathered through random interview coupled with open-ended and multiple choice questionnaires. For the present study, two circles were selected from each district (Lower Subansiri and Papum Pare) and almost 12 localities of rural, semi-urban and pure urban types were selected for the investigation purposively. A maximum number of 80 households were chosen from 12 localities surveyed, and from each household, at least two participants of either sex were selected for random interview. Thus, total participants (sample) interviewed was 160 (80 male and 80 female). The villages surveyed under Lower Subansiri District were: Hapoli, Raga and Godak, which fall under Raga and Ziro-I Circles; and the villages like Kamporijo and Milli falls under Kamporijo circle. The villages surveyed under Papum Pare Districts were: Doimukh, Yupia (Doimukh Circle) while, Poma, Ganga, Naharlagun, Itanagar and Chimpu are the villages fall under Itanagar circle.

The multiple choice and open ended task questionnaires were provided with the goal of capturing ethnobotanical, ethnozoological and food processing knowledge (boiled and Fried food) of the two age group populations (<16-50 years; <51-80 years) from rural, semi-urban and urban localities. The villages surveyed from rural localities were at least 20-50 km away from pure urban localities and villages surveyed from semi-urban localities were about 5-10 km away from the pure urban localities. The questionnaires were accompanied by audio-video recording of the knowledge gathered through oral interview to cross validate the data at the time of analysis. Out of 12 localities surveyed, 4 localities were entirely rural while another 4 localities were semi-urban and rest 4 localities were pure urban. The plant and animal parts used and the traditional conservation methods of food bioresources were recorded in both field notebook and questionnaire formats. The data generated through multiple-choice task were analyzed using Excel and SPSS computer software and the value were expressed in the SE mean, and percentage. The IUCN (2006) methods of categorizing the taxa were followed to assess the current status of the plants and animal species used in traditional food systems among the chosen tribe. In case of plant samples, herbarium sheets were prepared and identified at herbarium of Department of Botany, RGU and further crosschecked with Botanical Survey of India (BSI) herbaria and identified samples bearing botanical and vernacular name, area, and date of collection were deposited in the Herbarium of Department of Botany, Rajiv Gandhi University, Itanagar for future reference. The distribution and taxonomy of each plants were cross checked with standard Indian and Himalayan flora of 20th Centruy (J.D Hooker, 1888) and published regional floras.The author citation and botanical nomenclature were cited from IPNI checklist, ePIC (Kew garden) Index Kewinses of world flora and name in current use (NCU) as per ICBN rules (St. Louis Code 2000) were strictly followed. The animal species were collected in the form of live photography incase of domesticated animals, death samples in case of wild animals, bones and skins. The zoological names of animal species were identified at the museum of Zoological Survey of India (ZSI) Itanagar and State Forest Research Institute (SFRI) Itanagar and the author citation were confirmed with standard Indian animal taxonomic literatures. The authors sought Prior Informed Consent (PIC) of the informants of the study sites to confirm people's ownership over their preserved indigenous knowledge system and provide them with assured equitable benefit share in any future application.

Results

Ethnobotanical Resources and Traditional Food Systems

The present investigation has revealed almost 63 plants species belonging to 10 genera and 13 families which are commonly used as raw material sources in indigenous food processing technology (Table 1) prevalent among the *Nyishi* tribe of Lower Subansiri and Papum Pare District of Arunachal Pradesh. Among the wild plants, *Dendrocalamus gigantea, Dendrocalamus strictus, Bambusa pallida,* and *Phylostachys pubescens* are some of the important bamboo species used as wild vegetable foods during both summer and winter seasons. The young shoots *(Hiku hituk)* of the bamboos are consumed either in cooked, roasted, or fermented form. The freshly harvested young shoots are cooked and consumed during summer months and the fermented shoot *(hiku)* and fermented and sun dried shoots *(hiyup)* are preserved and consumed during the whole year. Our close observation also revealed that the fresh bamboo shoots are mostly cooked with flowers of *Enseta glauca* (*Kudum pupuk*) and leaves of *Piper pedicellatum (Riir oh)*. A few amount of preserved swine fat, salts, freshly crushed raw chilly *(Tiir)* and *Zingiber officinali (Taki)* are added and boiled till the preparation is consumable. The prepared vegetables are consumed with cooked local rice as main staple food. Among the wild food plants used as vegetable items are: *Pouzolzia benettiana, Clerodendrum colebrookianum, Centella asiatica, Oenanthe javanica, Houtuynia cordata, Pilea trinervia, Lactuca graciles, Gynura cusumbua, Diplazium esculentum, Fagopyrum esculentum, Gnaphalium affine, Polygonum molle, Sonchus arvensis, Spilanthes oleracea, Urtica zeylanica, Solanum nigrum, Solanum violaceum* and *Piper pedicellatum* (Figure 2). Few species such as *Sonchus arvensis, Oenanthe javanica, Houttuynia cordata, Centella asiatica* and *Gnaphalium affine* are consumed in raw along with maize whereas rest of the above mentioned species are cooked and consumed along with rice. The fruits of *Spondias pinnata, Zanthoxyllum rhetsa, Zanthoxylum armatum, Litsea cubeba, Citrus limonum* and its wild varieties are some of the wild fruits used as salad items, spice and condiments in addition to cultivated chili and ginger. Among the wild fruits, the hypanthium of *Ficus semicordata, Ficus glomerata,* the ripen fruits of *Saurauia armata, Rubus ellipticus, Rubus nevius, Castanopsis*

Table:1: List of cultivated and wild plants used as food and ethnomedicinal agents among the *Nyishi* of Lower Subansiri and Papum Pare District of Arunachal Pradesh.

Botanical name/family	Local name (*Nyishi*)	Part used	Ethnic food value	Ethnomedicinal uses
Allium hookeri Thwaites Fam: Liliaceae ES:H/Cult/ST/T/Tm/R	*Talap*	Leaves	Raw, paste used as salad consumed with Zinger and Apong	Stimulant, anti-inflammatory, wound healing, vermicide
Allium reballum M Beib. Fam: Liliaceae ES: H/Cult/ST/T/Tm/C	*Mud talap*	Leaves	Consumed in raw as salad	Skin allergy, wound and infection, brain numbness
Amaranthus spinosus L. Fam: Amaranthaceae ES: H/W/ST/T/Tm/C	*Puchu kinnyu*	Leaves	Boiled and consumed along with rice	Chest inflammation, trout pain, constipation
Amaranthus viridis L. Fam: Amaranthaceae ES: H/W/ST/T/Tm/C	*Yorko puchu kinyu*	Leaves	Boiled and consumed as vegetable	Rheumatism, chest pain, asthma, cough Boil/sores
Bambusa stricta Roxb. Fam: Poaceae ES: ES: Sr/W/ST/T/Tm/C	*Eh here*	Young shoot	Raw shoot cooked and consumed as vegetable item; fermented and dried one used as salad	Chest pain, indigestion, constipation, low blood pressure
Bauhinia variegata L. Fam: Ceasalpiniaceae ES: H/W/ST/T/Tm/C	*Pachaum*	Leaves	Tender leaves/tender shoots are boiled and consumed as vegetable	Liver disorder, chest pain, rheumatism
Begonia roxburghii A. DC. Fam: Begoniaceae ES: H/W/ST/T/Tm/C	*Boku yulu*	Leaf petiole	Consumed in raw as salad	Wound, boil and sores, stomache pain, indigestion
Cardamine hirsuta L. Fam: Brassicaceae ES: ES: H/W/ST/T/Tm/R	*Soram guyi*	Leaves	Consumed in raw as salad	Liver disorder, chest pain, cough, cut and wound, tootache
Centella asiatica L. Fam: Apiaceae ES: ES: H/W/ST/T/Tm/R	*Bodo*	Leaves/stem	Consumed in raw as salad	Brain numpness, unclear thinking, bodyache
Chenopodium album. L. Fam: Chenopodiaceae ES: H/Cult/ST/T/Tm/R	*Tai*	Seeds/Leaves	Seeds cooked and consumed as food; leaves boiled and consumed as vegetable	Indigestion, lack appetite, debility

(*Contd.*)

Botanical name/family	Local name (*Nyishi*)	Part used	Ethnic food value	Ethnomedicinal uses
Clerodendrum colebrookianum Walp.Fam: VerbenaceaeES: Sr/W/ST/T/Tm/R	*Potto*	Tender leaves	Boiled/Roasted and consumed along with rice	High blood pressure (BP), liver pain, insomnia, dysentery, diarrhea, cough
Coix-lacryma jobi L.Fam: Poaceae ES: Sr/Cult/T/ST/Tm/R	*Tangek*	Grain	Powdered/Cooked/Boiled /(*Dogom/Opo*)	Low vitamins, fats and carbohydrate, calcium supplements
Colocasia esculenta (L.) SchottFam:Araceae ES: H/Cult/ST/T/Tm/C	*Eng ngepop*	Leaves/Corm	Corm roasted/boiled and consumed as food; leaves consumed as vegetable	Indigestion, lack appetite, constipation
Colocasia indica (Lour.) KunthFam: Araceae ES: H/Cult/T/ST/Tm/C	*Enge*	Corm	Roasted/Boil/Fermented/consumed as food and beverage sources(*Ell-ash/Opo*)	Low stamina, laxative agent, constipation
Crassocephalum crepidioides Benth.) S. MooreFam: Asteraceae ES: ES: H/W/ST/T/Tm/C	*Yamen*	Leaves	Raw/Boiled and consumed as vegetable and salad	Constipation, chest pain, liver disorder, difficult delivery
Cucumis melo L.Fam: Cucurbitaceae ES: Cl/Cult/T/ST/Tm/C	*Meble*	Fruits/Leaves	Consumed in raw as fruit and vegetable items	Laxative, digestive thirst quencher
Cucumis sativa LFam: Cucurbitaceae ES: Cl/Cult/T/ST/Tm/C	*Muku*	Fruits/Leaves	Consumed in raw as fruits and vegetable items	Thirst quencher, laxative, constipation
Cucurbita maxima DutchFam: Cucurbitaceae ES: Cl/Cult/T/ST/Tm/C	*Tappe*	Fruits/Leaves	Boiled and consumed as vegetable sources, famine food	Seeds anti-inflammatory digestive, laxative
Dendrocalamus gigantea Wallich ex Munro Fam: PoaceaeES: Sr/W/ST/T/Tm/C	*Eeh Hiku*	Young shoot	Cooked and consumed as famine food; fermented and dried one used as salad	Chest pain, indigestion, constipation, low blood pressure

(*Contd.*)

Botanical name/family	Local name (Nyishi)	Part used	Ethnic food value	Ethnomedicinal uses
Dillenia indica L.Fam: DilleniaceaeES: Tr/W/ST/T/Tm/C	*Champak*	Calyx of fruit	Boiled and consumed as salad	Indigestion, stomachache, liver disorder
Dioscorea alata L.Fam: DioscoreaceaeES: Cl/W/ST/Tm/R	*Egin nginek*	Tuber	Roasted/cooked/fermentedFood and beverage(*Opo/Dogom*)	Laxative, asthma, stimulant
Dioscorea deltoidea Wall.Fam: DioscoreaceaeES: Cl/Cult/T/ST/Tm/R	*Egin nginte*	Tuber	Roasted/boil/fermented Food and beverage (*Dogom/Opo*)	Laxative agent, indigestion, fatigue
Diplazium esculantum Sw.Fam:Athyraceae ES: H/W/ST/T/Tm/C	*Taka peya*	Tender leaves	Boiled and consumed as vegetable	Liver disorder, alchohol addiction, indigestion
Elusine coracana (L.) Gaertn.Fam: Poaceae ES: H/Cult/T/Tm/C	*Teem*	Grain	Powdered/Cooked/ Roasted/Fermented local beer/(*Mirik dogom /Opo*)	Low vitamins, low stamina, iron deficiency
Ensete superbum CheesmanFam: Musaceae ES: Tr/W/ST/T/Tm/C	*Kodum*	Flower	Boiled and consumed as vegetable	Liver and chest pain, cut and wound, indigestion
Fagopyrum esculentum MoenchFam: PolygonaceaeES: H/W/ST/T/Tm/C	*Huku*	Leaves	Boiled and consumed as vegetable	Stomachache, headache, constipation, lack appetite
Ficus semi-cordata Buch.-Ham. ex Sm Fam: MoraceaeES: Tr/W/ST/T/Tm/C	*Tokuk*	Fruits	Consumed as carbo hydrate sources	Indigestion, constipation, asthma, brain stimulant
Gnaphalium affine D. DonFam: Asteraceae ES: H/W/ST/T/Tm/C	*Tikpeyingm*	Leaves	Consumed in raw along with maize and rice as salad item	Bodyache, urinary trouble, running stomach
Houttuynia cordata Thunb.Fam: Saururaceae ES:H/W/ST/T/Tm/C	*Hiya*	Stem/leaves	Boiled/Raw consumed as salad	Insomnia, high BP, diarrhea
Ipomoea batatas (L.) Lam.Fam: ConvolvulaceaeES: Cl/Cult/T/S/T/Tm/C	*Egin pegri*	Tuber	Boil/roasted/ Cooked/fermented food and beverage (*Dogom/Opo*)	Cure indigestion, lack appetite, asthma
Lactuca sativa L.Fam: Asteraceae ES: H/W/ST/T/Tm/C	*Rabjap*	Leaves	Consumed in raw along with maize	Cough, mild fever, stomach pain, gas trouble, liver disorder

(*Contd.*)

Biocultural Resources and Traditional Food Systems of *Nyishi*

Botanical name/family	Local name (*Nyishi*)	Part used	Ethnic food value	Ethnomedicinal uses
Lagenaria vulgaris Ser.Fam: Cucurbitaceae ES: Cl/Cult/T/ST/Tm/C	*Opum ojuk*	Leaves/Fruits	Boiled and consumed (*Oh*, Vegetable)	Asthma/chest suffocation/joint pain
Manihot esculenta Crantz.Fam: Euphorbiaceae ES: Sr/Cult/T/ST/Tm/C	*Sin Eegin*	Tuber	Boil/roasted/cooked/fermentedFood and beverage sources (*Dogom/Opo*)	Constipation, indigestion, raw juice emetic
Oenanthe javanica DC.Fam: Apiaceae ES: H/W/ST/T/Tm/C	*Bubu*	Tender leaves/tender stem	Consumed in raw in salad form	Insomnia, high blood pressure, chest pain, indigestion
Oryza sativa L Fam: Poaceae ES: H/Cult/T/ST/Tm/C	*Aam*	Grain	Cooked/Roasted/PowderFermented traditional beer (*Dogom/Etti/Opo*)	Asthma, diabetes, low stamina
Phaseolus vulgaris L.Fam: Papillionaceae	*Peren*	Pod/Seeds	Mature seeds/Young pod are cooked and consumed as vegetable protein source	Laxative, digestive constipation
Phyllostachys pubescens Mazel ex Lehaie f. Fam: PoaceaeES: Sr/C/ST/T/Tm/C	*Taab*	Young shoot	Raw shoot cooked and consumed as vegetable item.	Low vitality, indigestion, lack appetite, rheumatism
Piper pedicellatum C. DC.Fam: Piperaceae ES: Sr/W/ST/T/Tm/R	*Riir*	Leaves	Consumed in raw or boiled as vegetable	Insomnia, bodyache, chest pain, cough, lack appetite
Plantago major L.Fam: Plantaginaceae ES: H/W/ST/T/Tm/C	*Sot nyuru*	Leaves	Consumed in raw as salad	Lack appetite, cough, chest inflammation, joint pain
Polygonum molle WightFam: Polygonaceae ES:H/W/ST/T/Tm/C	*Yuru*	Tender leaves	Boiled/raw consumed as salad	Constipation, sensational urination, boil and sores
Pouzolzia bennettiana WightFam: Urticaceae ES:H/W/T/ST/Tm/C	*Huik*	Tender leaves	Cooked and consumed as vegetable items	Difficult delivery, laxative, constipation consumed in raw as fruits items
Prunus persica (L.) Batsch Beytr.Fam: RosaceaeES: Tr/Cult/ST/T/Tm/C	*Siikom*	Fruits	Constipation, cough	

(*Contd.*)

Botanical name/family	Local name (*Nyishi*)	Part used	Ethnic food value	Ethnomedicinal uses
Rubus ellipticus Sm.Fam: Rosaceae ES: Sr/W/ST/T/m/C	*Ngingek berek*	Berry	Consumed as carbohydrate sources	Indigestion, constipation, asthma, brain stimulant
Rubus lineatus Reinw.Fam: Rosaceae ES: Sr/W/Tm/C	*Ngintum bulum*	Berry	Consumed as carbohydrate sources	Indigestion, voice problem, constipation
Rubus niveus Thunb.Fam: Rosaceae ES: Sr/W/ST/T/m/C	*Kib-lukpum hench*	Berry	Consumed as carbohydrate sources	Indigestion, constipation, asthma
Saurauia nepaulensis DC.Fam: Sauruiaceae ES: Tr/W/Tm/C	*Sicho hench*	Berry	Consumed as carbohydrate sources	Indigestion, constipation, asthma, brain stimulant
Setaria italica (L.) P.Beauv. f.Fam: Poaceae ES: H/Cult/T/ST/Tm/C	*Tayak*	Grain	Boil/cooked/food fermented beverage (*Opo/Dogom*)	Diabetes, nullify excess blood sugar, chest suffocation
Solanum aculeatissimum Jacq.Fam: SolanaceaeES: H/Cult/ST/T/Tm/C	*Kasi biik*	Fruits	Cooked and consumed as vegetable/salad	Liver disorder, chest pain, fever, cough, stomach pain, indigestion
Solanum nigrum L.Fam: Solanaceae ES: H/W/ST/T/Tm/C	*Hoor*	Leaves	Consumed in raw and cooked form	Liver disorder, diabetes, stomache pain, cough
Solanum torvum Sw.Fam: Solanaceae ES: H/W/ST/T/Tm/C	*Sot biik*	Fruits	Roasted and consumed as salad	Liver disorder, chest pain, fever, cough, stomach pain, indigestion, toothache
Solanum viarum DunalFam: Solanaceae ES: H/W/ST/T/Tm/C	*Sibin biik*	Fruits	Roasted and consumed as salad	Liver disorder, chest pain, fever, cough, stomach pain, indigestion. toothache
Solanum violaceum OrtegaFam: Solanaceae ES: H/Cult/ST/T/Tm/C	*Biik*	Fruits	Raw/dried fruits consumed as salad	Liver disorder, chest pain, fever, cough, stomach pain, indigestion
Sonchus arvensis L.Fam: Asteraceae ES: ES: H/W/ST/T/Tm/C	*Tuku rubu*	Leaves	Raw and consumed along with maize	Diarhea, cough, liver disorder, skin inflammation
Spilanthes paniculata Wall ex DC.Fam: AsteraceaeES: H/W/ST/T/Tm/C	*Buud*	Leaves	Consumed in raw/cooked as salad	Toothache, constipation
Spondias axillaris Roxb.Fam: AnnacardiaceaeES: Tr/W/ST/T/Tm/C	*Paka kat*	Fruits	Consumed in raw as salad	Constipation, chest pain, circulation problem
Spondias pinnata (L.f.) KurzFam: AnnacardiaceaeES: Tr/W/ST/T/Tm/C	*Pakka*	Fruits	Consumed in raw as salad	Liver disorder, fever, cough, stomach pain, indigestion, skin inflammation

(*Contd.*)

Botanical name/family	Local name (*Nyishi*)	Part used	Ethnic food value	Ethnomedicinal uses
Trevesia palmata Vis.Fam: Araliaceae ES: Tr/W/ST/T/Tm/C	*Tago-meyo*	Flower	Cooked and consumed as vegetable	Asthma, indigestion, liver disorder
Urtica parviflora Roxb.Fam: Uriticaceae ES: Sr/W/ST/T/Tm/C	*Push pun*	Tender leaves	Boiled and consumed along with rice	Stimulant, numbness, constipation
Zanthoxyllum armatum DCFam: RutaceaeES:Sr/W/T/ST/Tm/R	*Honior*	Leaves	Leaves consumed in raw as salad	Laxative, stimulantmental retardation, Chest/liver pain
Zanthoxylum rhetsa DCFam: Rutaceae ES: Sr/W/T/ST/Tm/R	*Honior*	Seeds/Leaves	Powdered seeds used as salad, leaves are cooked and consumed as vegetable item	Stimulant, digestive, laxative, dumpness
Zea mays L.Fam: Poaceae ES: Sr/Cult/T/ST/Tm/C	*Toop*	Grain	Boil/roasted/cooked and consumed as food/fermented beverage (*Pumik/Opo*)	Nutrient supplement for the chronic patient, Lack stamina, debility
Zingiber officinale Roscoe Fam: ZingiberaceaeES:H/Cult/T/ST/Tm/C	*Taikke*	Rhizome	Raw paste used as ingredient in salad, consumed with *apong*	Vermicide, stimulant, debility, brain dampness

Legends: ES=Ecological Status; T=tropical; ST= Subtropical; Tm=Temperate; Alp=alpine; C=Common; R=Rare; E=Endangered; H= Herb; Sr=Shrub; Tr=tree; Cl=Climber; Cult.=Cultivated; W=Wild.; E=Endangered; Aq.=Aquatic; Agri=agricultural field; For.=forest; Ter,=terrestrial; Fam: Family *indica* and *Spondias pinnata* are some of the common species consumed during summer months as sources of carbohydrate and fats. These species are also used as revitalizing foods. The rootstocks of *Angiopteris evecta* and the young piths of *Wallichia oblongifolia* species are consumed as carbohydrate sources and as famine foods after necessary processing.

Among the wild tubers, *Dioscorea pentaphylla, Dioscorea deltoidea* and *Dioscorea alata* are some of the major wild species consumed in boiled or roasted form. They are famine food items consumed during the summer months when rice and other agricultural crops are not yet matured to be harvested. The village folk practices kitchen garden (*Bullu*) nearby home and also maintain small plots of agricultural land bit far away from home called *pollek*. They raise early growing crops during the month of March such as pumkin, cucumber, maize, sweet potato, *Colocasia, Dioscorea* and foxtail millet and hervated during the peak hour of famine during the month of July and August. They also practices shifting agriculture by clearing large hectare of forest in selected sites and crops are usually sown during the month of June and usually harvested during the month of October to December. Commonly cultivated vegetable species noticed in the shifting agricultural field are *Chonopodium album, Celosia cristata, Allium hookeri, Zingiber officinale, Lagenaria vulgaris, Cucumis sativa, Cucumis melo, Cucurbita maxima, Phaseolus vulgaris, Colocasia esculenta, Glycine max,* and indigenous varieties of legumes. Rice (*Oryza sativa*) is the main stable food crop cultivated by the subtropical and temperate inhabitants, and usually harvested during the month of November and December. Other commonly cultivated cereal crops are, *Oryza sativa (Aam), Zea maize (Toop), Setaria italica (Tayak), Eleusine coracana (Teem), Chenopodium album (Tai)* and *Coix-lacryma jobi (Tangek). Oryza sativa, Setaria italica* and *Coix-lacryma jobi* are cooked and consumed as major sources of protein, carbohydrates, and vitamins after removing the husk. While, the powdered finger millet and powdered maize grain are cooked in semi-liquid gel and consumed as major sources of carbohydrates and fatty acid. The grains of *Chenopodium album* are often cooked and consumed as substitute of rice. Among the above cereal crops, *Eleusine coracana, Setaria italica, Zea maize, Chenopodium album* and *Coix-lacryma jobi* are considered famine foods during the summer months and they are mostly cultivated along with rice in shifting agricultural field. Among the cultivated tubers, *Ipomoea batatas, Manihot esculenta, Colocasia esculenta* and *Dioscorea* species are some of the favourite tropical and temperate tubers food used as carbohydrate and protein sources during peak hour of rice scarcity in summer months.

As beverage items, apart from finger millet, other cultivated species such as foxtail millet, rice, banana fruit, tubers of tapioca (*Manihot esculenta*), sweet potato (*Ipomeoa batatus*), and Taro (*Colocasia esculenta*) are some of the selected species mostly used as principal raw material sources for the preparation of traditional beverage called *opo* (*apong*). The *opo* is considered sacred and integral parts of the traditional food habit and beverages of every Arunachalee tribes. The beverage is mostly used during socio-religious ceremony in their society and also widely found among the *Nyishis*. *Opo* is fermented with the help of yeast mixed in rice powder (*Opop*) and other fermenting plants materials such as *Lygodium flaxuosum*, *Leucas aspera* and *Piper bettle*. A yeast powder is usually prepared only by selected women with sound traditional knowledge about its ingredients. Preparation is usually done in a closed and confidential chamber free from dust and human sight, free from smell of acidic substance and menstrual women. The use of acidic products such as *Citrus limonum*, *Derris scandens* and fermented bamboo shoots are strictly prohibited at home during the time of yeast powder preparation, and it is believed that failure to observe proper taboos as per spiritual rules will spoil the entire effort. *Opo* (*apong*) is mostly served as energy booster during the hours of hard working in agricultural fields, at home as hallucination agent, during marriage ceremonies and in important social gathering. *Opo* is also offered to the guests and gods during socio-religious ceremonies to bring cheer and delight in every heart and soul.

Ethnozoological Resources and Traditional Food System

Our critical analysis of the collected animal species through field survey, literature studies, and museum consultation revealed that almost 41 animal species belonging to 7 genera and 6 families are widely employed as major protein sources in traditional food systems of the *Nyishi* of two districts (Papum Pare and Lower Subansiri) (Table 2). Among the category of wild animals used as protein sources, *Ursus tibetanus* (*Sudum*), *Bos gaurus* (*Uii sob*); *Macaki assamica* (*Seb bede*), and *Sus srofa* (*Sour*) and *Muntiacus muntjak* (*Sudum*) and these are some of the favorite wild larger mammals hunted by the *Nyishis* for meat bone and skin. Some members of the cat family such as *Panthera tigris* (*Hogla takar*), *Panthera leo* (*Ebi paat*), *Neofelis nebulosa* (*Nyori*), and *Felis benghalensis* (*Apa taas*) are some of the rare mammal which are not frequently hunted on account of mythological reasons. As per the *Abo-Tanii* myths and beliefs, the wildcat family was closely related to mankind in the past, but they possessed tremendous amount of both benevolent and malevolent energy. It is a religious believe among the village folk that when such animals are killed and if proper rituals are not performed to eliminate the evil spirits (malevolent), the lives of the whole community could be endangered. The *Elephas maximus* (*Soot*) were occasioanlly hunted for its meat and task in earlier decades, but elephant are now rarely seen in *Nyishi* belt.

Table 2: Animal species used as traditional food and cultural materials among the *Nyishi* tribe of Lower Subansiri and Papum Pare District of Arunachal Pradesh.

Zoological name (family)	Local name (*Nyishi*)	Part used	Ethnic food values	Ethnomedicinal/cultural uses
Aborichthys elongatus (Hora, 1921) Fam: Balitoridae ES: Aq./FWF/W/T/ST/Tm/R	*Ngui riib*	Whole body	Boiled and consumed as food along with newly harvested rice	Taboos food during *Yullo* ritual
Anguilla bengalensis (J,E, Gray, 1831) Fam: Anguillidae ES: Aq/FWF/W/T/ST/Tm/E	*Ngui Ngub*	Whole body	Consumed in boiled/raw	Taboos food during *Yullo* ritual; use to heal fire burn skin
Axis axis (Erxleben, 1777) Fam: Cervidae ES: For/W/T/ST/Tm/E	*Sudum dumtak*	Flesh/horn	Roasted/Smoke dry and consumed	Intestine used as digestive agent, cure constipation
Axis porcinus (Zimmermann, 1780) Fam: Cervidae ES: For/W/T/ST/Tm/E	*Sudum dumli*	Flesh/Horn	Roasted/Smoke dry and consumed	Intestine used as digestive agent
Bagarius bagarius (F. Hamilton, 1822) Fam: Sisoridae ES: Aq/FWF/W/T/ST/Tm/E	*Ngui ngurik*	Whole body	Boiled and consumed as food	Taboos food during *Yullo* ritual
Bandicota indica (Bechstein, 1800) Fam: Schistosomatidae ES: For/W/T/ST/Tm/E	*Mutum kubu*	Flesh	Roasted/boiled and consumed as food	Indigestion, lack appetite
Bos gaurus C.H. Smith, 1827 Fam: Bovidae ES: For/W/T/ST/Tm/E	*Uii sob*	Flesh/horn	Flesh is roasted/cooked and consumed as food	Taboos meat during *Yullo* ritual
Bubalus bubalis (Linnaeus, 1758) Fam: Bovidae ES: For/W/T/ST/Tm/E	*Mindik*	Flesh/horn	Roasted/Smoke dry and consumed	Taboos meat during *Yullo* ritual; child birth
Canis lupus Linnaeus, 1758 Fam: Canidae ES: Ter/M/W/T/ST/Tm/R	*Soccha*	Flesh/meat	Roasted/Boiled and consumed as food	Taboos food during *yullo* ritual and child birth
Cervus unicolor Kerr, 1792 Fam: Cervidae ES: For/W/Tm/E	*Sochar*	Flesh/horn	Roasted/Smoke dry and consumed	Taboos meat during *Yirne* ritual

(*Contd.*)

Zoological name (family)	Local name (*Nyishi*)	Part used	Ethnic food values	Ethnomedicinal/cultural uses
Capra hircus Linnaeus, 1758 Fam: Bovidae ES: Ter/M/W/T/ST/Tm/E	*Siib*	Whole body meat	Roasted/Boiled and consumed as food	Taboos food during *yullo* ritual; gall bladder is used to cure stomachache, low debility
Dremomys lokriah (Hodgson, 1836) Fam: Sciuridae ES: Arial/W/T/ST/Tm/R	*Takke kelli*	Flesh	Roasted/boiled and consumed as food	Intestine use to cure stomachache, cough
Elephas maximus Linnaeus, 1758 Fam: Elephantidae ES: For/W/T/ST/Tm/R	*Soot*	Flesh/Task	Roasted flesh is consumed as food item	Taboos meat during *Yullo* ritual
Felis bengalensis (Kerr, 1792) Fam: Felidae ES: Ter/M/W/T/ST/Tm/R	*Apa taas*	Whole body meat	Roasted/Boiled and consumed as food	Taboos food during *yullo* ritual and child conception and birth
Funambulus palmarum (Linnaeus, 1766) Fam: Sciuridae ES: Arial/W/T/ST/Tm/R	*Koche*	Flesh	Roasted/boiled and consumed as food	Intestine use to cure stomachache, cough
Funambulus pennanti (Wrouhgton, 1905) Fam: Sciuridae ES: Arial/W/T/ST/Tm/R	*Koche*	Flesh	Roasted/boiled and consumed as food	Intestine use to cure stomachache, cough
Garra amandalei Hora, 1921 Fam: Cyprinidae ES: Aq/FW/W/T/ST/Tm/R	*Ngui Ngoka*	Whole body	Boiled/Roasted and consumed as food	Taboos food during *Yullo* ritual
Hystrix indica Kerr, 1792 Fam: Hystricidae ES: Ter/W/T/ST/Tm/E	*Sush*	Flesh	Roasted/boiled and consumed as food	Intestine use to cure stomachache, cough
Lutra perspicilliata (I. Geoffroy, 1826) Fam: **Mustelidae** ES: Aq/W/T/ST/Tm/E	*Soram*	Flesh	Meat sources	Bone is used to asthma, digestive system is consumed to cure stomach pain.
Macaca assamensis McClelland, 1840 Fam: Cercopithecidae ES: Arial/M/W/T/ST/Tm/E	*Seb bede*	Whole body meat	Roasted/Boiled and consumed as food	Sacrifice animal during death; taboos food during child birth, and *Yullo* ritual

(*Contd.*)

Zoological name (family)	Local name (*Nyishi*)	Part used	Ethnic food values	Ethnomedicinal/cultural uses
Manis pentadactyla Linnaeus, 1758 Fam: Manidae ES: Ter/W/Tm/R/E	*Suchik*	Flesh/Digestive system	Meat sources	Stool of digestive system used to cure liver pain, stomachache
Moschus moschiferus Linnaeus, 1758 Fam: Moschidae ES: For/W/T/ST/Tm/E	*Sudum damchi*	Flesh/naval scent	Roasted/Smoke dry and consumed	Naval portion is used to cure malaria and diabetes
Muntiacus muntjak (Zimmermann, 1780) Fam: Cervidae ES: For/W/T/ST/Tm/R	*Sudum dump*	Flesh/horn	Roasted/Smoke dry and consumed	Intestine used as digestive agent, cure constipation
Mus booduga (Gray, 1837) Fam: Muridae ES: Agri/W/T/ST/Tm/E	*Nongo Kubu*	Flesh	Roasted/boiled and consumed as food	Indigestion, lack appetite
Neofelis nebulosa (Griffith, 1821) Fam: Felidae ES: Ter/M/W/T/ST/Tm/E	*Nyopak*	Whole body meat/teeth/jaw	Roasted/Boiled and consumed as food	Taboos food during *yullo* ritual and child conception and birth, death of person
Nycticebus coucang (Boddaert, 1785) Fam: Lorisidae ES: Arial/M/W/T/ST/Tm/E	*Socche*	Whole body meat	Roasted/Boiled and consumed as food	Intestine use to cure stomachache
Paguma larvata (Hamilton-Smith, 1827) Fam: Viverridae ES: Arial/M/W/T/ST/Tm/R	*Suiin*	Whole body meat	Roasted/Boiled and consumed as food	Taboos food during *yullo* ritual and child conception and birth
Panthera leo (Linnaeus, 1758) Fam: Felidae ES: Ter/M/W/T/ST/Tm/E	*Ebi paat*	Whole body meat/teeth/jaw	Roasted/Boiled and consumed as food	Taboos food during *yullo* ritual and child conception and birth, death of person
Panthera pardus (Linnaeus, 1758) Fam: Felidae ES: Ter/M/W/T/ST/Tm/E	*Nyorri*	Whole body meat/teeth/jaw	Roasted/Boiled and consumed as food	Taboos food during *yullo* ritual and child conception and birth, death of person
Panthera tigris (Linnaeus, 1758) Fam: Felidae ES: Ter/M/W/T/ST/Tm/E	*Hogla takkar*	Whole body meat/teeth/jaw	Roasted/Boiled and consumed as food	Taboos food during *yullo* ritual and child conception and birth, death of person

(*Contd.*)

Biocultural Resources and Traditional Food Systems of *Nyishi* 173

Zoological name (family)	Local name (*Nyishi*)	Part used	Ethnic food values	Ethnomedicinal/cultural uses
Petaurista petaurista (Pallas, 1766) Fam: Sciuridae ES: Arial/M/W/T/ST/Tm/E	*Sojo*	Whole body meat	Roasted/Boiled and consumed as food	Taboos food during *Yirne*, *Giir-talle* and *yullo* ritual; digestive system is consumed to cure stomachache, malarial fever
Presbytis pileatus (Blyth, 1843) Fam: Cercopithecidae ES: Arial/M/W/T/ST/Tm/E	*Seb besor*	Whole body meat /teeth/jaw	Roasted/Boiled and consumed as food/Skin used as daw cover	Intestine is consumed during dysentery and malarial fever; taboos food during child birth
Pteropus giganteus (Brunich, 1782) Fam: Pteropodidae ES: Arial/M/W/T/ST/Tm/E	*Tapen penga*	Whole body meat	Roasted/Boiled and consumed as food	Taboos food during *yullo* ritual and child conception and birth, death of person; cure dysentery, fever.
Ratufa bicolor (Sparrman, 1778) Fam: Sciuridae ES: Arial/W/T/ST/Tm/R	*Takke kera*	Flesh	Roasted/boiled and consumed as food	Intestine use to cure stomachache, cough
Rhinoceros unicornis Linnaeus, 1758 Fam: Rhinocerotidae ES: For/W/T/ST/Tm/E	*Soor*	Flesh/Task	Roasted flesh is consumed as food	Taboos meat during *Yullo* ritual
Rhinolophus luctus Temminck, 1835 Fam: Rhinolophidae ES: Arial/W/T/ST/Tm/R	*Tapen pinch*	Flesh	Roasted/Boiled and consumed as food	Taboos food during *yullo* ritual and child conception and birth, death of person; cure dysentery, fever.
Schizothorax progastus (McClelland, 1839) Fam: Cyprinidae ES: Aq/FW/W/T/ST/Tm/R	*Ngui Ngurii*	Whole body	Boiled/Roasted and consumed as food	Special aquatic food permitted for consumption during *Yullo* ritual
Ursus thibetanus (G. Cuvier, 1823) Fam: Ursidae ES: Ter/W/T/Tm/R	*Sutum*	Flesh/Gall baldder	Flesh as food malarial fever	Juice of gall bladder used as anti-diabetic,
Suncus murinus (Linnaeus, 1766) Fam: Soricidae ES: Ter/W/T/ST/Tm/R	*Pizz*	Flesh	Roasted whole body is consumed as food	Malarial fever, measles, viral fever, headache, cough
Sus scrofa Linnaeus, 1758 Fam: Suidae ES: For/W/T/ST/Tm/E	*Soor*	Flesh/skin	Roasted/Smoke dry and consumed	Gall bladder juice is used to cure dysentery
Talpa micrura Hodgson, 1841 Fam: Talpidae ES: Ter/W/T/ST/Tm/R	*Kurluk*	Flesh	Roasted and consumed as meat sources	Malarial fever, Viral fever, dysentery.

Among the bird species, golden fowl, wild pigeon, crow and *Aquilus chrysanthos* (*Petta peem*) are some of the species hunted for food. Wild pigeon and jungle fowl are the bird species used in food items, but the black crow and the golden eagle are rarely hunted on account of some mythological reasons and consumed for some specific pourposes. For example, the meat of black crows is consumed during sudden outbreak of epidemic diseases such as viral fever, dysentery and malaria. It is also consumed to prevent the effects of the evil spirits in individual lives. The *Aquilus chrysanthos* (*Golden Eagle*) is killed for its powerful sharp claws and beautiful stripped feathers. Feathers of Golden Eagle's are usually worn by the Priests of high status (Figure 1). The *Nyishi* folklore proclaimes that the eagle is the source of mystic power that enhance brilliant intellect, pride, courage and glory in individuals who wear it, and therefore Priests wear golden eagle's feathters during spiritual discourses, so that they can track the spiritual path within any hindrance.

Figure 1: Elderly *Nyishi* folk with traditional headgear during debate and discussion hour. Photo: Hui Tag

Among the rodents species (rat and squirrel), the *koche, takke, kojack, kobu, kurluck,* and *pizz* are some of the favourite species hunted for food. Those animals are mostly killed through stone and roof trap set in dense forest during winter months. Their fur and meat are roasted on fire and consumed with rice. The meat can also be consumed after smoke dry. The dead bodies of rodents are roasted on fire to remove the haires. It is usually dried on a traditional rack [*rakki*-lower shelf (half meter from fire heart), *ketten*-middle shelf (1 and half m from fire heart) and *Naka*-upper shelf (3 m high from fire heart)] made of bamboo constructed just half meter - 3 high above the traditional fire heart (*emik*). The dried rodents are packed in bamboo cages (*paka*) and preserved for over a year. It can be kept for self-consumption at home but are also used in barter trade, marriage and community festival festivals such as *Boori Boot* and *Nyokum*. As per the beliefs, accompaniment of rodent meat is a must during the inaugural eating of newly harvested rice. Taking newly harvested rice along with rodent meat, crab, prawn, and fish has been considered sacred and auspicious during the onset of crop harvesting in the villages of every *Abotani* tribe. The newly harvested rice is usually served to the guests along with rodent meat such as the meat of *Ratufa bicolor* (*takke*)-Malayan giant squirrel.

Apart from rodent meat, some of the villagers also consume the meat of snakes and lizards. Wild snakes meat, such as the meat of *Python molurus* (*Tib biram*), is cooked along with bamboo shoots, wild banana flowers and finely grinded raw ginger paste in closed bamboo vessel. However, the villagers knew that these reptilian species are very rare in natural habitat, and they believe that those species are possessed by powerful benevolent spirits *Doje-Yapom* and *Bur*. Villagers narrate mythology and folklore about rare fauna to younger generations for the protection and conservation.

The *Nyishis* are mostly settled near river valley and streams (200-1000m) in order to have easy access on fishery sources. They are one among the most efficient group of tribal fishermen found in the entire Eastern Himalayan region of India. Fishing is mostly done through various indigenous techniques such as *takom ganam* (indigenously crafted bamboo trap), *Esh menam* (damming of river), *Sibook penam* (river diversion), *Tom panam* (poisoning of river with fish poison plants) and *Lipum sinam* (aggregation of stones in slow moving river course). Among the common fish species, the *Tor*, *Catla catla*, and Indian major carp, among other species found in middle and lower elevations are consumed by villagers as major delicacies. Indigenous fishing techniques, gears, and knowledge applied by the villagers are comparable with earlier findings on *Adi* tribe of East Siang district of Arunachal Pradesh as reported by Tag et al. (2005) and Pallabi et al. (2005). Fishes are the all time favourite delicacies consumed by every households of *Nyishi* along with rice in both summer and winter. As mentioned before, prawns and crabs are cooked and consumed along with newly harvested rice as mark of the beginning of new season. It is believed that fish, prawns, and crabs are auspicious and they bring luck and fortune to family if they are consumed at the onset of winter seasons.

Among the domesticated animals, the *Bos frontalis* (*sob/sebbe*) is the most revered animal of the *Nyishi* tribe of Arunachal Pradesh. Apart from serving as a meat source, the *mithun* also serves *Abotani* society as status symbol of the family, exchange item during marriage, socio-religious ceremonies, and in their traditional barter trading systems. Marriage celebration in *Tani* society is considered incomplete without the exchange of live *mithun*. It is believed that the *mithun* is a sacred gift sent by the Almighty Donyi (the Sun God) to the humanity to be used as exchanged item during marriage ceremony, especially when the bride is to be sent to bridegroom's home. Such legendry tales has later evolved as inspirational sources of learning and also serve as an important source of customary laws in present day *Tani* society of Arunachal Pradesh. Apart from *mithun*, other domestic animals used as food and sacrificial items to be offered to gods during socio-religious ceremonies are the cow, goat, sheep, pig, chicken and dog. The meat of dog and cat are rarely and selectively consumed by only few male folk. Women and children are prohibited to consume dog and cat's meat on account of some mythological reason. The meat of cows, goats, chickens and pigs is consumed after sacrifice during socio-religious ceremonies such as marriage, y*ullo* rituals performed at home and done in the name of Sun God and in the name of other demi-Gods and Goddess. Domestic animals are also sacrificed during *Boori Boot* and *Nyokum yullo* (community festival) ceremony performed at community level for all round prosperity of mankind. During festive occasion the dishes are prepared according to the advice of senior folk in the villages. Younger generations are normally entrusted with the responsibility of preparing foods, beverages and collection of ceremonial items; while elderly folk guide them with their traditional wisdom and ideas.

Role of myths and legends in Conservation and Utilization of Food Bioresources

Our observation has further revealed that the *Nyishis* of Papum Pare and Subansiri have good knowledge relating to use of ethnobotanicals and ethnozoological resources in their daily diet apart from using them as medicines and cultural materials. Though the tribe does not have scripts to express their feelings, emotions and knowledge through writing, they have a powerful folklore and oral tradition that helps them in transmitting traditional knowledge from one generation to the next. Villagers are well versed in naming ecological zones in their local dialect such as *yorko* (tropical), *mopo* subtropical, *mud-moyam* (temperate) and snow-clad alpine zones as *tapam mud*. Villagers also expert in naming ecological habitats in their local dialect such as *liniek* for aquatic environment and *motum* for the terrestrial dense forest, swampy and marshy land as *sippa-sibla*, shifting agricultural field as *tump nongo*, wet-rice field as *sippa nongo or esh rig*, big river as *esh hoot,* and small stream as *esh koro*.

Nyishis are good in folklore and mythological story related to phytogeographical and ancient botanical knowledge. According to their myths and legends, the geographical area of mother Earth was belongs to the two sons of *Sichi* (the Mother Earth). The two sons born out of the contact between *Nyido* (rain God) and *Sichi* were known as *Tani* and *Doje*. *Tani* was the human-personified spirit and he loved to settle in village-like ecosystem. *Doje* was the elder brother and he loved astral existence and wilderness. The two mystic brothers often cheated each other by using their supernatural powers. They were staying with their wives and kids in the same wooden house, where the roof (Figure 2) was made of wild banana (*Kullu*) leaves (*Musa balbisiana)* and *taar rah (Calamus erectus*), floor and walls were made of bamboo culm, and the ropes were made of canes species –*taser and takek* (*Calamus flagellum* and *Calamus flaribundus).*

Tani and *Doje* were the first powerful godly spirit born and brought up on earth before human evolution. They had an enormous amount of gifted supernatural powers and knowledge related to cosmology and metaphysics. One day, a quarrel occurred between the two brothers due to differences over the matter related to safe existence of spiritual life on Earth and in the entire cosmos. The Sun

Figure 2: Traditional *Nyishi* House-roof made of *Calamus erectus* leaf (Arecaceae). Photo: Hui Tag

and Moon Gods (*Eji Donyi* and *Ath Pool*) along with other demi-Gods and Goddesses and seven sages (*Shi-Moro*) arrived at the home of *Tani* and *Doje* and requested them to end the quarrel immediately for the sake of safety of life on earth. With the instruction of Sun God and sages (*Shi-moro)*, all things on the Earth had to be equally divided between the two brothers. *Doje* was requested to take shelter in *Ficus religiosa* and *Altingia excelsa* as their abode since he loved wilderness. Further, *Doje* was further instructed to look after forest flora and fauna. However, the *Abo Tani* was instructed to take shelter in the village by living in a constructed house since he expressed willingness to lead village and community life by assuming human form. Finally, a consensus was arrived at *Si-Dony mirik* (the historic convention and gathering of Gods and Saints) with certain specific conditions. The matter debated during *Si-Doni mirik* has later evolved as heart touching folklore, mythological and historical source of divine knowledge through the medium of gifted priests of the human society. It is believed that the present day *Nyishi* customary laws is the offshoot of that historic *Si-Dony mirik* which is still regulating the social, cultural and moral life in present day *Tani* society of Arunachal Pradesh. Very recently, such cultural beliefs had been institutionalized, preached and practiced through *Central Nyedar Namlo* movement in *Nyishi* dominated belt of Arunachal Pradesh which is popularly known as *Aane Donyi & Jiwt Aane*, the sect of *Dony-Poloism* religious group.

Among the conditions granted during *Shi-Donyi mirik* to *Tani* for the utilization of biocultural resources, the Gods and Saints prohibited him not to go for ruthless harvesting of wild but rare and endangered plants and animal species such as cane, *Musa* species, *Ficus* species, lion, tiger, monkey, king cobra, rare bird species such as *Buceros bicornis* (Horn Bill), reptilian species including crocordile and snakes. These forest species can be harvested and hunted occasionally by *Tani* progeny, but with due propitiation of rituals. For example, propitiations are to be made beforehand to the benevolent God of wildlife –the *Doje -Yapom* as mark of respect and permission. As per *Shi-Donyi* convention, the soul authority of rare and endangered biodiversity wealth in wilderness is entirely under the custody of *Doje-Yapom*. Hence, it is still a belief among the villagers of rural localities that before collecting certain rare species like *Aconitum ferox, Musa valutina, Calamus* species, lion, tiger and leopard, special rituals has to be performed in the name of *Doje-Yapom* and *Nipo* (female spirit of river and streams). Such rituals involed sacrifices of white hen, fowl, dog, mithun, cow, goat and pig. Ritual and sacrifice is mainly done to propitiate *Doje-Yapom* and its subordinate spirits such as *Pomle-Jelli* and *Nipo* not to cause any harm on person concerned or on human society.

Our observations on cultural knowledge associated with biodiversity conservation among forest dwellers (*Nyishi*) in far-flung areas has revealed that their perception about biodiversity and ecosystem is based on their mythology and folklore which is, holistic approach, and a high degree of ecological ethics, which- some of them, are scientifically verifiable with the help of an advancing science, technology, morality and ethical laws of the 21st century. The Rio-Earth Summit of 1992, and the Promulgation of Kyoto Protocol (1997) for the conservation of environment and biodiversity resources can be seen as glaring example of the joint effort made at global level in 21st century, which can be well compared with the matter debated during *Si-Donyi* convention of the prehistoric time as per *Tani* mythology.

Another glaring example of mythological base conservation of biodiversity wealth was encountered during our ethnobiological field work with one of the elderly priest Shri *Temir Godak* of the Godak villages under Raga circle of Lower Subansiri District. He told us story about *Banyan* tree and its relation to human livelihood system in ecosystem. As per Shri *Temir version*:

"The people of their locality never select the plot for *jhum* agriculture where *Ficus religiosa, Ficus elastica* and its varieties, *Ficus benghalensis* and *Altingia excelsa* species are likely to be found".

They believe that these trees are the abode of benevolent spirit *Doje-Yapom* who once was the big brother of *Abotani*. Such mythological oral discourse is usually used to terrify the villagers not to fell banyan trees of the forest. Failure to compliment these age-old customary and spiritual laws would invite wrath and curse from the *Doje-Yapom*. It is believed that *Doje-Yapom* is the most powerful spirit capable of bringing terrible destruction to the natural ecosystem and human livelihood system. Here the modern day ecologist has proven the fact that the *Ficus* and *Altingia* species are the keystone species that provide shelter to millions of insects, thousands of birds, reptiles and mammals beside releasing fresh oxygen and moisture to the atmosphere by absorbing carbon dioxide that keep the surrounding environment cool and comfortable (Odum 1969). *Ficus'* root system is extensive enough to prevent landslides and soil erosion and it is also capable enough to absorb and retain water and moisture in the soil that enable proliferation of other biodiversity nearby. Similarly, animal species such as lion are not hunted for food without any cause. It is believed that during the times of human evolution in ancient days, two brothers went to the jungle for collecting rats for food that were being killed in trap (*Ewd*) made of flat stone. The younger brother discovered with surprise that the big brother ate up in raw at least half part of rat body which he collected from the stone trap (*Ewd*). The younger brother with humanly personified asked his elder brother as to why he was eating the rat meat in raw. As soon as younger brother (*Nyia*) came to know about the wild food habit of the elder brother (*Ebi Path*), the elder brother requested *Nyia* not to reveal the unpalatable secret to the family at home. He further instructed his younger brother to slab on his face, nose and mouth, and also asked him to message whole body with his sharp nail. The younger brother did accordingly and to his surprise, the entire human body of his big brother suddenly turns into a ferocious lion body. The lion embodied big brother *(Ebi Path)* requested *Nyia* to go back to the village with assurance that he will keep on hunting animal for *Nyia* and his family for food because his body had already turn out to be unsuitable for practice shifting agriculture anymore after assuming lion's body. *Ebi Path* further promised his brother that he would never harm or hunt human being for food as long as their livelihood security and status quo of territorial jurisdiction is properly observed and respected. The moral behind the story is that faunal diversity is an integral part of our ecosystem and every organism of the natural ecosystem is interdependent. Each organism has its own ethics, and encroachment of the territory by either group would cause displacement and annihilation of the others. Hence, a balance approach is required to safeguard biodiversity of the forest for the sustainable existence of biodiversity and human livelihood system.

Lion being a ferocious carnivorous animal kills the *mithun,* the most revered animal of *Nyishi* (the progeny of *Aath Nyia,* the ultimate descendant of *Abotani*) occasionally when their natural habitat and food security are threatened due to the encroachment of *Nyia's* progeny. In such circumstances, group hunting is organized to kill the lion and after killing, the head portion of lion is buried in the soil after removing its teeth and jaws. An special ritual *(gir-nirom)* is performed at the entry gate of the village in order to prevent entry of the evil spirits associated with the lion. Teeth and jaws are dried with smoke and only the priest with supernatural attributes can wear by attaching them to the string attached to the cover of long *Nyishi* daw (sword). It is believed that the lions' jawbone and teeth boost courage, enhance velour and fighting spirit, improve clear mental faculty, and bring luck, wealth and happiness in individual and community life. Lions are never killed for food, but they are killed for their teeths and jaw.

Hence, the modern days conservationist can afford to do a job by entrusting responsibility to traditional knowledge holders at village or community level to protect the degrading Earth's environment and bioresources through traditional approaches which is simple, inexpensive, most holistic and sustainable process to be followed.

Role of Rituals and Taboos in Conservation and Utilization of Food Bioresources

Sacrificial rituals are integral part of *Nyishi's* culture. For certain specific events, rituals are performed before the event for its success. The rituals and taboos practiced among the *Nyishis* can be broadly categorized into two types: i.e. rituals and taboos associated with birth, marriage and death (life cycle); and rituals and taboos associated with agricultural seasons and animal husbandry, fishing, and hunting (economic activities).

Rituals, Taboos and Traditional Foods on Various Sociocultural Occasions

Under this category lie all rituals and taboos performed in the name of almighty *Donyi and AathPolo* Gods and Goddesses for prosperity in livelihood system, communal harmony, and blessing. During the birth of a child, a special ritual is performed to protect the child from the eyes of evil spirits. According to *Nyishi* mythology, the behaviour and types of activity performed by the parents during prenatal and postnatal period influence the behaviour of the child. Therefore, certain *taboos* are observed during this period. The parents are not supposed to fell banana trees, dig and fil holes on earth. Father are also not suppose to eat monkey, tiger, lion, cow, squirrel, and pangolin meat because of the existing belief that the behaviour of such animals would influence child's behaviour, food habit, and physical appearances. Some part of this rationale belief and taboos though one may be perceived as superstitious belief, can be compared with modern law of genetical inheritance as proved by Dr. Gregor Mendel in late 19th Century.

Certain plants products are also encouraged as well as restricted during the pre-natal and post-natal periods. Plants such as *Pouzolzia bennettiana, Gynura cusumbua* and the powdered seeds of *Sesamum indicum* are encouraged to be consumed by the mother during the last quarter of pregnancy. It is believed that these plants have a laxative effect on the womb and enhance easy delivery of the child. Other food items such as *Dioscorea*, *Colocasia*, sugar cane, *Clerodendrum colebrookianum*, papaya fruits, and sticky wild fruits rich in concentrated sugar and bitter alkaloids substances are strictly prohibited for the expectant mother because of the belief that those food items generally cause difficult delivery and some fruits such as papaya are restricted for abortive reason. Some of the above mentioned beliefs, food habits and taboos, cultural practices during prenatal and post natal period are common among Indian and South East Asian families.

In rituals related to birth, marriage, and death, green leafy vegetables are not brought to home for consumption for few days as per the instruction given by the priest who performed the particular rituals. The taboos observed are ritual-specific and each ritual involves taboos of different length ranging from 3 to 10, 15 days or up to one month and even one year.

Men folk never take food from the hands of menstrual women when they are planning for fishing, hunting, or when performing special rituals. After the death of a person, the rituals are performed to purify the soul of every individual involved during cremation of the death body. Vegetable items such as *Allium hookeri*, ginger, and *sesamum* oil seeds are strictly prohibited for consumption after the death of some member of the family because of the existing belief that consumption of such vegetable items induce hair graying, chest suffocation, and voice retardation. Dog meat is restricted for women folk. It is believed that women who ate dog meat in their present birth will remain childless in their next birth. Patients are asked to refrain from dog, *mithun,* cow, goat, pig, rodent and other wild animal meat during specific ritual performed in the name of most holistic Gods and Goddesses responsible for sweet oration, meticulous thinking, power of rationality, soul purification and brilliant intellects, and such taboos length vary from 1 to 12 months. *Nyishi* believes that violation of spiritual taboos would cause dull intellect and have adverse side effect to the patient. Animals like monkeys, Chinese pangolins, dogs, *mithun* and cows are sacrificed before the crematorium ground (*Nyiblu*) with chanting of hyms by the priest so that the soul of those animal would act as a faithful servant to the departed human soul in their next world. Among the domesticated animals, dog and *mithun* are usually prepared for sacrifice before the crematorium ground. It is a popular belief that these animals would remain as faithful servant to the departed human soul. Among the wild animal, monkeys are usually chosen for sacrifices. It is believed that the origin of monkeys and men are the same and they remain as faithful friends to each other in their next world. Such taboos observed are in accordance with the spiritual laws as told by elderly priest, orators, and knowledgeable persons of the community.

Strict taboos are observed during and after performance of special ritual for pacifying evil spirits who caused death by accident, food poisoning, suicidal case, brutal murder and through snake bite. The family members are asked to restrict from consumption of hot and fermented food items such as *Glycine max*, and *Colocasia*. According to Tani philosophy, these food items are usually a precious possession of evil and angry spirits *(Gir and Talle)*. Failure to observe proper taboos on these food items would cause another fatal accident and injury within the family in immediate future.

The snake bite patient is allowed to stay out in separate hut near the main house. They are provided with separate utensil and food items but sometimes the prepared food is served without touching the patient body. Irritant food items such as *Citrus* fruits, *Colocasia,* and *Dioscorea* are restricted for the patient till the wound is healed. The *Nyishi* community of Arunachal Himalayan Region consider snake bite patient as most unholy and somewhat untouchable in earlier decades. Similarly, a person who practice traditional surgical to remove the testis from the scrotum of male pig are also considered unholy. Such people are never allowed to enter into the home during community festivals, marriage and home rituals. Certain animals are not allowed to be consumed during taboos period after special rituals performed in the name of benevolent spirit. They are cat family, *mithun,* birds, reptiles, rodents and other animal meat except chicken meat along with rice in their food items. *Apong,* the fermented beverage made from finger millet and rice is usually allowed to drink before and after rituals which is important part of their ceremony.

Rituals and Taboos associated with ethnic foods and biocultural resources in relation to Agricultural Practices, Animal Husbandry, Fishing and Hunting

Shifting agriculture is the mainstay of the tribes of Arunachal Pradesh. It is their primary occupation besides fishing, hunting, and animal husbandry. The traditional knowledge associated with agricultural seasons, crop cultivation, and harvesting methods and their nexus with socio-religious taboos have not been well explored by the earlier researchers who worked in rural parts of Subansiri and Papum Pare districts of Arunachal Pradesh. Our present investigation revealed that the ethnic agricultural practices followed by the rural farmers are closely linked with seasonalty. As per *Nyishi* local calendar, the summer month (May to September) is known as *Dhir* and the winter months (October to April) as *Dhira. Dhir* is the time of sowing seedlings of paddy and other agricultural crops. After July, weeding starts and it would continue till the month of September. Summer months are hard for the rural *Nyishi* who do not have access to modern amenities like motorable road transport, foodstock such as fair price shop and medical facilities, because during this season all agricultural products of the previous year usually get exhausted. Farmers relay on wild fruits, young pith *of Wallichia oblongifolia*, bamboo shoots, fish, and wild tubers such as *Dioscorea* to suppliment their foodgrain difficiency.

For bumber crop harvesting, a special ritual is performed at community level during the months of January and February called *Nyokum Yullo* (Figure 3) and *Boori Boot yullo*. This is the biggest public festival and ritual celebrated in the name of almighty *Donyi-Polo* Gods for bumper harvesting, protection from epidemic diseases, and for the promotion of social harmony throuhout the year. Such important rituals performed at community level is a part of thanks giving ceremony where domesticated animals, and fermented wine are offer to Gods, and pray the almighty gods for further blessing to humanity for better agricultural production, proliferation of domesticated animals, and for ensuring social and communal harmony and overall ecological security. The villagers for the sake of prosperous existence also perform separate *yullo* rituals at home depending upon the expenditure one can bear for such grand rituals. Usually, such large-scale rituals are done by the rich family of the village who can afford to manage the incidental expenditure involved.

Figure 3: *Nyishi* folk of Arunachal Pradesh in their traditional attire during *Nyokum Yullo* Celebration at community level. Photo: Hui Tag

The precious animals such as *Sob* (*Bos frontalis),* chicken, pig, coupled with other food and beverage items such as rice powder and fermented rice and millet beer (*apo)* are offered to the Gods, demi-gods and goddesses, and associated spirits as token of gift with puritan heart and mind. During and after accomplishment of *Yullo Uii*, strict taboos are observed. Certain fruits items such as *Citrus, Lageneria vulgaris*, and certain green leafy vegetable items such as *Allium hookeri*, *Allium cepha,* and *Allium sativa*, brinjal and *Glycine max*, and rice powder of other clan and family members of the villages are restricted for atleast 3-6 months. Similarly, animal species such as red stripped fish, crocodile, and scale less fish *(Ngurik and Ngub)*, giant lizard, tiger meats, *Felis benghalensis,* and fox meat are also restricted for the member of the household for 2-6 months with belief that consumption of such animal meat would lead to the enlargement of tyroid gland in neck (goiter), graying of hair, susceptible to epidemic diseases, dull intellect, numbness, and broken tone and invite bad luck in individual and family life. The side effect due to improper observance of taboos may not show its immediate effect, however, it would certainly show their effect in long run with belief that the angels of Sun Gods closely observe each and every activity of the member of the concern family who performed rituals in their name. The angels convey the message regarding violation of divine rules to the almighty Gods and consequently the punishment will be awarded to the persons concerned who violate the ritual laws and spiritual taboos. It is also believed that complete observation of taboos would please the Almighty and the blessing deserved through patience observation and faithfulness usually yield enormous luck and happiness for the entire family.

Nyishis are good in animal husbandry and veterinary, poultry and livestock. *Mithun* is their semi-domesticated animal allowed roam freely in jungle mostly in secondary forest for meat and as cultural symbol of the family. It is also used to symbolize individual status, and represents pride and glory of the family. For simple marriage ceremony a minimum number of 2 mithun along with filtered millet beer (*apo*) are used as ritual items to fulfill the formalities. One *mithun* cost around 50,000 INR.

Apart from *mithun,* sheep, goat, cow, pig, dog and chickens are some of the animal species domesticated by the *Nyishis* as source of raw materials for ritual ceremonies, foods, and nutrition. During outbreak of epidemic diseases like foot and mouth diseases, infection of anthrax in cow and *mithun*, the village folk usually perform special rituals to prevent further outbreak of the epidemic diseases. All susceptible porous pass of the jungle are closed through wooden fencing and priest chant specific hyms to invite the spirit who check the outbreak of epidemics diseases in the neighboring villages and message are also passed to the neighbouring villages to check the infected *mithun* from passing through their reserved forest area. The domesticated fowl and chicken, goat and pig are some time killed by the epidemic diseases. In such situations, the village customary law enforces rules of prohibition to check the use of such infected meat brought from the neighboring villages and culprits are usually levied with heavy fine amounting to INR 2000-5000 or the he/she will have to mortgage his *mithun* or Tibetan made mystic bell as alternative mode of payment to defend his stand before village folk.

Fishing remains as fascinating source of recreation as well as an important source of food in order to support livelihood of the rural *Nyishi* since time immemorial. The major rivers available for fishing in *Nyishis* dominated area extended to an area of almost 1100 km length, which account for 55% of the total river length in Arunachal Pradesh, which serves as major center for ethnofisheries activities (Tag et al. 2005). During the fishing operation, menstrual women are not allowed to enter into the site with belief that their presence would render fruitless.

Similar taboos are observed during hunting. Men folk never sleep or receive food from the hands of menstruating women because of the belief that such act is unethical and it would render men folk powerless, lack enthusiasm and dull intellectuality, and also render the important initiative, works and events unsuccessful. Taboos related to menstruation are strictly observed during hunting of wild animal for food. Meats of domesticated animal such as *mithun*, goat, cow, and dog are not allowed to roast at home when the hunting related ritual is in progress before and during individual and group hunting in the jungle for a day. Another method of hunting wild animal in the places far away from their villages is known as *Dha*. Usually, the menfolk operate *dha* (trapping methods) individually or with two to three persons of the same family in pristine forest. Each clan has its own sacred virgin forest area acquired by their ancestors during their migration time and kept undisturbed and reserved for operating agriculture, fishing and hunting of wild animal such as rodents, cat family, pangolin, wild bor, deer, mask deer, bear, and wild pigeon. The persons involved in *dha* operation usually stay in the jungle for a month together. They survive with food items such as rice and maize they carry from the home which is also a part of strict taboos and formalities. They never go back to their home until all the trapping work is finished. It is believed that going back to village would pollute the entire effort. Special rituals are performed with white hen and egg to propitiate the ancestral gods and spirits who are thought to be the owners of those wild animals of the chosen jungle where they operate hunting. For attracting luck, confident and protection of life of the hunters during trapping and group methods of hunting, a special ritual is performed at home in the name of powerful spirit responsible for hunting which is popularly known as *rator*. Male dog and adult fowl are usually sacrificed to propitiate the *rator* spirit and requested him to bring luck, self confidence, charm and safety in hunter's lives. Such associated cultural belief among the *Nyishis* are in accordance with the age-old mythology, folklore and oral literature, which is basically a divine method of viewing the systems of trapping and hunting that support their livelihood issues.

During their departure for hunting and trapping, men never reveal their destination and intention to other fellow villagers. They maintain strict confidentiality until setting of the whole hunting and trapping methods in appropriate location is over. They consider such move as sacred and believe that lack of confidentiality would render fruitless in their efforts. The animals killed during hunting and trappings are roasted over the burning flame to remove haires, fur and feathers. The digestive system along with blood, heart, kidney and liver are cooked in cylindrical bamboo vessels (*Udu*) along with a pinch of salt and wild banana flowers. Such delicacies are thought to be tasty and nutritious which is consumed by the hunters during their stay in the jungle. The remaining is packed in the leaves of *Phrynim pubinerve* (*Kokam ook*), and brought to home as gift to their beloved wife and children. The meat of the animal body/flesh is cut into narrow strip and sharply pointed bamboo stick is inserted from one end passing through another end of the stripped meat. Bundles of such stripped meat are kept over the temporary rack built around 0.5m high above the burning fire heart at jungle. The dried meat is packed in a cage made of bamboo roof (*paka*) and is carried to the home. The dried animal meat is popularly called as *Edin dingko* and use as exchange item during trade and commerce, marriage and other socio-religious ceremony apart from self-consumption at home. Special salad is prepared from *dingko* mixed with chilli powder, powdered fruits of *Zanthoxyllum rhetsa* (Honior), paste of *Zingiber officinale* (*taikke*), *Allium hookeri* (*talap*) and dried bamboo shoot (*hiyup*) mixed with pinch of salt which is served with wine (*apo*) to the guests and family members during special gathering like *nyelle* and *lippi sukor* (a special large wooden plateform constructed for social gathering at village field to discuss some important issues) village level to boost morality, charm and delight in the minds of young and old alike. Prepared salad is also consumed along with rice to boost up appetite in chronic patients, and also consume during stomach indigestion and constipation. The *dingko* is usually preserved in *Naka* (a 7-8 fit high platform raised above which act like ceiling) for a long period. It remains free from microbial infecrion due to the continuous supply of smoke from the *emik* (traditional fire heart). In case of deer meat, it is taken in semi-roasted form or cooked in bamboo vessel along with wild banana flowers and served with cooked rice and *apo*. Wild banana flowers are frequently used because they are considered nutritious, boost taste and flavor in the main dishes. Wild banana flowers are found available in the forest of Eastern Himalayas. The meat preserved in the form of *dingko* plays a significant role especially during marriage ceremony. It was found to be customary to offer one *paka* of *dingko* to bride's family from boy's parents as important gift items besides offering *apo*, pig and *mithun* meat. From economic point of view, one bundles of *dingkos* containing 10 strip *dingko* and 10 bundles of *dingko* means 100 strip *dingko* which is equivalent to Rs 5000 – 8000/- in current market price if sold in urban area, which is seen as most demanding food items and commercially lucrative in both rural and urban areas. The villagers perceived it as profitable business to generate economic wealth in rural areas but such practices are not very sustainable and also not a common and frequent practices of every villagers because in each village only few professional hunters of the village who kill large number of animals could afford to go for such business.

Indigenous Food Biocultural Resources: Alternative Source of Ethnomedicine

Plant and animal species of both cultivated and wild nature are conserved and used as both food and ethnomedicinal agents. The *Nyishis* have their own rationale and hypothesis behind each plant and animal species used as food and medicines. They believe that the plant and animal species consumed as food can also have curative properties if consumed in the right season and occasion either alone or through compound formulation. Their food preparations are mostly boiled, and cooked without oil. Spices such as ginger, and the leaf of *Zanthoxyllum rhetsa* (Honior) are added to the cooked green leafy vegetables in order to enhance taste and flavor (Figure 4). Such spices are consumed as refreshing, stimulating, digestive, and antidepressant agents by adding local varieties of Chilli. Similarly *riir* (*Piper pedicellatum*) after mixing with tender bamboo shoots is consumed (Figure 5). This species is reported to be used in body pain, and stimulent as well (Saha and Sundriyal 2013). The paste of *taike* (*Zingiber officinale*) and *talab* (*Allium hookeri*) is considered sacred.

Figure 4: *Zanthoxylum rhetsa* (Honior) –Fam: Rutaceae leaf consumed by *Nyishi* Folk as vegetable as well as spice and medicine to cure hypertension. Photo: Hui Tag

Figure 5: *Piper pedicellatum (Riir) Fam: Piperaceae –* Leaf consume by *Nyishi* folk as vegetable which is usually cooked with young bamboo shoots. Photo: Hui Tag

Those food items are believed to have anti-microbial, blood purifier, and immune system stabilizing properties. They are also believed to have power to dispel evil spirits and also capable of purify individual sin. The prepared paste is mixed well and served to the guests along with *apo* during *yullo* ritual performed at both individual and community level with belief that the god of grain and wealth possesses such spice items. The medicinal and nutritional value of the cultivated traditional rice varieties have been proven by the modern biochemists in different laboratories of the world during late 19th to 21st Century (Panda, 2001) and they come to the conclusion that majority of the wild food plants and animal species used in traditional food systems are yet to be ascertained in biochemical and pharmacological level to prove its bioactive principles. However, the *Nyishi* have their traditional rice germplasm carried during their migration time from the north (present Tibet and beyond). The local farmer cultivates number of rice varieties ranging from 60-70 whose nutritional analysis and medicinal value are yet to be varified in scientific laboratories. For example, the brownish gel (*mirik*) prepared from powdered finger millet seeds are encouraged to consumed during low vitamins, physical weakness and aneamia while maize powder (*top pomik*) is encourage to consume during constipation, general weakness and indigestion.

The *apo* prepared through fermentation of rice and finger millet is also use to fight cool, and mental depression. It is also used as energy booster during hard labour and ritual ceremonies. Tobacco (*Nicotianum tabaccum*) is usually smoked and then the burnt residue is consumed as anti-depressant and brain stimulating agent. Tobacco is mostly consumed by older generation of the localities and children and young people are told to restrain from chewing and smoking tobacco. The older folk usually use tobacco during hard work and at leisure periods. However, to nullify tobacco's adverse effects, they eat boiled leaves of *Clerodendrum coolebrokianum* (*potto ooh*) and also eat a lot of foods on regular basis coupled with physical laour in agricultural field. Perhaps it is their perception through physical labour prevents them from adverse side effect of tobacco. Still, the local folk lived for about 70-90 years as told by Mr. Magra Teen, one of the older folk of Kicho village of Raga circle. However, their average life expectancy is 72 years for both men and women (SHDR 2005). Tobacco is also used as anti-inflammatory and wound healing agent during insect bite, scorpion bites, cuts and wound. The tender frond of *Diplazium esculentum* (*taka peya*) is another important fern species used as medicinal agents to cure asthma, diabetes, insomnia, and liver cirrhosis and also given to the alcohol-addicted person to relieve toxicity in their liver and nervous system. The young frond is boiled along with chilli and salt, which is consumed along with rice. Apart from *Diplazium,* the species like *Solanum etiopicum, Solanum torvum, Solanum violaceum, Solanum nigrum, Clerodenrum colebrrokianum* and animal products like bile juice of the gall bladder of mithun and Himalayan Black Bear (*Sutum Eppi*) are also used as digestive, stimulant, anti-malaria, anti-diarrhoea and anti-diabetic agents. Furthermore, it was found that the whole plants of *Houttuynia cordata*, stem wood of *Trichosanthes tricuspidata*, fruits of *Solanum torvum, Solanum violaceum, Solanum viarum;* leaves of *Sonchus arvensis* and whole plant of *Centella asiatica* are the well known species commonly used as anti-diarrhoea, anti-dysentery, and brain stimulating agents consumed in the form of salad. The leaves of *Gynura cusumbua, Fagopyrum esculentum* and *Pouzolzia benettiana* are some of the wild vegetable species which are cooked and consumed along with rice during severe constipation, headache and indigestion. During severe constipation, the tubers of *Dioscorea, Colocasia esculenta,* and *Ipomoea batatas* are usually consumed after being roasted in fire in order to enhance laxativity. Wild fruits such as hypanthium of *Ficus semi-cordata,* berries of *Rubus nevius, Saurauia armata* and fruit peel of *Spondias pinnata* are some of the important fruits consumed in raw as stimulant, refreshing and purgative agent during summer month apart from using them as major source of carbohydrate, protein, vitamins and mineral sources. Pith of *tach* (*Wallichia oblongifolia*) and swollen stem of *taab* (*Angiopteris evecta*) are used to cure asthma and joint pain apart from used them as source of carbohydrate during peak hour of famine. Raw leaves of *Spilanthes oleraceae* and roasted fruits of *Solanum viarum* are used to cure toothache, headache, viral fever, while the juice extract of leaves of *Bidens pilosa* var. *minor* is applied to cure cut and wound, earache and sinusitis.

Certain vegetable items are usually avoided by the villagers because they know the side effect of such food items on their physiology due to the long term experience they got through repeated uses. For example, the fruits of *Kullu* – the wild banana (*Musa balbisiana*), *Baiik* (*Solanum violaceum), taka peya* (*Diplazium esculentum)* and indigenous varieties of *Baiyom* - the tomato (*Lycopersicum esculentum*) and leaves of *potto oh* (*Clerodenrum colebrookianum)* are avoided during summer months. It is belief that these vegetable items are though very tasty and delicious but have adverse side effect on health by reducing stamina and vitality if consumed frequently. Such belief is evident through their cultural proverb *"baiik ngo nikchi, baiyom ngo kotar, kullu ngo hibu"*. The proverb says that a fruit of *baiik* is responsible for causing deep sleep and formation of debris on the eyes; *baiyom* is responsible for lethargy, *Kullu* is responsible for reducing plump face to sunken and dried one. The proverb is obviously having deep meaning and developed through repeated observation and experimentation due to their close association with nature since time immamorial. The modern biochemist may claim that these plants might have some moderate to high concentration of active phyto-consitituents such as group of alkaloid, steroid, anthroquinone and triterpenoid. Experimentation at pharmacological level could yield some important compound for the future application as alternative herbal drugs for reducing high cholesterol level, high fatty acid deposition, high blood pressure, and insomnia and hypervitaminous conditions. They could also be used as anti-diabetic and hypolidemideamic agent to cure alcoholic patients suffering from liver disorder.

Few animal species such as *kurluk, piiz, sojo,* giant lizard, black bear, *Seb besor* (long tailed monkey), black crow, and dog are most often consumed as medicine. For example, the bile juice squized out from gall bladder of himalayan black bear is used to cure dysentery, diarrhoea and chronic malaria. The meat of black crow (bird), *kurluk* and *pizz* (both rodents) is roasted and consumed during malarial fever, dysentery and to boost immune system. The village folks believe that these animals are capable of dispelling evil spirit from the human surrounding. Priests and orators usually avoid such meat. Digestive parts and stools of *seb besor*, meat and stool of *sojo*, are also used to treat malaria, viral fever and dysentery. Digestive systems of aerial wild birds (*petta pik, pisik, pung, podur, rass, poblek*) are roasted and consumed along with stools. It is their perception that these aerials birds mostly feeds on wild edible fruits of *Cinnamomum*, *Macranga denticulata* and other species are rich source of nutrient and act as medicinal agents to cure bodyache, stomach complain, and chest pain.

Besides wild vegetable plants, the *Nyishi* of remote villages were found to be heavily relaying on cultivated food plants and animal products as another major source of ethnomedicinal agents. However, milking of animal and use of milk product was found to absent in food habit of *Nyishi*. Impacts of modern Indian food habit were seen in some villages located near semi-urban area where semi-fried foods are consumed by the few houshold.

Biodiversity and Traditional Food: Learning on Knowledge Gap

The rating of the ethnobotanical knowledge of the participants interviewed through multiple choice task questionnaire methods has revealed that 8% informants represent poor; 52% represent average; 40% informants represent good category under age group <16-50 years. Whereas, in age group <51-80 years, 5% informants are poor, 34% are average and 61% informants are good in ethnobotanical knowledge (Figure 6). The rating of the ethnozoological knowledge of the participants interviewed revealed that 16% informants represent poor, 59% represent average; 25% informants represent good category under age group <16-50 years. Whereas, in age group <51-80 years, 10% informants are poor, 44% are average and 46% informants are good in ethnozoological knowledge. It was observed that ethnbobotanical and ethnozoological knowledge increase with increasing age of the informants which is in consensus with the finding of earlier workers such as Martin (1994), Phillips (1996) and Reyes Garcia (2003).

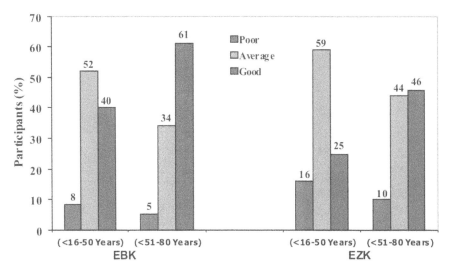

Figure 6: Comparison of Ethnobotanical and Ethnozoological Knowledge of the Informants of two age group [Total Sample (n = 160); age group <16-50 (n=105), age group <51-80 (n=55)] [Legend: EBK=Ethnobotanical Knowledge; EZK=Ethnozoological Knowledge].

People living in high altitude zone (1500-2000 m) like Talley Valley, Kemliko and Godak mostly relayed on boil, roasted and fermented food whereas the participants interviewed from township area such as Boa Simla, Raga, Tamen (1200-1800 m), Doimukh and Itanagar (150 m) usually relayed on both boil and fried food. Statistical analysis on dependency rate of informants on boiled and fried food revealed that almost 97% informants consumed boiled food in rural localities (Figure 7) and almost 68% in semi-urban localities while 12% informants in pure urban localities consumed boiled food. Further, it was also observed that participants dependant on fried food in rural sites is just 3% while the figure is 32% in semi-urban localities. In pure urban localities, fried food dependant is almost 78%. This implies that the food habits of rural localities are almost boiled. The fried item is hardly seen except during specific occasion such as VIP or officer's visit. However, in semi-urban and pure urban localities, people usually opt for both boiled and fried food. It can be inferred from the present finding that people's attitudinal change and modern outlook due to the influence of alien culture, rapid religious conversion and acculturation could be a few important driving forces that contribute to such gradual change in traditional food habit among the urban and semi-urban *Nyishis* of Arunachal Pradesh.

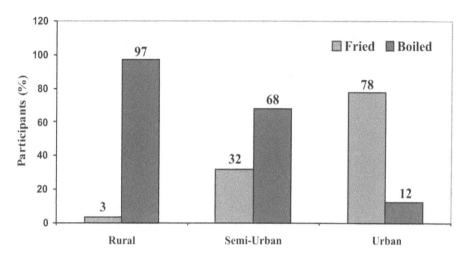

Figure 7: Participants Dependancy rate on Boiled and Fried Food in Rural, Semi-Urban and Urban Localities [Total Sample (n)=160].

Our interaction with local informants revealed that the participants of the semi-urban and urban localities were found to be gradually tilting towards the philosophy of Christianity, and Hinduism in recent decade. It is well evident that the cultural transitions in *Nyishi* society has manifested itself in the form of gradual change in their traditional food habit, and ritual customs, ideology, and dressing sense among the younger generation, and they are more pro-towards alien culture. However, some glaring example of traditional food habit has been witnessed among the older generation of the semi-urban and urbans *Nyishis* residing near Itanagar, Naharlagun, Ziro, Doimukh, Raga and Tamen. The older folk of these villages even after their converted into Christianity and also having modern educated kids at their home are still relaying on traditional boiled diets. But such fried food habit is mostly opted by the younger generation in the above mentioned semi-urban and urban localities. The villagers collect wild vegetable items from the nearby village and jungle and sold it in the market of Ganga, Doimukh, Raga, Boa Simla and Tamen. Such business are mostly run by the women and only few men were encountered during our ethnobotanical market survey except retailers of imported vegetable items mostly marketed by male folk but non-Arunachalee communities such as *Assamese, Nepali, Bengali,* and Bihari residing in these mini-towns.

Wide gap in traditional biodiversity knowledge and drastic changes in cultural food habit among the younger generation *Nyishi* in recent decades can be perceived as manifestation of their changed attitude and outlook due to the impact of modern culture on traditional institution and in their livelihood system. Such situation could certainly lead to social chaos, cultural erosion, identity crises and civilization confusion in near future. Most importantly, the traditional food item is being considered nutritious, health friendly on account of compatibility with ecosystems because of the fact that they are herbal and animal based and such food knowledge under threat could be lost forever if the knowledge gap continue to persist among the younger at current rate.

Hence, urgent initiative is require to document and train the younger generation on traditional food and biocultural knowledge least they lost under the powerful influence of alien culture. The techniques like indigenous agricultural practice, traditional fishing and hunting method, food and beverage processing techniques if found unique and resource conserving capability, should be encouraged to use, and must be preserved as the part of community biodiversity register and must be patented in the name of the community in order to prevent piracy of these vital knowledge system by the other organization and multinational company who intended to developed industrial products using these indigenous knowledge system in commercial scale without prior permission from the grass root level traditional knowledge holders.

Discussion and Conclusion

Traditional institutions in the *Nyishi* dominated regionsuch as *nyelle* and *lippi sikor* (*Kebang* like panchayat body as seen among the *Adis*) play an important role in conservation of whole ecosystem through traditional clan forest and *tump-no-ngo* (*jhum* agriculture) system coupled with holistic concept based on age-old mythology and folklore. Priest and orators, elderly generation of the village usually play a significant role in disseminating traditional ethics and morality in their society for conserving bioresources for the sustainable co-existence of human and nature. In view of degrading environment due to the population pressure in present century, the policy makers can take a noble initiative by educating people about environmental ethics emulating indigenous approach in conserving biodiversity based ethnic food resources in mountain ecosystem of Eastern Himalayan Region. People transmigration in recent decade has already created a cultural chaos in mountain ecosystem of Arunachal Himalayan Region of India and other states of Northeast India and such chaos has finally manifested itself in the form of deforestation, ruthless exploitation of natural resources, and habitat destruction where survival of many rare plants and animals species have been pushed to the edge of vulnerability and extinction. For example, the *Nyishis* of Kurung Kame district migrating towards the semi-urban area of Lower Subansiri and Papum Pare district in the form of daily wage labourer do not have the sufficient knowledge and skills about the environmental ethics of newly found places since their practical experience about *jhumming*, fishing and hunting season they had in their original native region were totally different in certain level such as selection of fallow period, *jhum* cycle, sowing season of the agricultural crops. As soon as they migrates to new area, they would adopting their rudimentary knowledge and skills to ensure food security by cultivating a plot of *jhum* agriculture but without proper consultation and without having traditional knowledge and wisdom about fallow period and seed sowing seasons which the farmers of Lower Subansiri and Papum Pare usually followed. Consequently, their whole effort do not yield much productivity, thus remain hunted by the famine for whole years. Moreover, the indigenous weeding techniques followed among the *Nyishis* are location specific thus each group of *Nyishis* residing in different geographical location in 5 districts of Arunachal Pradesh with their own indigenous agricultural, fishing and hunting techniques adopted for food security vary widely from place to place. Thus, in order to ensure their food security, a proper training about season, climatic and soil types coupled with traditional beliefs, wisdom, ethics and morality should be taught to the people who have just migrated near urban or semi-urban sites. Such initiative and awareness campaign can help restoration of both traditional food knowledge systems in particular and the overall health of the mountain and subtropical ecosystem in general.

From the present ethnobiological investigation, it may be concluded that the *Nyishis* of Arunachal Pradesh are good in ethnobotanical and ethnozoological knowledge for using them as both food and cultural materials. The traditional food bioresources they used are season and location specific, which is only harvested in right time couple with traditional wisdom, ethics, beliefs and practices such as rites, rituals, animal sacrifice and propitiation offer to the malevolent and benevolent gods and goddesses of the nature. Majority of their food processing methods are boil, which is health friendly as they used a good admixture of both plants and animal products rich in essential nutritious elements, which could be beneficial for the both physical and mental health. Social and religious taboos observed among the *Nyishis* play a significant role in conservation and restoration of food bioresources in particular and overall biodiversity wealth of natural ecosystem in general. Investigation should be made at molecular and biochemistry level to assess the targeted gene responsible for cure target ailments, anti-oxidant property, nutritional status, cytotoxicity effect, and long-term reliability of these natural products. Famine most often hit the villagers during middle parts of the year that is during the month of July and August. During this period, *bullu* (the kitchen garden) they natured in their home backyard and *pollek* has been found to be best alternative to combate famine and in ensuring food security among the rural *Nyishis*. The similar situation prevails with other tribes of Arunachal Pradesh. The *pollek* is a term use to denote a plot of land cultivated in the area surrounded by dense forest where early ripening crop such as pumpkin, cucumber, sweet potato, Colocasia, tapioca and maize are sown during month of April in temperate and March in tropical. It has been witnessed that *pollek* play a significant role in combating famine during July and August when previous year food stock get exhausted and wild harvesting is nearly impossible due to natural calamity, heavy downpour of rainfall and harsh weather conditions. However, the *bullu* and *pollek* system practiced among the *Nyishis* has been perceived as highly inconsistent as per our present finding. The majority of the families (60%) in the villages do not adopt this dual agricultural system on regular basis in right season. Thus, to fight famine and to ensure food security in the remote villages, the farmers should be encouraged to cultivate *pollek* as important part of their main agricultural system to ensure availability of food and vegetable during peak hour of famine. However, gap is wide enough related to traditional food habit and knowledge system among the younger generation of urban and semi-urban localities. Traditional knowledge campaign coupled with well thought strategy could help in rescuing traditional food resources and food habit from further degradation.

Acknowledgements

The first, second and fourth author are thankful to Department of Science and Technology, GOI, New Delhi for funding support through Women Scientist (WOS-B) project. The authors are thankful to all the traditional knowledge holders and villagers who rendered valuable service and cooperation during the field work at chosen localities. The authors also like to thank district authority of Papum Pare and Lower Subansiri District of Arunachal Pradesh for their guidance and support. The participants who shared traditional food knowledge with us during our field survey and also acted as local guide of the respective localities is highly solicited. Cooperation rendered by Dr. R.C.Srivastava, Joint Director, BSI, Itanagar in taxonomic work on plants is highly solicited.

References

Albert LS, Kuldip G (2006) Traditional use of medicinal plants by the *Jaintia* tribes in North Cachar Hills district of Assam, northeast India. *Journal of Ethnobiology and Ethnomedicines* 2:33 doi:10.1186/1746-4269-2-33. URL Available at http://www.ethnobiomed.com/content/2/1/33[accessed on 14th July 2006]

Angami A, Gajurel PR, Rethy P, Singh B, Kalita SK (2006) Status and potential of wild edible plants of Arunachal Pradesh. *Indian Journal of Traditional Knowledge*, 5(4):541-550

Annonymous (1960-2000) Index Kewensis (in series). Royal Botanic Garden, Kew, Richmond Surrey, UK

Annonymous (2001) Statistical handbook of lower subansiri district, Arunachal Pradesh, 3-58

Annonymous (2001) Statistical handbook of Papum Pare district, Arunachal Pradesh, 6-40

Aparajit D (2004) An overview of Hornbills: Distribution and conservation in Arunachal Pradesh. *Arunachal Forest News,* 20:40-64

Arora RK (1980) Nature food plants of northeastern tribals. *In:* Jain sk (ed) Glimpses of Indian ethnobotany. Oxford and IBH Publication, New Delhi, 91-136

Bhaskar S, Tag H, Das AK (2007) Ethnobotany of foods and beverages among the rural farmers of *Tai Ahom* of North Lakhimpur district, Asom. *Indian Journal of Tradational Knowledge*, 6 (1): 126-132

Borang A, Thapaliyal GS (1993) Natural distribution and ecological aspects of non-human primates in Arunachal Pradesh. *Indian Forester,* 119(4):834-844

Borang A (1996) Studies on certain Ethnozoological aspects of *Adi* tribes of Siang district, Arunachal Pradesh, India. *Arunachal Forest News*, 14(4):1-5

Champion HG, Seth SK (1968) Revised survey of forest types of India, Delhi

Das AK (1986) Ethnobotany of East Siang District of Arunachal Pradesh. A PhD thesis submitted to the Guwahati University (Unpublished)

Dutta BK, Dutta PK (2005) Potential of ethnobotanical studies in Northeast India: An overview. *Indian Journal of Traditional Knowledge,* 4(1):7-14

Electronic Plant Information Centre (ePIC), hosted by Royal Botanic Garden, Kew (UK) – available at http://www.kew.org/epic, accessed on 13th June 2004

Gajural PR, Rethy P, Singh B (2003) Preliminary studies on wild edible plants resources of Dihang Debang Biosphere Reserve. *In:* Abstract and Souvenir of 13th Annual Conference of IAAT and International Symposium on plant taxonomy: Advances & Relevance. Held at Department of Botany, TM Bhagalpur University (Bhagalpur) Bihar, 14-15th Nov.2003

Greuter W *et al* (2000) International Code of Botanical Nomenclature (St. Louis Code) adoted in 16th International Botanical Congress, St. Louis, 1999, USA. Available at http://www.bgbm.org/iapt/nomenclature/code/SaintLouis/0001ICSL.Accessed on 16th September 2002

Greuter W, Brummitt RK, Farr E, Kilian N, Kirik PM, Silva PC (1993) Names in current use for the extant plant genera. Koelt, Konigstein, Germany. *Regunum Vegetable*, 129:1464

Greuter W. Brummitt RK, Farr E, Kilian N, Kirik PM, Silva PC (1994) International code of botanical nomenclature (Tokyo Code) adopted by the 15th International Botanical Congress, Yokohama, 1994. Regunum Veg. 131

Groombridge B (1992) Global biodiversity: Status of Earth's living resources, WCMC, Cambridge, UK

Haridasan K, Bhuyan RL and Deori ML (1990) Wild Edible Plants of Arunachal Pradesh. *Arunachal Forest News*, 8(1&2): 1-8

Haridasan K, Bhuyan RL (2002) Medicinal Plants distribution and farming potential in Arunachal Pradesh. *SBRDT Workshop cum Training Bulletin,* 26, SFRI, Itanagar

Hegde SN, Haridasan K (2002) Mishmi teeta – an endemic and endangered medicinal plant in Arunachal Pradesh. *SBRDT Workshop Cum Training Bulletin*, p- 25, SFRI Itanagar

Hooker JD (1862-1889) *Flora of British India* Vol. 1-7

International Plant Nomenclatural Index (IPNI) Hosted by International Association for Plant Taxonomy (IAPT) Vienna – available at http://www.ipni.org. – accessed on 3rd Dec. 2003

IUCN (2006) Guidelines for Using the IUCN Red List Categories and Criteria Version 6.2 (December 2006). Prepared by the Standards and Petitions Working group of the IUCN-SSC Biodiversity Assessments Sub-committee, Huntingdon Road, Cambridge, UK. (available at http://app.iucn.org/webfiles/doc/SSC/RedListGuidlines.pdf. accessed on 3rd Jan. 2007)

Jain SK, Rao RR (1980) Glimpses of Indian Ethnobotany, 84-91. Oxford & IBH Publishing Co. New Delhi

Kaul KN, Haridasan K (1987) Forest types of Arunachal Pradesh: A preliminary study. *Journal of Economic and Taxonomic Botany,* 9(2):379-389

Kohli YP (1996) Antibacterial and antifungal activity from the extract of leaves of *Clerodendrum colebrookianum. Arunachal Forest News*, 14 (2):17-19

Mao AA, Hyniewta TM (2000) Floristic diversity of Northeast India. *Journal Assam Science Society*, 41(4):255-266

Martin G (1995) Ethnobotany: a methods manual. Chapman & Hall Co. London, 120-270

Milburn MP (2004) Indigenious nutrition using traditional food knowledge to solve contemporary health problems. *The American Indian Quarterly*, 28(3&4): 411-434

Murtem G (2005) Ethnobotanical studies of upper Subansiri District of Arunachal Pradesh. A PhD thesis submitted to the Department of Botany, Rajiv Gandhi University (Arunachal University) (unpublished), 15-208

Murtem G (2000) Wild vegetables of *Nyishi* tribe of Arunachal Pradesh. *Arunachal Forest News*, 18(1&2):66-77

Myers N, Muttermeier RA, Muttermeier CA, da Fonseca, GAB, Kent J (2000) Biodiversity hotspots for conservation priorities. *Nature*, 403:853-858

Nowak M (1991) Walker's Mammals of the World. Fifth Edition. Johns Hopkins University Press, Baltimore, 563-564

Odum EP (1971) Fundamentals of Ecology. W.B. Saunders Company, Philadelphia, PA (USA).

Pal GD (1984) Observation on the ethnobotany of the tribals of Subansiri, Arunachal Pradesh. *Bulletin of the Botanical Survey of India*, 26(1-2): 26-37

Pal GD (1990) Flora of Lower Subansiri, Arunachal Pradesh. PhD thesis submitted to the University of Culcutta (unpublished)

Panda H (2001) Herbal Foods and its medicinal values. National Institute of Industrial Research, 106-E, Kamla Nagar, Delhi-110007

Pallabi K, Tag H, Mukhopadyay PK, Das AK, Mukerjee AK (2004) Indigenous fishing techniques practiced by the tribes of Arunachal Pradesh (North East India). *Aqua Asia Journal*, 10(2):35-37

Panigrahi G, Naik VN (1961) A botanical tour to Subansiri Frontier Division (NEFA). *Bulletin of the Botanical Survey of India*, 3(3-4):361-388

Phillips OL (1996) Some quantitative methods for analyzing ethnobotanical knowledge in ethnobotanical research. In: Alexiades MN and Sheldon JA (ed) Field manual, pp-171-197. New York Botanical Garden (NY)

Prater SH (1948) The book of Indian Animals. Bombay Natural History Society Publication, Bombay

Rao RR, Jain SK (1977) A hand book of field and herbarium methods. Today and Tomorrow's Printers and Publushers, New Delhi-110 001

Rawat MS, Chowdhury S (1998) Ethnomedicobotany of Arunachal Pradesh (*Nyishi* and *Apatani* tribes), pp-1-206. Bishen Singh Mahindra Pal Singh, New Conaught Place, Dehra Dun

Reyes-Garcia V, Byron E, Godoy R, Vadez V, Apaza L, Perez E, Leonard W, Wilkie D (2003) Ethnobotanical knowledge shared widely among Tsimane Amerindians, Bolivia. *Science*, 299:1707

Saha D, Sundriyal RC (2013) Perspectives of tribal communities on NTFP resource use in a global hotspot: Implications for adaptive management. *Journal of Natural Sciences Research*, 3(4):125-169.www.iiste.org

SDHR (2005) State human development report. Published by Department of Planning, Government of Arunachal Pradesh, India. http://www.undp.org/content/dam/india/doc/human_develop_report_arunachal_pradesh_2005_full_report.pdf.Accessed on 10-01-2014

Singh RK (2004) Using diversified ethnic fruits for food security and sustainable livelihoods, IK practice detail-2 (Available at http://www4.worldbank.org/afr/ikdb/search.cfm; accessed on 8th Jan. 2007

Singh RK, Singh A, Sureja AK (2007) Traditional foods of *Monpa* tribe of West Kameng, Arunachal Pradesh. *Indian Journal* of *Traditional Knowledge*, 6(1):25-36

Solanki GS, Chongpi B, Awadesh K (2004) Ethnology of the *Nyishi* tribes and wildlife of Arunachal Pradesh. *Arunachal Forest News,* 20:74-85

Tag H, Das AK, Pallabi K (2005) Plants used by the Hill Miri tribe of Arunachal Pradesh in Ethnofisheries. *Indian Journal* of *Traditional Knowledge,* 4 (1): 57-64

Tag H, Das AK (2003) Significant plants among the *Nyishi* tribe of Arunachal Pradesh (North East India), an ethnobotanical view. In: Abstract & Souvenir of 13th Annual Conference of IAAT and International Symposium on plant taxonomy: Advances & Relevance. Held at Department of Botany, TM Bhagalpur University (Bhagalpur) Bihar, 14-15th Nov.2003

Tag H, Das, AK (2004) Ethnobotanical notes on Hill Miri tribes of Arunachal Pradesh. *Indian Journal of Traditional Knowledge,* 3(1): 80-85

Tag H, Das AK, Kalita P (2005) Plants used by the Hill Miri tribe of Arunachal Pradesh in ethnofisheries. *Indian Journal* of *Traditional Knowledge,* 4(1): 57-64

Turner NJ (2001) "Doing it right": Issues and practices of sustainable harvesting of non-timber forest products relating to First Peoples in British Columbia. *British Columbia's Ecosystems and their Management,* 1(1):1-11

Turner NJ (2005) Earth's Blanket: Traditional teaching for sustainable living. Douglas & McIntyre Ltd, Quebec Street Suite, Vancouver, British Columbia, Canada

Turner NJ, Turner SE (2004) Food, forage and medicinal resources of forests. Encyclopedia of Life Support Systems. EOLSS Publishers, 1-41

Chapter – 8

New Shoots, Old Roots — the Incorporation of Alien Weeds into Traditional Food Systems

Michelle Cocks*, Tony Dold and Madeleen Husselman

Abstract

Leafy vegetables have largely been portrayed as fulfilling an important nutritional supplement in local people's diets, as a safety net function to rural households during periods when agricultural production is low and as a food supplement for the poverty stricken. In this paper we present evidence to show that the consumption of wild leafy vegetables (pot-herbs) represents more then just an important source of nutrition and/or a safety net function for poor households, but that they also fulfil an important cultural function for both poor and wealthy Xhosa speaking people in the Eastern Cape, South Africa. We show that the consumption of food needs to be studied as a bio-cultural phenomenon which includes both nutritional and anthropological understanding. The consumption of traditional leafy vegetables, called *imifino*, remains stable across peri-urban areas and urban centres, at 49% and 42% respectively. Their consumption is not restricted to the rural poor as these greens are also regularly consumed by wealthy households in both peri-urban and urban centres. Traditionally the preparation and consumption of *imifino* dishes is the domain of women and children as a social activity. Recently, thanks to media

---

*m.cocks@ru.ac.za

education programmes and despite traditional taboo, men are increasingly also eating *imifino*. The empirical results clearly demonstrate the need to give more attention to the social processes impacting the use of wild plant products in order to fully appreciate the scope of conserving indigenous cultural heritage and biodiversity in ever-changing times. The findings also point to the need for those studying agricultural biodiversity systems to give attention to alien weeds and plant species which are not actively managed but which may have a significant role in food systems.

Keywords - Xhosa, weeds, edible wild greens, *imifino*, South Africa, traditional foods.

Introduction

Humans have always depended on wild edible plants and it is reported that approximately 7,000 plant species have been collected as food throughout history (Moore and Tymowski 2005). Since the beginning of agriculture some 12,000 years ago many of these edible plants have been domesticated. There are an estimated 45 000 species of plants in Sub-Saharan Africa, of which about 1 000 can be eaten as green leafy vegetables (Etkin 2000). However, a rapidly growing global population and changing food consumption patterns have resulted in a shift away from use of these plants towards modern, commercialized agriculture. In many parts of Africa the introduction of European species such as *Brassica* (i.e. cabbage and its relatives) has replaced the collection and consumption of traditional vegetables, often to the nutritional detriment of local people (Ndoye et al. 1997; Ngwerume and Mvere 2000; Gockowski et al. 2003). This, however, is not the case in all parts of Africa. For example, Gockowski et al. (2003) describe how the introduction of exotic agricultural crops appears to have had little effect on the sales and consumption of traditional leafy vegetables in Cameroon.

Wild or weedy leafy vegetables have been acknowledged as providing an important nutritional supplement in local people's diets (Fleuret 1979; Wehmeyer and Rose 1983; Gockowski et al. 2003, Ogoye-Ndegwa and Aagaard-Hansen 2003; Moore and Tymowski 2005), and as serving a 'safety net' function to rural households during periods when agricultural production is low (Guinand and Lemessa 2000; Shackleton 2006). For example, in Ethiopia, wild leafy vegetables have been described as 'famine-foods' that are consumed only during times of drought (Guinand and Lemessa 2000). It has also been documented that local people, particularly women, are involved in the trade of these vegetables, making an important contribution to households' economy in

Tanzania and parts of southern Africa (Fleuret 1979, Shackleton 2006, Nguni and Mwila 2007). Recently the Food and Agriculture Organization of the United Nations (FAO) and the International Plant Genetic Resources Institute (IPGRI) are leading an international initiative to promote the sustainable use of biodiversity in programmes contributing to food security and human nutrition, and to thereby raise awareness of the importance of this link for sustainable development. A call has thus been made for further research to gather data on composition and consumption of edible wild-growing greens to increase the evidence base of linkages between biodiversity and food security and human nutrition (Toledo and Burlingame 2006).

In this paper we present evidence to show that the consumption of wild leafy vegetables (*imifino*) in South Africa represents more than just an important source of nutrition and/or a 'safety net' function for poor households, but also fulfils an important cultural function. This is despite massive urbanisation and rapid changes occurring in many regions, resulting in livelihood strategies becoming increasingly diversified (Ellis 1998; Kepe 2008). Notwithstanding these changes in southern Africa, many households are still reliant on wild resources (Shackleton et al. 2001; Shackleton and Shackleton 2004; Wiersum et al. 2004 and Cocks 2006).

Furthermore, we provide evidence to support the argument that the consumption of food needs to be studied as a bio-cultural phenomenon, which includes both a nutritional and anthropological understanding. For example, the way in which food is collected, distributed and shared within a community is guided by local customs, practices and traditions. By focusing only on the health aspect and nutritional contributions of food, one fails to acknowledge the significance of cultural elements (Ogoye-Ndegwa and Aagaard-Hansen 2003; Kepe 2008). It has therefore been suggested that when one studies communities' food habits, one needs to "invite the curious eyes of historians, geographers, sociologists and folklorists" into the analysis (Fieldhouse 1998). This study attempts to provide some of those insights by focusing on the social aspects of the consumption of traditional leafy vegetables in the Eastern Cape Province of South Africa.

Research Methodology

The socio-political history of the region, indeed of the whole of southern Africa, has had a major impact on the peoples of the region. The Eastern Cape Province is inhabited predominantly by *isiXhosa* speaking ethnic groups who are descendants of the Bantu groups who migrated from Central Africa some 2000 years ago and settled in the region. In the late 1700s Dutch-speaking pastoralists entered this area with disastrous results for the *amaXhosa* who were banished from their land, economically and politically subjugated and brought under the heel of the colonial state (Ainslie 2002). For over a century, the consolidation of racially based subjugation entailed what may be regarded as an unbridled assault on Xhosa identity and culture (Beinart 1980). The South African apartheid government established two Homelands within the borders of the Eastern Cape Province, the Republic of Ciskei, were the peri-urban survey presented in this study was undertaken, and the Republic of Transkei. Homelands were established under the apartheid vision that saw South Africa's population divided into a number of ethnic groups, each with its own territory and inherent potential that would be developed into separate sovereign nations (Sharp 1988). Under this imposed system, more than 3.5 million people were shifted to mostly infertile lands in the Ciskei (Beinart 1994). African people living in the Homelands mostly worked as labourers in the 'white' industries and urban areas in South Africa. Today post-apartheid rural households continue to draw heavily on livelihoods generated from urban areas and/or State benefits such as pensions, rather than revenue flows from farming practices or biodiversity use (Turner 2004), although land-based livelihood strategies are still important (Shackleton et al. 2001).

Despite the onslaught of the apartheid regime and the ongoing impact of global economic influences and changes in the area, cultural practices and traditions are still adhered to by many households in the now inclusive Province of the Eastern Cape. For example, a study revealed that in the peri-urban communities, 40% of the total natural resources collected and utilised at a household level per annum are used for cultural purposes and in the urban areas this was as high as 85% of the total natural resources used (Cocks 2006).

The information presented in this chapter has been drawn from three sets of data, two quantitative surveys and one qualitative study. The quantitative household survey was conducted in 2003 and aimed to document the use of wild plant resources in six peri-urban communities and two urban communities (Cocks 2006). The six peri-urban communities surveyed, included: Pirie Mission, Chata, Woodlands, Ntloko, Benton and Crossroads (Figure 1), which are located in the Peddie and King William's Town districts of the former Ciskei homeland, now Eastern Cape Province, in South Africa (Cocks et al. 2006). These villages are characterised by poor infrastructure, high population densities, and high poverty levels (De Wet and Whisson 1997, Palmer 1997). The urban surveys were conducted in King William's Town and East London in the same region (Cocks 2006). A 100% questionnaire household survey (n=1 011) was conducted in the peri-urban villages to document household composition, wealth, and use of wild plant material. The questions on wild plant consumption focused on determining the species, use categories and quantities of wild plant material collected. In the two urban centres a stratified sampling methods was used to ensure a range of households across different socio-economic conditions, were interviewed. This included sampling in suburbs which were perceived to represent high income, middle income and low income classes. Within each suburb, households were selected through random sampling technique. A total of 302 households were interviewed; 151 questionnaires in each urban centre (Cocks 2006).

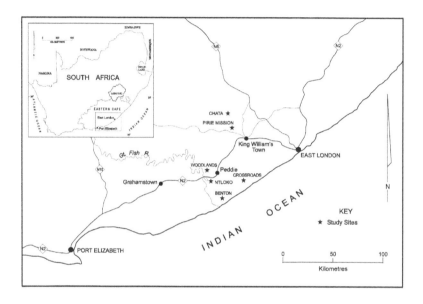

Figure 1: Map showing locality of the peri-urban and urban study sites.

Information for the qualitative study was based on 25 in-depth interviews and these were carried out amongst household members living in middle class and wealthy suburbs of Zwelitsha (King William's Town) and Mdantsane (East London) with Similar interviews were conducted amongst 35 households in four of the peri-urban areas, included in the survey (Chata, Woodlands, Ntloko and Benton). Households were selected randomly. In order to identify plants correctly, voucher specimens were collected and are housed in the Selmar Schonland Herbarium (GRA) at the Albany Museum in Grahamstown.

A third set of quantitative interviews were carried out, with customers of a street vendor selling cooked *imifino* at a busy Highway market in Mdantsane. A total of fiffty-nine customers were asked questions regarding how often they purchase and consume *imifino*, their knowledge of these species and their preferred species.

Results

Imfino Plants

In the quantitative household survey nineteen species were reported as being collected (Cocks et al. 2006). All but one of these *imfino* plant species are cosmopolitan weeds and it is therefore difficult to trace the history of the use of *imifino*. How and when these plants were brought to South Africa remains uncertain, but by the 19th century most of them were already considered part of the traditional diets of African people. Written records show us that certain species were already eaten here in the late 17th century, for example *isiqwashumbe* (*Raphanus raphanistrum*) was cultivated by Jan van Riebeeck at the Cape of Good Hope in 1652 and *utyuthu* (*Amaranthus hybridus*) documented as "very frequently stewed instead of spinach", as early as 1680 (Van Wijk 1986).

All *imifino* plants in the study area are collected from home gardens, disturbed land or fallow agricultural fields. Of the 10 commonly cited species, the dominant three are *Amaranthus hybridus*, *Chenopodium murale* (Figure 2) and *Sonchus oleraceus* (Figure 3). Almost all the popular species originate from other parts of the world (Table 1), but are common weeds in South Africa, where they grow abundantly in fields, gardens and disturbed areas after rain. The fact that almost all the popular species eaten as *imifino* are alien weeds is not unique to our study sites. For example, Ogoye-Ndegwa and Aagaard-Hansen (2003) describe how most of the traditional vegetables eaten by the Luo in Kenya are weeds that are collected from previously cultivated farmlands.

Figure 2: Example of one of the ten most commonly consumed species, *Chenopodium murale*. Photo: Michelle Cocks

Figure 3: Example of one of the ten most commonly consumed species, *Sonchus oleraceus*. Photo: Michelle Cocks

Table 1: The ten most commonly cited imifino species arranged in alphabetical order by family.

Family	Botanical name	Voucher	Xhosa name	Origin South Africa	No. of respondents	% of respondents
Aizoaceae	*Aizoon glinoides* L.f.	Dold and Husselman DH10	Ingcethe	(Germishuizen and Meyer 2003).	4	6
Amaranthaceae	*Amaranthus hybridus* L.	Dold and Husselman DH8	Utyuthu	Central and South America (Van Wyk and Gericke 2000).	57	86
Asteraceae	*Bidens pilosa* L.	Dold and Husselman DH9	Umhlabangulo	South America (Wells et al. 1986).	6	9
Asteraceae	*Sonchus oleraceus* L.	Dold and Husselman DH4	Ihlaba	Europe, Asia and Mediterranean origin (Van Wyk 2005).	42	64
Brassicaceae	*Raphanus raphanistrum* L.	Dold and Husselman DH7	Isiqwashumbe	Europe, Asia (Wells et al. 1986)	6	9
Chenopodiaceae	*Chenopodium murale* L.	Dold and Husselman DH6	Imbikicane	Europe, Asia (Wells et al. 1986).	20	30
Cucurbitaceae	*Cucurbita pepo*	Dold and Husselman DH5	Imitwane	Central America (Van Wyk 2005).	11	17
Solanaceae	*Solanum nigrum* L.	Dold & Husselman DH1	Umsobo	European origin (Van Wyk 2005).	18	27
Urticaceae	*Urtica dioica* L.	Dold and Husselman DH2	Irhawu	European and Asian origin (Van Wyk 2005).	8	12
Urticaceae	*Urtica urens* L.	Dold and Husselman DH 3	Urhaljani	Northern temperate regions (Grubben and Denton 2004).	7	11

While *imifino* plants are reported to be highly nutritious by some authors (Grubben and Denton 2004; Van Wyk 2005), others treat them as weeds and problem plants with respect to commercial agriculture (Wells et al. 1986). There has been no attempt to cultivate *imifino* in the study area and the general perception is that the species are "wild" (indigenous).

Data collected from the in-depth interviews in the urban centres revealed that a total of 17 species were consumed, of which seven were named by more than 10% of the respondents (Table 1). An average of three species were regularly collected per household. *Amaranthus hybridus* was considered the tastiest, followed by *Solanum nigrum* (Figure 4).

Figure 4: *Solanum nigrum* was considered as one of the tastiest *imifino* species.
Photo: Michelle Cocks

The most popular *imifino* species in our study area are commonly used in the region (Rose and Guillarmond 1974; Bhat and Rubuluza 2002; Shackleton et al. 2002). However, Shackleton's study in three South African provinces (2003) indicated that the relative importance of species differs from region to region. Although use and popularity is linked to local availability, the species recorded in our study are evidently typical ingredients for traditional *imifino*, their popularity being influenced by cultural preferences. The influence of cultural preference on selecting *imifino* species is also suggested by Voster and van Rensburg (unpubl. doc.).

Consumption of imifino

The peri-urban household surveys found that just less than 50% of the households collected and consumed *imifino* species. The mean quantities collected amounted to 35.1 ± 30.3 kg per annum per household (Cocks et al. 2006). Wealth was found not to significantly influence either the consumption or the quantity of *imifino* collected at a household level (Cocks et al. 2006).

In the urban areas the percentage of households collecting and consuming *imifino* dropped to only 42%. (Cocks 2006) The wealth of the household in the urban areas, however, did influence significantly the amount of material collected (H = 9.65, df = 2, p = 0.008). Poor cluster households collected the highest mean quantity per user household, 41 kg ($6) per annum, and the middle and rich clusters collected similar quantities, 25 kg ($3) and 29 kg ($4) respectively, per annum (Table 2) (Cocks 2006).

The two major reasons given for households' not consuming *imifino* were: 1) no person in the household was able to recognise the species, and 2) scarcity of these plants. Wild leafy vegetables are called *imifino* in the study area, but the term also refers to a particular dish that consists of several species mixed together and cooked with maize flour. Often other ingredients such as animal fat, onions and spices are added to the dish. Most people prepare the dish according to a specific recipe (Husselman and Sizane 2006). Ten different recipes were recorded, which vary considerably in the supplementary ingredients used, though the method of preparation is usually the same.

Participants in the study revealed that in the past, *imifino* was consumed primarily by women who collected, prepared and ate *imifino* socially in the late afternoon (Figure 5). It was generally believed that if men consumed *imifino* they would become weak and prone to gossip "like women". This belief is also mentioned by Mtuze (2004). Today, it is still

Figure 5: A group of women who have collected *imifino*. Photo: A. Dold

predominantly eaten by women and children as a social pastime with family and friends. Collecting is done in pairs or groups, and the preparation and eating is shared with extended family and friends who often sit outside on the ground on reed mats (Figure 6).

Table 2: Use and quantity of *imifino* collected according to wealth status of households in urban households

Resource use		Overall use			Wealth Cluster							
					Poor cluster			Middle cluster			Wealthy cluster	
	n	Mean± St. Dev	Median ±MAD	n	Mean ±St. Dev	Median ±MAD	n	Mean ±St. Dev	Median ±MAD	n	Mean ±St. Dev	Median ±MAD	Test Statistic
Imifino ($)	127	4.23±3.72	2.80±3.11	37	5.65±4.26	4.94±3.17	51	3.42±2.97	2.80±2.08	39	3.94±3.76	2.80±3.11	H = 9.65, df = 2, p = 0.008

Resource use		Overall use			Wealth Cluster							
					Poor cluster			Middle cluster			Wealthy cluster	
	n	Mean± St. Dev	Median ±MAD	n	Mean ±St. Dev	Median ±MAD	n	Mean ±St. Dev	Median ±MAD	n	Mean ±St. Dev	Median ±MAD	Test Statistic
Imifino (kg)	127	31±27	36±23	37	41±31	36±23	51	25±22	20±15	39	29±27	20±23	H = 9.65, df =2, p = 0.008

Wealth		Cluster 1 ('poorest')	Cluster 2 ('richest')	Test statistic		
		N	n			
Imifino	Using	214	179	43%	χ^2= 2.74	df=1	p = 0.0974
	Mean ± SD	36%					
		37.1 ± 36.1	33.1 ± 26.0				
	Median ± MAD	110	76		W=4 399.5		p = 0.5279
		22.8 ± 17.3	22.8 ± 16.9				

Leaves and soft stalks are removed from the plant and soaked for ±30 minutes in cold water to remove dust and soil. The material is then chopped finely on a plate or chopping board and boiled for about five minutes in clean water with a pinch of salt added. The cooked vegetables are then mixed with maize rice (a product made from maize starch), or maize flour and boiled together to create a thick porridge, known as *isigwampa*. Occasionally *imifino* leaves are eaten on their own as a side dish with *pap* (maize porridge) or rice. However, this is uncommon when it is eaten as a social snack in a group setting.

Figure 6: A group of women in a peri-urban village preparing *imifino* outside in the afternoon sun. Photo: A. Dold

In urban areas, women generally congregate in the sitting room once the dish has been prepared (Figure 7). The *imifino* dish is placed in the centre of the room and eaten using the fingers of the right hand to take a pinch of *imifino* off the palm of the left hand. It is considered inappropriate to eat *imifino* off crockery plates, or to use cutlery. One middle-class urban family reported serving a small portion of *imifino* with the regular European-style Sunday roast, showing a persistent urban link with traditional customs.

Figure 7: A group of women sharing and tasting the *imifino*. Photo: Michelle Cocks

Imifino was considered by the majority of female respondents to be healthier than commercially available vegetables, such as cabbage and spinach. The latter are also regularly prepared in a similar way, especially when the wild species are scarce, but the taste of the wild *imifino* was generally preferred by the respondents above 30 years old. The younger generation either preferred the commercial vegetables (50%) or were oblivious of the difference (19%).

Imifino is considered to be an important food for women, because, as mentioned by respondents "we come and sit together and talk about family problems". *Imifino* is usually eaten socially as a snack and only in 26% of the urban households is it eaten as a meal. In the rural areas it is not considered part of a meal as household members will rarely eat it alone as they may do with other foods. In the urban areas approximately half of the respondents eat *imifino* with the whole family and 20% of all women always eat it with friends, taking turns to prepare the dish. In less than 10% of the households the children did not like *imifino*.

A significant proportion of the households interviewed in the in-depth study indicated that the male members of the family increasingly also partake of the *imifino* dish. Of the 26 households interviewed in the in-depth study in the two urban locations only three families indicated that male members of their families did not enjoy eating *imifino* at all. Moreover, then half of the 59 customers purchasing cooked *imifino* on the market in Mdantsane were men and more than 60% of these men had not eaten *imifino* when they were young. Media (radio) campaigns and advice given at State clinics were mentioned as the main reasons for men beginning to appreciate *imifino*. It is reported that men in the former Transkei region also consume *imifino* (Kepe 2008).

The majority of peri-urban households prepared *imifino* at least twice a week in season. A smaller number consumed it every day, while only a few consumed it once a month. The majority in the former two categories were between the ages of 51 and 66, followed by those between 36 to 50 years old. Those who consumed *imfino* only once a month were from middle class families and represented the 36 to 50 years age category.

Availability of Imifino Species

All *imifino* plants are collected opportunistically and no cultivation of the species has been recorded in the study area. *Imifino* plants growing along roadsides and wastelands are generally considered to be "dirty", as "men and dogs relieve themselves" in these areas. Nevertheless, people may maintain and encourage the plants to survive in their gardens, by letting them set and disperse seeds. In the peri-urban villages, cultivated fields are good sources of *imifino*, and therefore these species are more abundant and consumed more often in these areas. A few urban respondents said that while visiting relatives in the countryside they would collect *imifino* and bring it back with them when they returned back to the city. *Imifino* species are commonly collected after Spring and Summer rainfall and are scarce in the dry winter season. They have a short shelf life and can be kept for only two or three days.

The collection of *imifino* species was for the most part an activity of the older generation. The respondents below 30 years old knew fewer species (i.e. <6) than their elders (i.e. 6-15). Men rarely collect *imifino* and most of them could not recognise any species. Respondents in the peri-urban areas were confident that *imifino* consumption would continue in the future, because of the nutritional importance. The main reason given by urban respondents for continued use was the taste as well as the fact that it was traditional food. Most informants indicated that they preferred to collect *imifino* themselves rather than purchase from hawkers without knowing the source of the plants. Yet, 65% indicated that they would buy *imifino* if it was sold in supermarkets or vegetable shops where it would have implicit quality assurance.

Very few of the female respondents were aware of the possibility of drying *imifino* to increase its shelf life as was done traditionally in the past. The drying process has been recorded as preserving leafy vegetables for months (Voster van Rensburg unpubl. doc., Nguni and Mwila 2007, Bhat and Rubuluza 2002). Only a few elderly women in the peri-urban areas knew of the tradition. Different methods for drying were described by these respondents. Species considered suitable for drying were *Sonchus oleraceus, Amaranthus hybridus* and *Solanum nigrum*. The most common method is to place freshly harvested leaves on a flat rock or sheet of corrugated iron to dry in the sun for 3 to 10 days, depending on the weather. Alternatively the leaves can be dried inside the house, which gives them a different flavour. Another method includes a maturation process. Fresh leaves are chopped and placed in a paper bag or grain sack for 4 days. The leaves are then placed in the sun to dry. Only one respondent in Crossroads village was recorded as using this method, saying that she preferred the flavour to fresh leaves.

Trade in Imifino

Despite the preference for wild *imifino* vegetables amongst the urban respondents and the scarcity of material in the dry winter months there is almost no trade in fresh leaves in the study area. Occasionally, women from the rural areas vend fresh *imifino* from door to door in urban centres. Two respondents in the urban centre of King William's Town said that they had asked a friend to collect for them, in return for cash.

Only one stall at a busy informal market in East London was observed selling freshly cooked *imifino*. The trader had been doing so for the last five years. She collected *imifino* from her own garden when available. Both men and women representing all age groups were observed purchasing *imifino* from her.

Very few of the peri-urban respondents interviewed were aware of the trade in either raw or cooked *imifino* in the area, and those who were had only seen trade in the larger city centres such as Port Elizabeth and King William's Town.

Discussion and Conclusion

Despite the social and political transitional changes that the isiXhosa speaking people of the Eastern Cape have undergone over the last 100 years, the consumption of traditional foods such as *imifino* has not radically declined, as has often been recorded in other parts of Africa (Ogoye-Ndegwa and Aagaard-Hansen 2003). This is supported by the high prevalence of collection and consumption of *imifino*, even within the urban centres, in this study.

The case study has clearly illustrated that the consumption of *imifino* does not simply represent a source of nutrition for families, but its use is guided by local customs, practices and traditions. As a result, the consumption of *imifino* is not restricted to a poor person's diet; nor is it simply 'famine-food'. Our findings show that it is consumed regularly in both wealthy and poor urban households as a traditional dish, prepared according to a recipe typical for the region. This confirms the hypothesis that the use of natural resources is not restricted to utilitarian subsistence livelihood and 'poor man's safety-net' requirements, but also fulfils an important role in local people's cultural practices and traditions, and contributes towards a sense of community (Cocks 2006). For example, *imifino* is collected and eaten predominantly by women in an important social context. In the past it was taboo for men to participate in these activities, indicating a strong gender bias similar to that reported in other parts of Africa, for example amongst the Luo of Kenya (Ogoye-Ndegwa and Aagaard-Hansen 2003).

The empirical results thus clearly demonstrate that we should not assume that the use of wild plants is restricted to specific socio-economic conditions. They suggest that more attention needs to be given to the social processes impacting on the use of wild plant products and biodiversity (Cocks 2006) and that the connections between food, culture, and identity are inseparable (Bergquist 2006). These sentiments are in line with the call made by UNESCO to ensure that biodiversity finds a central place in the policy agendas so as to ensure that the social and economic benefits to society of the links between biological and cultural diversity, go beyond mere economic values (Persic and Martin 2008).

These sentiments also feed directly into the need for the food habits of a community to be studied as a bio-cultural phenomenon. This is because, despite periods of massive political upheaval, globalization and modernization, indigenous communities do not simply "enter or leave modernity" but they rather enter a dynamic process of trans-cultural exchange, which results in the persistence of certain cultural norms and activities (Cocks 2006). Canclini (1995), describes this process as occurring not as a simple cultural syncretism but rather as a dynamic process, whereby the 'modern' fails to 'substitute' for the 'traditional,' but rather results in constant rearticulations of tradition. Our study illustrates this dynamism of tradition: whereas *imifino* consumption was considered inappropriate for men in the past, increased awareness of its nutritional values has changed this perception. A comprehensive understanding of these processes is crucial to fully appreciate the scope of conserving indigenous cultural heritage and biodiversity in an ever-changing world.

Interestingly, these leafy vegetables are not commonly traded, nor are they cultivated, although demand prevails in our urban study areas, especially during the dry seasons. This finding demonstrates the need for those documenting agricultural biodiversity systems to also give attention to alien weeds and plant species which are not actively planted or managed (Cromwell et al. 2001). This is in line with a call made by Cromwell et al. (2001) for a new approach to be adopted by agricultural research and development, an approach based on developing strategies aimed at farming systems rather than particular crops, as the former are often also less reliant on external inputs. Attention to the cultivation of alien weeds/wild plants alongside food crops would provide such a strategy. It would also provide a greater appreciation of the multiple goods and services that could be offered by diversity in agricultural ecosystems which are perceived as being the way forward to feed people, alleviate poverty and protect the environment (Norris 2008).

Acknowledgements

This study was sponsored by SANPAD (South African Netherlands Programme on Alternatives in Development) and Rhodes University Joint Research Council (JRC). We thank Lindsey Bangay for her assistance with the statistical analyses, and the household members who welcomed us into their homes to be interviewed.

References

Ainslie A (2002) Introduction: setting the scene. *In:* Ainslie A (ed), Cattle ownership and production in the communal areas of the Eastern Cape, South Africa. Programme for Land and Agrarian Studies, Research Report No. 10

Bhat RB, Rubuluza T (2002) The biodiversity of traditional vegetables of the Transkei region in the Eastern Cape of South Africa. *South African Journal of Botany,* 68:94-99

Beinart W (1980) Labour migrancy and rural production: Pondoland c.1900-1950. *In:* Mayer P (ed), Black villagers in an industrial society. Oxford University Press, Cape Town

Beinart W (1994) Twentieth-century South Africa. Oxford University Press, Oxford

Bergquist KJS (2006) From kim chee to moon cakes: feeding Asian adoptees' imaginings of culture and self. *Food, Culture and Society,* 9(2):141-153

Canclini G (1995) Hybrid cultures: Strategies for entering and leaving modernity. University of Minnesota Press, Minneapolis

Cocks ML (2006) Wild resources and practices in rural and urban households in South Africa: Implications for bio-cultural diversity conservation. PhD Dissertation, Wageningen University, The Netherlands

Cocks ML, Bangay L, Wiersum KF, Dold AP (2006) Seeing the wood for the trees: The role of woody resources for the construction of gender specific household cultural artefacts in non-traditional communities in the Eastern Cape, South Africa. *Environment, Development and Sustainability,* 8:519-533

Cromwell E, Cooper D, Mulvany P (2001) Agriculture, biodiversity and livelihoods: Issues and entry points. In: Koziell I and Saunders J (eds), Living off biodiversity: Exploring livelihoods and biodiversity issues in natural resource management. IIED, London

De Wet C, Whisson M (1997) From reserve to region. Apartheid and social change in the Keiskammahoek District of (former) Ciskei: 1950-1990. Occasional Paper No. 35. Institute for Social and Economic Research (ISER), Rhodes University, Grahamstown, South Africa.

Ellis F (1998) Household strategies and rural livelihood diversification. *Journal of Development Studies,* 35(1):1-38

Etkin N (2000) The cull of the wild. In: Etkin N. (ed) Eating on the wild side: The pharmacologic, ecologic and social implications of using noncultigens, pp1-28. The University of Arizona Press, Arizona

Fieldhouse P (1998) Food and nutrition. Stanley Thornes Ltd, Cheltenham

Fleuret A (1979) Methods for evaluation of the role of fruits and wild greens in Shamba diet: A case study. *Medical Anthropology,* 3: 244-269

Gockowski J, Mbazo'o J, Mbah G, Moulende TF (2003) African traditional leafy vegetables and the urban and peri-urban poor. *Food Policy,* 28:221-235

Grubben GJH, Denton OA (2004) Plant resources of tropical Africa 2: Vegetable. PROTA Foundation, Backhuys Publishers, Netherlands

Guinand Y, Lemessa D (2000) Wild-food plants in Ethiopia: Reflections on the role of 'wild-foods' and 'famine-foods' at a time of drought. University of Pennsylvania, African Studies Center, http://www.africa.upenn.edu/Hornet/famp0300.html. Cited 03/2000

Husselman H, Sizane N (2006) *Imifino*: A guide to the use of wild leafy vegetables in the Eastern Cape. Netherlands Institute for Southern Africa (NIZA). ISBN: 0-86810-426-4. Dupli Print, Grahamstown.

Kepe T (2008) Social dynamics of the value of wild edible leaves (*imifino*) in a South African rural area. *Ecology of Food and Nutrition,* 47:531-558

Moore G, Tymowski W (2005) Explanatory guide to the international treaty on plant genetic resources for food and agriculture. IUCN, Gland, Switzerland and Cambridge, UK in collaboration with IUCN Environmental Law Centre, Bonn, Germany

Mtuze PT (2004) Introduction to Xhosa culture. Lovedale Press, Alice

Ndoye O, Ruiz Perez M, Eyebe A (1997) The markets of non-timber forest products in the humid forest zone of Cameroon. Rural Development Forestry Network Paper 22c

Ngwerume FC, Mvere B (2000) Report on the findings of a socio-economic survey on the marketing, consumption and production of traditional vegetables in the urban and peri-urban areas of Harare, Zimbabwe. DR&SS-HRC/NRI Project A0892

Nguni D, Mwila G (2007) Opportunities for increased production, utilization and income generation from African Leafy Vegetables in Zambia. *African Journal of Food Agriculture, Nutrition and Development,* 7(4):1-20

Norris K (2008) Agriculture and biodiversity conservation: opportunity knocks. *Conservation Letters* 1(1): 2-11 doi:10.1111/j.1755-263X.2008.00007.x

Ogoye-Ndegwa C, Aagaard-Hansen J (2003) Traditional gathering of wild vegetables among the Luo of western Kenya - A nutritional anthropology project. *Ecology of Food and Nutrition,* 42:60-89

Palmer R (1997) Rural adaptations in the Eastern Cape, South Africa. Working paper 11, Institute of Social and Economic Research (ISER), Rhodes University, Grahamstown, South Africa

Persic A, Martin G (eds.) (2008) Links between biological and cultural diversity. Report of the International Workshop, UNESCO and The Christensen Fund. UNESCO HQ Paris

Rose EF, Guillarmond AJ (1974) Plants gathered as foodstuff by the Transkeian peoples. *South African Medical Journal,* 48:1688-1690

Shackleton CM (2003) The prevalence of use and value of wild edible herbs in South Africa. *South Africa Journal of Science,* 99:23-25

Shackleton SE (2006) The significance of local trade in natural resource products for livelihoods and poverty alleviation in South Africa. Dissertation, Rhodes University, South Africa

Shackleton CM, Shackleton SE (2004) Use of woodland resources for direct household provisioning. In: Lawes MJ, Eeley HAC, Shackleton CM and Geach BS (eds), Indigenous forests and woodlands in South Africa: policy, people and practices, pp195–197. University of KwaZulu-Natal Press, Scottsville, South Africa

Shackleton CM, Shackleton SE, Cousins B (2001) The role of land-based strategies in rural livelihoods: the contribution of arable production, animal husbandry and natural resource harvesting. *Development Southern Africa,* 18:581-604

Shackleton CM, Shackleton SE, Ntshudu M, Ntzebeza J (2002) The role and value of savanna non-timber forest products to rural households in the Kat River valley, *South Africa Journal of Tropical Forest Product,* 8:45–65

Sharp J (1988) Ethnic group and nation: The apartheid vision in South Africa. *In:* Boonzaier E and Sharp J (eds), South African keywords: The uses and abuses of political concept, pp79-100. David Philip, Cape Town

Toledo A, Burlingame B (2006) Biodiversity and nutrition: A common path toward global food security and sustainable development. *Journal of Food Composition and Analysis,* 19(6-7):477-483

Turner S (2004) Community-based natural resource management and rural livelihoods. In: C Fabricus, and E Koch with H Magome and S Turner (eds), Rights, resources and rural development: community-based natural resource management in southern Africa. Earthscan, London and Sterling, pp44-66

Van Wijk Y (1986) The practical book of herbs. Chameleon Press, Cape Town

Voster HJ, Jansen van Rensburg W. The utilisation of African vegetables. Unpublished document.

Van Wyk BE (2005) Food plants of the world. Briza Publications, Pretoria

Van Wyk BE, Gericke N (2000) People's plants. Briza Publications, Pretoria

Wehmeyer AS, Rose EF (1983) Important indigenous plants used in the Transkei as food supplements. *Bothalia,* 14:613-615

Wells MJ, Balsinhas AA, Joffe H, Engelbrecht VM, Harding G, Stirton CH (1986) A catalogue of problem plants in southern Africa. Memoirs of the Botanical Survey of South Africa No. 53

Wiersum KF, Singhal R, Benneker C (2004) Common property and collaborative forest management; rural dynamics and evolution in community forestry regimes. *Forests, Trees and Livelihoods,* 14:281-293

Chapter – 9

Edible Fungi in Mesoamerican Lowlands: A Barely Studied Resource

Felipe Ruan-Soto* and Joaquín Cifuentes

Abstract

It's relatively frequent to hear about the customary consumption and commercialization of wild and cultivated mushrooms in European, North American and Asian tempered zones. In contrast, such a phenomenon has been widely unrecognized in Mesoamerican lowlands. The scarce interest for leading ethnomicological studies in these zones may be caused by a preconception according to which peoples inhabiting them do not use macroscopic fungi. Yet, archaeological evidence shows a relation between fungi and Mayan peoples existed during pre-Hispanic times.

Recently, the consumption of 25 fungi species has been documented; of these, *Schizophyllum commune* and *Pleurotus djamor* are the most appreciated. An interesting finding is people's association of edible mushrooms with meat in terms of flavour and nutritional properties. A set of documented cultural practices related to mushroom gathering, commercialization and other informal economic activities permit sustainable advantages providing both nutrimental

*ruansoto@yahoo.com.mx

and monetary benefits. Although consumption of wild mushrooms in Mesoamerican lowlands appears to be a desirable practice, ongoing transculturation attempts the endurance of these millenarian practices.

Keywords: Fungi, traditional commercialization, local mycological science, ethnobiology, ethnomycology, edible fungi.

Introduction

Mesoamerica is more than a concept designed to geographically delimit middle America; it was constructed to characterize a region that comprises shared cultural features that differentiate it from other American societies (Kirchhoff 1960) (Figure1).

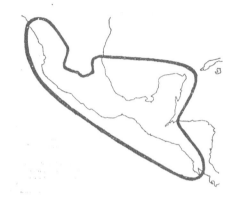

Figure 1: Map that shows the mesoamerican limits in the middle of XVI century (taken and modified from Kirchoff 1960).

Mesoamerica is an ethnically diverse region. On a global scale, Mexico (approximately half of contemporary Mexico's territory corresponds to Mesoamerica) ranks 6th on a 12-country list of ethnic diversity (56) (Toledo 2001). Additionally, Guatemala comprises 21 different ethnic groups (Morales-Esquivel 2001). In spite of such diversity, Mesoamerican peoples share a common history (Kirchhoff 1960), similar land utilization, domestication and plant usage (Casas 2001), common construction strategies, pottery patterns and a similar cosmogony as well as similar forms of understanding the nature.

Furthermore, this region displays a very special feature: it contains a great biological diversity. In Mesoamerican lowlands, there are tropical rainforests and tropical dry forests, biomes considered among the richest in terms of the existing number of species (Raven et al. 1999). According to certain estimations, humid rainforests maintain up to 50-70% of the total of the planet's species (Fox 1992; Gurevitch et al. 2002).

Such factors have influenced Mesoamerican lowland peoples so, that throughout time they have constructed different conceptions of nature as well as different ways of social organization and appropriation of natural resources. These interacting elements remain in a constant adaptation process.

Local environmental knowledge includes, among other aspects, the gathering practices as well as the usages and handling of diverse organisms in order to satisfy the populators' needs. Among these, dressing, transportation, recreation, healing, spiritual needs and feeding can be mentioned. Thus, human societies progressively cease to conceive organisms as alien elements and conceptualize them as resources.

Even though each Mesoamerican lowland community has developed different perceptions about available natural resources and consequently different ways to name and classify them (Lazos and Paré 2000), all of them share a traditional pattern of environment utilization and handling. Although these groups base their diet on maize, they also utilize a great diversity of wild and planted resources to complement it (Williams 1990; Casas 2001; Levy-Tacher et al. 2002). According to Caballero, in 1984 about 5000 plant species used by Mesoamerican natives, most of which were obtained by gathering, were registered (Caballero 1984). An illustrative case are the Lacandon people in Chiapas, in the southeast of Mexico. In their traditional *milpas* (Nahuatl name for Mesoamerican agricultural fields), maize and bean along with other 50 species are grown mainly for self-consumption (Nations and Night 1980; Levy-Tacher 2000). During off-season, people make use of other resources such as edible plants and land or aquatic animals.

Within this diversified resource utilization pattern, wild mushrooms become an element that can potentially complement people's diet. Ethnomycology studies both the relations between fungi and human societies, and the local mycological knowledge of these societies. These interactions have been widely studied in the highlands of central Mexico, but in Mesoamerican lowlands, the situation is quite different (Ruan-Soto et al. 2004; Ruan-Soto et al. 2006).

In this chapter, we offer a review of the state-of-the-art in the study of wild edible fungi in Mesoamerican lowlands: challenges, potentialities and results; the related cultural practices found and perspectives for intensive utilization of fungal resources. Firstly, we will explain the importance of edible fungi in subsistence strategies of Mesoamerican highlands' ethnic groups. Secondly, we present an account of the different ways in which human groups related to fungi in Mesoamerican lowlands before the arrival of the Spaniards at the end of the sixteenth century. Finally, we offer a review of current uses of mushrooms by inhabitants of Mesoamerican lowlands.

For the writing of this essay a bibliographical revision of diverse sources, such as scientific articles, grade thesis, books and technical reports, was done. This, together with the authors' fieldwork experience in the region allowed for a meta-analysis of these studies.

Edible Fungi in Survival Patterns of Mesoamerican Highlands Societies

In Mesoamerican highlands societies, a broadly documented mushroom consumption tradition along with deeply rooted mycological knowledge exists. Through such knowledge, these societies identify edible fungi species and diverse aspects of the biology of such organisms: *i.e.* origins, climatic and environmental factors necessary for their appearance, physical characteristics and structure, the time of the year in which they may be found, the substrate on which they appear, the types of vegetation propitious for their appearance and some existing relations with other organisms (Estrada-Torres and Aroche 1987; Reygadas-Prado 1991; Montoya 1992; Mariaca et al. 2001; Montoya et al. 2001; Garibay-Orijel et al. 2006).

Thus, wild fungi become a nutritional resource to which countryside inhabitants can accede during the rain season. Throughout this season, different fungi species appear in the forest according to their phenology: *Liophyllum* and *Agaricus* species appear in May and June, during the first rains; *Ramaria* spp, *Amanita* spp., *Lactarius* spp., *Helvella* spp. and *Laccaria* spp. among others appear between July and September. By the end of the rain season between October and November *Morchella* species can be found (Reygadas-Prado, 1991).

During the fungi-gathering season, private properties and other forest sections are freely trespassed (Mariaca et al. 2001). Since fungi are perceived as free-access resources, impoverished families may utilize them. These families make long walks throughout the nearby forest early in the morning in order to gather a considerable amount of mushrooms; all family members take part in this activity regardless age or gender.

Generally, most of the edible fungi species are gathered for self-consumption in family households. According to Villareal and Pérez-Moreno (1989), by the early 1990's, about 200 edible fungi species were registered in Mexico and most of them in the Mesoamerican region. Additionally, more than 1000 local names used to refer to different edible fungi species in the same region have been documented (Guzmán 1997; Garibay-Orijel 2009). Considering further ethnomycology researches carried out in this region during the last 15 years, the number of edible fungi species easily increases to 300.

Since edible fungi are seasonal resources, Mesoamerican highlands' societies have devised methods to conserve mushrooms when they are no longer available in the forests. The most common way to do so is by drying them, either in the sun or by hanging them near the fireplace (Montoya 1992).

In addition, the ways in which fungi are cooked vary. They can be boiled, roasted or fried; they sometimes constitute the main ingredient in some traditional and elaborated dishes, or else are mixed with different ingredients like beans, eggs or different meats.

The fact that edible fungi are a highly valued food is not casual. Several researches have documented the great nutrimental value of many traditionally consumed fungi species in Mesoamerican highlands. Some species such as *Boletus edulis* Bull., *Clitocybe gibba* (Pers.) P. Kumm., *Morchella* spp., *Lyophyllum decastes* (Fr.) Singer, *Gomphus floccosus* (Schwein.) and *Cantharellus cibarius* (Fr.), contain between 34,7 % and 21,4 % of crude protein (Aguilar-Pascual 1988). Moreover, many of the traditionally consumed species contain essential amino acids like lysine, methionine, threonine, valine, leucine and tryptophan. The average content of fats is 4%. Furthermore, they constitute an important source of thiamin, riboflavin, niacin, biotin and ascorbic acid (Martinez-Guerrero 2000).

Many edible species are collected and sold in local markets. This activity provides monetary benefits for many families, which allows them to buy goods they are not able to produce. In fact, certain central Mexico highlands communities conceive themselves as being "fungi gatherer-eater societies" due to the great amount of families dedicated to collecting and selling edible fungi during the rain season (Montoya 1992; Ruan-Soto et al. 2004). Fungi commercialization varies from door-to-door and local market sales to a complex transportation, sale and re-sale process from rural communities to bigger population centers (Ruan-Soto 2002).

Given the benefits and great value of edible fungi in Mesoamerican highlands, an important question can be put forward: could intensive fungi gathering cause extinction of such species? Eventhough researches on this subject are few, a couple of studies carried out in Finland and the Netherlands (Readhead 1997) and another in Switzerland (Egli et al. 2005) seem clear doubts: intensive fruiting body gathering does not have adverse effects in vegetative mycelium, the number of mycelium nor the mycorrhiza; long-term and systematic harvesting reduces neither the future yields of fruit bodies nor the species richness of forest fungi, irrespective of whether the harvesting technique was picking or cutting (Egli et al. 2005). The reason for this is that the only fungi structure used is precisely the fruiting body, whereas the body (mycelium) itself remains in the substrate without suffering any damage. Thus, far from endangering fungi species, intensive utilization facilitates their conservation and avoids deforestation.

Utilization of Fungi in Mesoamerican Lowlands During pre-Hispanic Times

Very little is known about the relation between human societies inhabiting Mesoamerican lowlands and fungi before the arrival of Spaniards, nevertheless, available data sheds some light on the subject. Such data can be grouped in three categories: sculptures, codices and chronicles.

Regarding sculptures, in diverse sites of Guatemala and Chiapas (Mexico) many stone figurines have been found (Lowy 1971; Mayer 1977; Wasson 1983; Guzmán 2003). Since the late 19th century nearly 300 fungus-shaped statuettes have been registered in Mayan areas (Mata 1987).

Yet, the significance of these stone statuettes remains unclear since it's people without any archaeological training or the necessary skills to interpret their possible meaning who have extracted many of them. Wasson (1983) suggests an existing relation between these stone figurines and fungi-related cults accounted for by friars in the 16th century. He argues that anthropomorphous figures carved on the base of these fungi-shaped sculptures may represent women grinding entheogen (hallucinogenic) fungi on a *metate*, a grinding stone tool used for processing grains. This is a prevalent practice in ceremonies involving mushroom ingestion in Oaxaca, Mexico.

Moreover, based on a 17th century list of Mayan words in the Kaqchikel linguistic zone near the City of Guatemala and Antigua Guatemala, Wasson argues that highland Mayans knew entheogen fungi. It is thus possible that the referred sculptures had a ritual significance associated to mycolatry, a practice that has vanished from the Mayan territory (Lowy 1971).

Few are the evidences of fungic representations in pottery, particularly in the Mayan region. Nevertheless, published field notes of archaeologist Matthew Stirling show fungus-shaped pottery censers found in a cave in the Zoque region in Chiapas (Pailles and Beutelspacher 1989).

Concerning codices, both the Madrid codex and the Dresden codex contain figures that some authors have interpreted as fungi representations as they are held by standing human figures in an offering position (Lowy 1972; Guzmán 2003) (Figure 2).

Regarding written evidence, in the sacred Mayan book Popol vuh a ritual use of fungi is mentioned: "... *and when the blood had been drunk by the Gods, thus spoke the stone, when the priests and executioners arrived and took their offerings to them, and so they did in front of the symbols, burning pericón* (Tagetes lucida) *and Holom Ocox...*" (Recinos 1960) *Holom* is a K'iche' word meaning "head", whereas *Ocox* means "muhsroom" (Mayer 1977; Guzmán 1997).

Figure 2: Fragments of the Madrid and Dresden codex (taken and modified from Lowy 1974).

However, relations between inhabitants of Mesoamerican lowlands and fungi in those times are still unclear. In fact, no evidence exists that conclusively shows the consumption of edible fungi. In contrast, among contemporary Mesoamerican societies such practice can effectively be witnessed.

Utilization of Edible Fungi in Mesoamerican Lowlands

Relations between fungi and Mesoamerican lowlands' inhabitants, including the focus on edible fungi, have seldom been studied. This situation could be explained by an extended belief that Amazonian peoples and similar societies do not use fungi and display either non mycophlylic attitudes (neither attraction nor repulsion towards fungi) or mycophobic attitudes (aversion towards fungi)

(Fidalgo 1965; Guzmán 1987; Mapes et al. 2002; Goes-Neto and Bandeira 2003; Ruan-Soto et al. 2006; Ruan-Soto et al. 2007). Nevertheless, recent studies have demonstrated that many tropical lowlands' inhabitants utilize fungi and even display mycophilic attitudes (Ruan-Soto et al. 2006; Ruan-Soto et al. 2007; 2009).

Currently, several studies have tackled the subject at issue in Mexico: among the Totonac people in Veracruz (Chacón 1988); the Maya people in Yucatan (Mata 1987); the Chinantec, Chol, Zoque and Chontal people in the coastal plain of the Gulf of Mexico (Ruan-Soto 2002; Ruan-Soto et al. 2004; Ruan-Soto et al. 2006); the Lacandon people and mestizos in the Lancadon Rain Forest in Chiapas, (Ruan-Soto et al. 2007; 2009); the Zoque people in northern Chiapas (Alvarado-Rodriguez 2006) and the Mam people in the Tacana volcano at the southern Mexican border (Medina-Arias 2007); there also is one general revision of the fungi in the maya culture (Guzmán 2003). Additionally diverse ethnic groups have been studied in Guatemala (Sommerkamp 1990; Morales-Esquivel 2001) and the consumption of certain fungi species has been documented in El Salvador (Guzmán 1987).

Edible species utilized

In tropical Mesoamerican lowlands, 25 edible fungi species have been registered (Table 1). Mesoamerican societies recognize an average of 5 to 11 species (with 23 out of the 25 edible species reffered overall) almost all of which share a common feature: they live upon lignicolous substrate and they have a more rubbery consistency than many terrestrial and/or ectomycorrhizal fungi. A possible explanation for the preference of lignicolous species by local folk is that producing resistant fruiting bodies keeps them from quickly rotting in the hot and humid rainforest environment. Unlike these species, terrestrial and/or ectomycorrhizal fungi generally produce fleshy fruiting bodies, which rot more easily (Ruan-Soto et al. 2007; 2009). This is how Van Dijk et al. (2003) explain the preference for lignicolous species among inhabitants of humid forests in southern Cameroon: they prefer non easily rotting fungi. Besides, other Mesoamerican lowlands societies like the Chontal and Chinacanteco people from the coastal plains of the Gulf of Mexico, or the mestizo people from the southern Lacandon rainforest in Chiapas, perceive terrestrial fungi as poisonous. Therefore, this type of fungi is not included in their diet (Ruan-Soto et al. 2004; Ruan-Soto et al. 2006; Ruan-Soto et al. 2007; 2009).

Table 1: Edible fungi species registered in Mesoamerican lowlands.

1.	*Auricularia polytricha* (Mont.) Sacc.
2.	*Auricularia fuscosuccinea* (Mont.) Farl.
3.	*Auricularia mesenterica* Pers.
4.	*Auricularia delicada* (Fr.) Henn.
5.	*Auricularia cornea* (Ehrenb) Ehrenb ex. Endl
6.	*Schizophyllum commune* Fr.
7.	*Schizophylum fasciatum* Pat.
8.	*Favolus tenuiculus* (Beauv.) Fr
9.	*Pleurotus djamor* (Fr.)Boedijn
10.	*Pseudofistulina radicata* (Schw.) Burdsall
11.	*Armillariella tabescens* (Scop. Ex Fr.) Sing.
12.	*Armillariella* sp
13.	*Cookeina sulcipes* (Mont.) Berk.
14.	*Cookein tricholoma* (Mont.) Kuntze
15.	*Hohenbuhelia petaloides* (Bull. Ex Fr.) Schul.
16.	*Ustilago maydis* (DC.) Corda
17.	*Polyporus alveolaris* (DC. Fr.) Bond
18.	*Lentinus crinitus* (L.:Fr.)Singer
19.	*Lentinus velutinus* Fr.
20.	*Lentinus strigosus* (Schw.) Fr.
21.	*Calvatia cyathiformis* (Bosc.) Morgan
22.	*Pluteus harrisii* Murr.
23.	*Tremella fimbriata* Fr.
24.	*Oudemansiella aff. stefendii* (Rick) Sing
25.	*Oudemansiella canarii* (Jungh.) Höhn.

Although there are no quantitative studies on the preferences of certain species, while analyzing the most referred species it can be seen that some of the preferred species in this region are: *Schizophyllum commune* Fr., registered in six researches (Figure 3); *Pleurotus djamor* (Rumph. ex Fr.) *Boedijn*, mentioned in five researches (Figure 4); *Auricularia polytricha* (Mont.) Sacc., *A. delicata* (Fr.) Henn. and *Favolus tenuiculus* (Beauv.) Fr., mentioned in four researches; and *Auricularia fuscosuccinea* (Mont.) Farl. and *Ustilago maydis* (DC.) Corda mentioned in three researches (Table 2).

Figure 3: *Schizophyllum commune*.
Photo: Julio Cesar Ruiz-Velasco

The ways in which these species are cooked are equally diverse. For instance, the *Auricularia* species are generally mixed with beans and cooked in broth, boiled (Alvarado-Rodriguez 2006) or roasted on a *comal* (an iron flat grill placed over an open fire) (Figure 5). *Schizophyllum commune* and *Favolus tenuiculus* are generally scrambled with eggs or tomato (*Lycopersicum esculentum*), onion (*Allium* spp.), red pepper (*Capsicum* spp.), oregano (*Leppia* sp.), acuyo (*Piper* sp.) and garlic (*Allium sativum*); also, several different fungi can be found in elaborated dishes like *mone*, a traditional dish in some sites of Tabasco, Mexico. *Mone* is prepared with spring onion (*Allium* spp.), sweet pepper (Capsicum spp.), leaves of *momo* (*Piper auritum*), tomato (*Lycopersicum esculentum*) and oregano (*Leppia* sp); it is stuffed in banana leaves (*Musa paradisiaca*) and cooked in a *comal* or steamed (Ruan-Soto et al. 2004). It can be

Figure 4: *Pleurotus djamor.*
Photo: Amaranta Ramírez-Terrazo

Figure 5: *Auricularia delicata* roasted in a *comal.* Photo: Felipe Ruan-Soto

stated that the great variety of traditional dishes that include fungi, along with the fact that they remain well positioned in local traditional feeding patterns, reflects their relevance in daily life of Mesoamerican peoples

Though in some zones of the coastal plain of the Gulf of Mexico *Schizophyllum commune* are sun dried and kept in bags for later consumption once the rain season is over (Ruan-Soto et al. 2004), fungi conservation practices are not common in Mesoamerican lowlands; it's generally preferred to consume them when they are still fresh.

Edible Fungi in Mesoamerican Lowlands

Table 2: Places where wild fungi consumption has been registered in Mesoamerican lowlands.

Species / Place	a	b	c	d	e	f	g	h	Total
Auricularia polytricha (Mont.) Sacc.	✓	✓	✓	✓					4
Auricularia fuscosuccinea (Mont.) Farl.	✓			✓	✓				3
Auricularia mesenterica Pers.	✓								1
Auricularia delicata (Fr.) Henn.			✓	✓	✓	✓			4
Auricularia cornea (Ehrenb) Ehrenb ex. Endl					✓				1
Schizophyllum commune Fr.	✓		✓	✓	✓	✓	✓	✓	6
Schizophylum fasciatum Pat.	✓								1
Favolus tenuiculus (Beauv.) Fr			✓	✓	✓		✓		4
Pleurotus djamor (Fr.)Boedijn	✓		✓	✓	✓	✓			5
Pseudofistulina radicata (Schw.) Burdsall							✓	✓	2
Armillariella tabescens (Scop. Ex Fr.) Sing.	✓								1
Armillariella sp					✓				1
Cookeina sulcipes (Mont.) Berk.	✓								1
Cookein tricholoma (Mont.) Kuntze	✓								1
Hohenbuhelia petaloides (Bull. Ex Fr.) Schul.	✓								1
Ustilago maydis (DC.) Corda	✓			✓	✓				3
Polyporus alveolaris (DC. Fr.) Bond					✓				1
Lentinus crinitus (L.:Fr.)Singer	✓								1
Lentinus velutinus Fr.					✓				1
Lentinus strigosus (Schw.) Fr.					✓				1
Calvatia cyathiformis (Bosc.) Morgan					✓				1
Pluteus harrisii Murr.					✓				1
Tremella fimbriata Fr.					✓				1
Oudemansiella aff. stefendii (Rick) Sing					✓				1
Oudemansilla canarii (Jungh.) Höhn.					✓				1

a.- Veracruz, México (Chacón 1988); b.- Yucatán, Mexico (Mata 1987); c.- Coastal plain in the Gulf of Mexico (Ruan-Soto 2002; Ruan-Soto et al. 2004; Ruan-Soto et al. 2006); d.- Lacandona rain forest, Chiapas, Mexico (Ruan-Soto et al. 2007; Ruan-Soto et al. 2009); e.- Selvas del Norte, Chiapas, México (Alvarado-Rodríguez 2006); f.- Tacana volcano (Medina-Arias 2007); g.- Guatemala (Sommerkamp 1990; Morales-Esquivel 2001); h.- El Salvador (Guzmán 1987).

The associations inhabitants make between edible fungi and meat are worth highlighting; to them, fungi have a very similar taste to meat and may even substitute it (Guzmán 1987; Ruan-Soto et al. 2004; Ruan-Soto et al. 2007; 2009). This perception is not exclusive of Mesoamerican societies; some authors e.g. Prance (1984) and Van Dijk et al. (2003) have observed this phenomenon; in Amazonian societies in Brazil and in Cameroon respectively.

Popular fungi perceptions

Among the diverse studied Mesoamerican lowlands' towns, rain is related to fungi appearance (Ruan-Soto et al. 2004; Medina-Arias 2007), and in some stories divinities are conceived to be the creators of fungi (Ruan-Soto et al. 2007).

Although in humid rainforests the rain season stretches throughout a great part of the year (about eight months form June to January), and consequently it may seem quite obvious that the presence of many fungi species be permanent, people indicate specific periods (generally July and August) as propitious for their appearance.

Regarding ecological aspects, edible fungi are thought to be organisms that emerge from rotten logs. In general, some tree species that constitute edible fungi substrata are well identified. *Bursera simaruba, Heliocarpus donnellsmithii, Gliricidia sepium* and *Hampea nutricia* are some examples of trees on which different edible fungi species grow (Chacón 1988; Ruan-Soto et al. 2004). Additionally, agricultural fields, cattle ranches and *acahuales* (previously harvested spaces left for their natural regeneration) are thought to be propitious spots for the appearance of great amounts of edible fungi. The reason for such conceptions has to do with the prevailing slash-and-burn agricultural system. In it, trunks of trees are not totally burned and remain within agricultural fields, becoming an available substrate for edible fungi species and concentrating a greater amount of it in a single zone than do some of the best-conserved patches of rainforest.

Resource handling

Apparently, in Mesoamerican lowlands gathering of edible fungi are not a deliberate activity as is the case in central Mexico highlands, but occurs only when fungi are found by chance in great amounts while working on the fields or along the footpaths (Chacón 1988; Ruan-Soto et al. 2006). As far as the access to this resource is concerned, although edible fungi are supposed to be a freely accessible resource (as they are not a product of someone's work) people only gather fungi pertaining to their own property. This is a generalized practice since agricultural fields and cattle ranches are private properties. Thus, when fungi appear in greater amounts within agricultural fields, they are accessible only to their owners (Ruan-Soto et al. 2006; Ruan-Soto et al. 2009).

However, it is also possible to witness informal economic activities concerning edible fungi, such as the interchange and reciprocity among inhabitants of different communities. Frequently, when someone finds a great amount of edible fungi he gathers them and presents them, either fresh or cooked, to his neighbors and friends. For some authors, such practices reinforce social ties among the members of any given community (Casaverde 1981; Ruan-Soto et al. 2009; Ruan Soto et al. 2006).

Fungi sale

Fungi sale is one of the least studied subjects in Mesoamerican lowlands, and yet available data is quite abundant to evaluate their place in family economy, since fungi clearly offer monetary advantages.

In the coastal plain of the Gulf of Mexico, the sale of *Schizophyllum commune* and *Favolus tenuiculus* has been registered (Ruan-Soto et al. 2006). Furthermore, in Guatemalan lowlands the sale of the above mentioned in addition to *Pseudofistulina radicata* has also been registered (Sommerkamp 1990).

In these sites, it's mainly women who sell fungi in local markets (Chacón 1988; Ruan-Soto et al. 2006). A traditional work division, in which men are in charge of agricultural fields, while women dedicate themselves to the gathering of edible fungi for both self-consumption and selling, causes this phenomenon.

Generally, fungi are sold along with many other regional goods produced within households; *Schizophyllum commune* is widely sold in Mexican and Guatemalan lowlands. One may observe small to great plastic bags filled with great amounts of it waiting to be sold (Sommerkamp 1990; Ruan-Soto et al. 2006). When it comes to *Favolus tenuiculus* and *Pseudofistulina radicata,* the sale dynamics is similar although quantities are usually much smaller. Another selling strategy is offering the gathered produce door-to-door among neighbors or in nearby towns.

Issues on Edible Fungi Consumption in Mesoamerican Lowlands

Although edible fungi consumption is a deeply rooted feeding practice with no environmental impact, some factors may still cause it's abandonment.

Among Mesoamerican inhabitants, it's frequently mentioned that fungi gathering was more widely practiced in past times (Ruan-Soto et al. 2009). Formerly, since Mesoamerican lowlands' settlers inhabited isolated regions, they consumed a greater variety of wild resources, such as snakes, toads, worms and different fungi species. Nevertheless, thanks to the opening of new pathways, highways and the appearance of other means of transportation, these communities were no longer isolated. Such conditions allowed people to buy industrialized food as new ideas and ways of conceiving natural resources broke through their culture. Nowadays, Mesoamerican local knowledge about natural resources and all related traditional practices are progressively disregarded, positioning western science as the only valid knowledge. In this uneven relation, hegemonic ways of thinking and living are imposed in different social contexts.

Among young Mesoamerican people, negative perceptions about traditional resources begin to be constructed, which results in conceiving wild edible fungi (and many other traditional foods) like "emergency food". As Fidalgo poses it (1965) these are food resources used only when other preferable food is not available. Consequently, industrialized food from the cities is thought to be "better" whereas wild fungi are perceived as "food for poor people", a reminiscence of an undesirable past.

The hegemonic western way of living leads young Mesoamericans to neglect their traditional feeding practices thus losing interest for edible fungi species and all other cultural knowledge that allows taking advantage of this natural nutritional resource. This makes traditional knowledge vulnerable before ongoing acculturation processes; as old people die, all the notions contained in oral history are in danger of being lost forever.

In the end, the result is a change in the feeding patterns: unhealthy, industrialized and expensive food coming from large cities substitutes highly nutritious local products, which additionally enjoy a broad cultural acknowledgment.

Perspectives in Wild Fungi Use in Mesoamerican Iowlands

Facing such an unfavorable situation, it's necessary that Mesoamerican ethnomycology takes on the promotion of an intensive utilization of edible fungi in local contexts, along with its permanent task of explaining the existing relations between fungi and Mesoamerican inhabitants.

To accomplish this, it's necessary to implement specific research tasks: intensify ethnomycography studies in order to characterize local mycologycal knowledge, register popular perceptions about fungi, ways in which such organisms are named and classified, their utilization and indeed the different edible species. In addition, it's important to characterize nutrimental properties of edible fungi to recognize the potential nutritional income in communities that commonly eat them. With all this information at hand, it will be possible to identify species that could be cultivated, and with the help of different biotechnological techniques, rural production spaces could be constructed in order to take better advantage of agricultural residues usable to grow culturally significant fungi species.

Although these actions are necessary and constitute the main objectives of ethnomycology, it's also relevant to carry out participative actions that promote the utilization of fungic resources. One of these is the promotion of edible fungi in different local spaces such as fairs, exhibitions, communitarian workshops and gastronomical shows (Figure 6). Communitarian workshops, for instance, are crucial for rural production and training; as such, they could give people an alternative opportunity to utilize fungi species even off-season for both self-consumption and commercialization.

However, the most important task for Meso-american lowlands ethno-mycology is, in our opinion, the rescue of mycological knowledge bound to disappear facing the prevailing tras-culturation processes in Mesoamerican societies. Through ethno-mycology studies, a written registry of the collective memory may be developed thus preserving it and avoiding its extinction. Furthermore, it's urgent to work for the recognition of traditional local knowledge as a valid way to perceive and name fungi (and other natural resources) in order to place it in a fair position before scientific knowledge. These tasks could eventually constitute the basis of culturally relevant educative processes, thus promoting a sustainable utilization of traditional nutritional resources.

Figure 6: Abimael, a boy form Playón de la Gloria community in the Lacandona Rain Forest of Chiapas, México; handling a *Pluteus harrisii*.
Photo: Felipe Ruan-Soto

It's also important to develop activities that contribute to the utilization of fungi within comprehensive forest manangent plans as is proposed by Garibay-Orijel et al. (2009): awareness-raising focused on local authorities and forestry technicians on the ecological importance of mushrooms; illustrated catalogues on useful species accessible to communities (which would also allow for the taxonomic identification of species) built from a local cultural perspective and in native languages; training of local people in diversified use of fungi as a result of a dialogue of knowledges; local technical training to obtain and disseminate strains of fungi and obtention of resources to develop infrastructure to cultivate mushrooms are good examples.

Conclusion

As can be appreciated throughout the chapter, in Mesoamerican lowlands a series of characteristics which may favour the intelligent use of fungal resources are conjugated:

a) A high richness of potentially usable macroscopic fungi species .

b) Vestiges of the existence of a long tradition in use and management of these resources from prehispanic times.

c) A developed local micological science constructed through generations of observation and experimentation.

d) Interest of local cultural groups to develop productive use projects of their forest resources.

It's vital that government instances involved in environmental and productive decision-making take into account macrsoscopic fungi as an important resource in forests. Thus they may incorporate micological knowledge and local practices concerning fungi into local management plans and general public politics.

The main thing is that sustainable development projects, concerning fungi or not, must be designed according to the cultural knowledge of the region, respond to the real needs of peoples inhabiting it and have the dual goal of preserving the environment and ensuring the wellbeing of people.

Acknowledgments

The authors wish to thank Dr. Ranjay K. Singh for his invitation to collaborate in this book. We would also like to thank Biologists Ruth Alvarado-Rodriguez and Freija Guadalupe Medina-Arias for generating some of the data presented here. The first author would like to thank the Posgrado en Ciencias Biológicas of the Universidad Nacional Autónoma de México. For the translation and correction of the English version we thank anthropologist Eduardo González Muñiz and biologist Marisa Ordaz Velazquez. Finally we would like to give a special thanks to all the people in mesoamerican rural communties who lived below and to the left of the earth, and who have worked alongside us in search for hope.

References

Alvarado-Rodríguez R (2006) Aproximación a la etnomicología zoque en la localidad de Rayón, Chiapas, México. BA Thesis, Universidad de Ciencias y Artes de Chiapas, Tuxtla Gutiérrez
Caballero J (1984) Recursos comestibles potenciales. *In:* Reyna T (ed) Seminario sobre la alimentación en México, 1a edn. Instituto de Geografía-UNAM, México D.F
Casas A (2001) Silvicultura y domesticación de plantas en mesoamérica. *In:* Rendon B, Rebollar S, Caballero J, Martinez-Alfaro M A (eds), 1st edn. UAM-SEMARNAP, México D.F
Casaverde J (1981) El trueque en la economía pastoril. In: Llobera J (ed) Antropología económica, estudios etnográficos, 1st edn. Anagrama, Barcelona
Chacón S (1988) Conocimiento etnoecológico de los hongos en Plan de Palmar, Municipio de Papantla, Veracruz, México. *Micologia Neotropical Aplicada,* 1:45–54

Egli S, Petera M, Buserb C, Stahelb W, Ayera F (2005) Mushroom picking does not impair future harvests: results of a long-term study in Switzerland. *Biological conservation*, 129:271–276

Estrada-Torres A, Aroche RM (1987) Acervo etnomicológico en tres localidades del Municipio de Acambay, Estado de México. *Revista Mexicana de Micología*, 3:109-131

Fidalgo O (1965) Conhecimento micológico dos indios brasileiros. *Rickia*, 2:1-10

Fox FM (1992) Tropical Fungi: Their comercial potencial. In: Isaac S (ed) Aspects of Tropical Micology, 1st edn. Cambridge University Press, Cambridge

Garibay-Orijel R (2009) Los nombres Zapotecos de los hongos. *Revista Mexicana de Micología*, 30:43–61

Garibay-Orijel R, Córdova J, Cifuentes J, Valenzuela R, Estrada-Torres A, Kong A (2009) Integrating wild mushrooms use into a model of sustainable management for indigenous community forests. *Forest Ecology and Management*, 258:122-131

Garibay-Orijel R, Cifuentes J, Estrada-Torres A, Caballero J (2006) People using macrofungi in Oaxaca, México. *Fungal Diversity*, 21:41-67

Goes-Neto A, Bandeira FP (2003) A Review of the etnomycology of indigenous people in Brazil and its relevance to ethnomycologycal investigation in Latin America. *Revista Mexicana de Micología*, 17:11-16

Gurevitch J, Scheiner S, Fox AG (2002) The Ecology of Plants. Sinauer Associates Inc., Sunderland

Guzmán G (2003) Fungi in the Maya Culture: Past, Present and Future. In: Gomez-Pompa A, Allen MF, Fedick SL, Jimenez-Osornio JJ (eds.) The Lowland Maya Area: Three millennia at the human wildland interface, 1st edn. Food Products Press. New York

Guzmán G (1997) Los nombres de los hongos y lo relacionado con ellos en América Latina. Introducción a la etnomicología aplicada de la región. CONABIO-Instituto de Ecología A.C., Xalapa

Guzmán G (1987) Distribución y etnomicología de Pseudofistulina radicata en mesoamérica, con nuevas localidades en México y su primer registro en Guatemala. *Revista Mexicana de Micología*, 3:29–38

Kirchoff P (1960) Mesoamerica: Sus límites geográficos, composición étnica y caracteres culturales. *Tlatoani*, 3:1–15

Lazos E Paré L (2000) Miradas indígenas sobre una naturaleza entristecida: percepciones del deterioro ambiental entre nahuas del sur de Veracruz. Plaza y Valdes editores-Universidad Nacional Autónoma de México, México D.F

Levy-Tacher S (2000) Sucesión causada por la roza–tumba–quema en las selvas de Lacanhá, Chiapas. PhD Thesis, Colegio de Postgraduados, Texcoco

Levy-Tacher SI, Aguirre JR, Martínez MM, Durán A (2002) Caracterización del uso tradicional de la flora espontánea en la comunidad lacandona de Lacanhá Chiapas Mexico. *Interciencia*, 27(10):512–520

Lowy B (1971) New Records of mushrooms stones from Guatemala. *Mycologia*, 63:983-993.

Lowy B (1972) Mushroom symbolism in maya codices. *Mycologia*, 64:816-821

Mapes C, Bandeira FP, Caballero J, Goes-Neto A (2002) Mycophobic or Mycophilic? a comparative Etnomycological study between Amazonia and Mesoamerica. In: Stepp J R, Wyndham F S, Zarger R K (eds) Ethnobiology and Biocultural Diversity. Proceedings of the Seventh International Congress of Ethnobiology, University of Georgia Press, Athens, 23-27 October 2000

Mariaca R, Silva LC, Castaños CA (2001) Proceso de recolección y comercialización de hongos comestibles silvestres en el Valle de Toluca, México. *Ciencia Ergo Sum*, 8(1):30-40

Martínez-Gerrero MA (2000) Desarrollo tecnológico para la producción Intensiva de Lentinus edodes, Neolentinus lepideus y Ganoderma sp., haciendo uso de materiales orgánicos regionales de la actividad agrícola y forestal. MSc thesis, Universidad Autónoma de Puebla, Puebla

Mata G (1987) Introducción a la etnomicología maya de Yucatán: El conocimiento de los hongos en Pixoy, Valladolid. *Revista Mexicana de Micología,* 3:175–188

Mayer KH (1977) The mushroom stones of mesoamerica. Acoma Books, New York

Medina-Arias FG (2007) Etnomicología Mam en el Volcán Tacaná Chiapas México. BA thesis, Universidad de Ciencias y Artes de Chiapas, Tuxtla Gutiérrez

Montoya A (1992) Análisis comparativo de la etnomicología de tres comunidades ubicadas en las faldas del Volcán La Malintzi, Estado de Tlaxcala. BA thesis, Universidad Nacional Autónoma de México, México D.F

Montoya A, Estrada-Torres A, Kong A, Juárez-Sánchez L (2001) Commercialization of wild mushrooms during market days of Tlaxcala, Mexico. *Micologia Aplicada International,* 13(1):31-41

Morales-Esquivel O I (2001) Estudio Etnomicológico de la cabecera municipal de Tecpán Chimaltenango, Guatemala. BA thesis, Universidad de San Carlos de Guatemala, Guatemala

Nations JD, Night RB (1980) The evolutionary potential of lacandon maya sustained-yiel tropical rain forest agriculture. *Journal Anthropology Research,* 36(1):1–33

Pailles M, Beutelspacher L (1989) Cuevas de la región Zoque de Ocozocuautla y el Rio La Venta. El Diario de campo de 1945 de Matthew W. Stirling con notas arqueológicas. Notes of the New World Archaeological Foundation. No. 6. Brigham Young University. Provo

Prance GT (1984) The use of edible fungi by amazonian Indians. In: Prance GT, Kallunki M (eds) Ethnobotany in the neotropics, 1st edn. Kansas City Allen Press, New York

Raven PH, Everet RF, Eichhorn S (1999) Biology of Plants. W. H. Freeman and Company, New York

Readhead SA (1997) The pine mushrooms industry in Canada and the United States: Why it exist and where it is going. *In:* Palm M, Chapela I (eds) Mycology in sustainable development: expanding concepts, vanishing borders, 1st edn. Parkway Publishers Inc. Boone

Recinos A (1960) Popol Vuh. Fondo de Cultura Económica, México D.F

Reygadas-Prado GF (1991) Estudio etnomicológico de la subcuenca Arroyo El Zorrillo, D.F. BA thesis, Universidad Nacional Autónoma de México, México D.F

Ruan-Soto F (2002) Aproximación al conocimiento micológico tradicional en tres regiones tropicales del sureste mexicano, a través de un estudio en mercados. BA thesis, Universidad Nacional Autónoma de México, México D.F

Ruan-Soto F, Garibay-Orijel R, Cifuentes J (2004) Conocimiento Micológico Tradicional en la Planicie Costera del Golfo de México. *Revista Mexiacana de Micología,* 19:57–70

Ruan-Soto F, Garibay-Orijel R, Cifuentes J (2006) Process and dynamics of traditional selling wild edible mushrooms in tropical Mexico. *Journal of Ethnobiology and Ethnomedicine.* doi:10.1186/1746-4269-2-3

Ruan-Soto F, Mariaca R, Cifuentes J, Limon F, Pérez-Ramírez L, Sierra-Galván S (2007) Nomenclatura, clasificación y percepciones locales acerca de los hongos en dos comunidades de la selva lacandona, Chiapas, México. *Etnobiologia,* 5:1–20

Ruan-Soto F, Cifuentes J, Mariaca R, Limón F, Pérez-Ramírez L, Sierra-Galván S (2009) Uso y manejo de hongos silvestres en dos comunidades de la Selva Lacandona, Chiapas, México. *Revista Mexicana de Micología,* 29: 61-72

Sommerkamp Y (1990) Hongos comestibles en los mercados de Guatemala. Universidad de San Carlos de Guatemala, Guatemala

Toledo VM (2001) Indigenous peoples and biodiversity. *In:* Asher LS (ed) Encyclopedia of biodiversity, Vol. 3. Academic Press, New York

Van Dijk H, Awana-Onguene N, Kuyper TW (2003) Knowledge and Utilization of Edible Mushrooms by Local Populations of the Rain Forest of South Cameroon. *AMBIO* 32(1):19–23

Villarreal L, Pérez-Moreno J (1989) Los hongos comestibles silvestres de México, un enfoque integral. *Micologia Neotropical Aplicada,* 2:77–114

Wasson RG (1983) El hongo maravilloso: Teonanacatl. Micolatría en mesoamérica. Fondo de Cultura Económica, México D.F

Williams T (1990) A review of sources for the study of nahuatl plant classification. *Advances Economic Botany,* 8:249–270

Chapter – 10

Menu for Survival: Plants, Architecture, and Stories of the Nisga'a Oolichan Fishery

Nancy Mackin* and Deanna Nyce

Abstract

Oolichan (*Thaleichthys pacificus*), a small anadromous smelt also known as "savior fish", were a necessity of life for Pacific Northwest Coastal and adjacent interior indigenous peoples, as they were the first fish to return to the rivers after winter and produced not only tasty flesh but also a nutritious oil, or "grease." A focus of the traditional oolichan culture is the Nass River of northwestern British Columbia, Canada. The valleys and mountains surrounding the Nass River are the traditional homelands of the Nisga'a people, whose technological innovations including implements, buildings, and landscape structures facilitated fishing, production, and preservation of oolichan, as well as salmon. Over thousands of years, the Nisga'a developed plant-based technologies for use in the oolichan fishery and for preparation trade of rendered oolichan "grease". In turn, structures and implements associated with the oolichan fishery facilitated knowledge exchange across generations, since the oolichan processing areas and fishing camps were the places where elders conveyed their cultural and ecological knowledge to younger people. Structures fashioned from plants helped people to recall the stories, traditions,

---

*nma@telus.net

spirituality, and respect associated with the Nisga'a oolichan fishery. Technological knowledge, gathered and adapted over countless generations, continues to be important in the Nass Valley oolichan fishery, which remains strongly rooted in traditions that retain their value within an increasingly technological world.

Keywords: Oolichan (*Thaleichthys pacificus*), fishing technology, Nisga'a, NorthWest Coast of North America.

Introduction: Oolichan[1], plant technologies, and survival

Traditional ecological knowledge (TEK) can be thought of as "a never-ending choreography of spatial, temporal, and cultural dynamics" (N. J. Turner pers. comm. 2008). The dynamic, place-based knowledge conveyed orally across generations often holds keys to human survival as well as to the well-being of cultures and ecosystems (Turner 2005). An excellent example of traditional knowledge linking the survival of individuals, cultures, and even entire ecosystems is indigenous peoples' customs surrounding procurement, preservation, and preparation of oolichan[1] (*Thaleichthys pacificus*) – small anadromous fish living along the Pacific Coast of North America from California to Alaska. A particularly rich body of traditional knowledge surrounds the oolichan fishery of the Nisga'a, the indigenous people of the Nass River Valley, an expanse of about 20,000 square kilometers of mountainous and forested land at the northern tip of Canada's Pacific Coastline. This chapter examines how the Nisga'a oolichan fishery employs technologies that are rooted in ancient tradition yet flexible enough to adapt to landscape, cultural, and climatic changes. Further, since ecosystems also are sustained within a dynamic, non-linear balance (Capra 1982; Berkes 1993), we will examine how the changing yet constant traditions surrounding oolichan production may contribute to the well-being of ecosystems.

In the Nass Valley and other cold northern regions, oolichan, or *saak*, were relied upon to prevent starvation in the early spring when food stores were often nearly depleted. Because the oolichan meant an end to winter privations and an affirmation of life, the Nisga'a and other northern peoples developed a complex and detailed culture surrounded the procurement, preparation, and preservation of the fish and its nutritious oil.

The complex culture associated with oolichan arose at least partially from necessity. Not only did oolichan help keep human communities from starvation, but they are also a key component of the food chain for other species that provide people with alternative sources of food and materials for clothing and shelter. From their diet of marine plankton, fish eggs, insect larvae, and small crustaceans, oolichan store up to fifteen percent of their body weight as fat. Their high fat content makes them a rich food source for ocean and shore predators, particularly seals and sea lions, which in turn are eaten by Nisga'a and other indigenous peoples of the Northwest Coast.

Oolichan runs are both variable and unpredictable. Like other adronomous fish, oolichan smolts (young fish) swim to the ocean from their home river or stream, where they feed and live as adults, returning to their freshwater birthplace to spawn. The number of offspring within a given oolichan population varies considerably; one stream may have many oolichan while a neighboring stream may produce few or no oolichan. Moreover, a stream or river may host fairly regular and sizeable runs of oolichan and then, without warning, have few oolichan for one or more years. To secure a good supply of oolichan, Nisga'a and other Indigenous fishers have maintained oral records of fish runs over countless generations, thereby assembling an unwritten geographic and temporal database detailing oolichan abundance patterns. Elders and other knowledge-holders are able to employ these long-term records to determine possible reasons for large or small runs in a given year and to make decisions about quantities of fish that should be harvested that year.

As a part of their oolichan monitoring and management strategies, the Nisga'a people utilized a range of plant materials to construct implements, containers and buildings associated with different stages of oolichan harvesting and production. Many of these technological and architectural innovations continue to be in use today. The structures may be seen as a form of constructed memory, helping people to retain the knowledge invested in thousands of years of oolichan production in the Nass Valley.

Memories about oolichan were sustained within a complex web comprised of architecture, plants, and stories. These three interconnected components of traditional knowledge helped the Nisga'a people remember the codifications associated with oolichan fishing, preparation, and preservation. Stories were retold at feasts, but in times of disaster or stress there was a risk that storytellers would not have the time or place to share ancient oral wisdom and practices. During ecological turmoil, such as times of volcanoes or floods, or during social upheaval, such as during the period of residential schools imposed on Canada's aboriginal peoples, architectural works often endured as reminders of resource management technologies and controls as well as the wisdom, respect, and spirituality associated with oolichan production. A third ingredient in the mnemonic recipe – plants employed in buildings and implements – also returned to the valley as characters in stories and as reminders of how to build. Ecological succession and human activities such as planting (Turner 2005) were two processes that enabled successive generations to find the materials they needed to keep oolichan fishery traditions in active use. The buildings kept the stories alive (Nisga'a Elder Emma Nyce pers. comm. 2003) and the stories reminded people of how to employ plants to create the buildings and manage the important oolichan fishery.

Oral history: Oolichan Arrive in the Nass Valley

There are many stories about oolichan, and there are specific traditions about what plant-derived materials are used in the curing of oolichan and rendering of oolichan grease and the way the rendering buildings are assembled. The stories tell of places, plants, and the origin of traditions. One of the oldest and most widely-shared stories tells of **Txeemsim**, one of the great supernatural beings in Nisga'a narrative tradition.

Txeemsim (also known as Wiigat) went down the coast to a supernatural place known today as Hlgu Ul Cliff near Quinamas Bay, also known as Ten-mile. This occurred at about the time of the year when winter is nearly over and the people begin to walk out in search of food supplies. It was the month of Xsaak (March) when Wiigat set out on this part of his journey. He was going to go and gamble at a taboo place and so he wore his black bear hide on this occasion. He had with him his andaxan, which is a two-foot by three-foot mat made of woven ash tree saplings [Note: the traditional story refers to ash tree, but since there are no ash tress in the Nass, Allison Nyce talked to her grandmother Emma Nyce to clarify the story. Emma says the andaxan were made from the bark of red cedar (*Thuja plicata, simgan* in Nisga'a). She remembers her father talking about that story and the mat would be cured with mountain goat oil.]. *Almost any design and dye could be found on these wooden mats, and Txeemsim's was very attractive indeed. He also had with him his haxsan, which are small round bones about the length and diameter of a ball point pen, and are used in gambling much the same way as people use playing cards today.*

Txeemsim carried on down to Hlgu Ul Cliff as planned. Once there, he made his canoe submerge beneath that cliff. This was in the month of March. He made his canoe submerge so he could gamble with others beneath the sea under the cliff. When he reached the bottom, he parked his canoe near the entrance of the House of Rockcliff and entered.

He shook his bear-hide coat off and said, "The slime of and oolichan-spawn of Txeemsim's dip-nets splashed on me when I was unloading the canoe. And I hardly got any sleep all night, as a certain person caught three canoe loads of oolichan further up the river."

"Hah!" said one chief. "How could that be? It is months before the oolichan arrive yet."

Another said, "You are a liar." No-one in the sbi naxnok (House of Supernatural Beings) believed Txeemsim's claim.

But Txeemsim was not deterred. After all, he had planned his gambling foray quite far in advance. Txeemsim had made preparations well before reaching the cliff of Hglu Ul. He had stopped off at a place called Wil Ukwsbaxhl K'ask'oos (where the heron runs out towards the river), a pleasant ango'oskwhl (traditional territory) belonging to the Heron and situated a short distance up Gits'oohl (Observatory Inlet). He had then waived down a passing gull with a herring in his beak on the pretext that he wanted to talk.

Txeemsim had told the Sea Gull that standing out on the point not too far ahead there was a long-legged person who said that the oolichan run would begin in a few days' time. The seagull replied to Txeemsim that it was not even close to the time of oolichan arrival. "That person you speak of it absolutely crazy."

"He is standing right over there", Wiigat interjected. To prove his point, the gull said, "I will go over there and straighten him out."

When the gull returned to his favorite roost, Txeemsim was strategically located immediately alongside the roost. While the seagull was landing, Txeemsim quickly grabbed him and flung him to the ground, exclaiming, "You should not talk about y friend like that" as he was squeezing the poor gull in the stomach. Txeemsim cried, "Nothing will stop me from squeezing you, since you called my best friend a liar." All of a sudden the herring that the gull had just swallowed popped back up...and Txeemsim grabbed it and walked away, pleased with himself.

Txeemsim went back to his beached canoe and squeezed out the spawn from some herring into the bottom of his canoe. He spread the spawn around and rubbed it in a bit. His idea was to make the herring spawn look like oolichan spawn. He also slid the remains and the tail just under the stern sheets, to make the ruse look even more authentic. He did all this before submerging his canoe to visit the gambling house beneath the ocean cliff. Consequently, when the others in the gambling house called him a liar, the notorious Txeemsim said, "Perhaps you should look in my canoe then, if you don't believe me."

The chief sent several people down to check. Sure enough, they returned and had to admit, "He told the truth."

Deanna Nyce, president and CEO of Wilp Wilxo'oskwhl Nisga'a Institute, (WWNI), the Nisga'a University, brought to this work the translation of the preceding story about the oolichan fishery. The story is owned by the Nisga'a people as a whole and can be shared, unlike many of the sacred stories or the stories that are owned by individual families.

Methodology and consent

The primary research that informs this paper was undertaken with the assistance and consent of Wilp Wilxo'oskwhl Nisga'a Institute. Under the Nisga'a Treaty ratified in May 2000 by the Nisga'a, British Columbian, and Canadian governments, WWNI has jurisdiction over research undertaken in the Nass Valley and concerning Nisga'a language, culture, and resources. Working closely with WWNI, we talked with elders from the Nisga'a Nation. We also obtained consent from each interviewee to respectfully employ their knowledge. In addition, each March for many years we visited Fishery Bay, fourteen miles upriver from Nass River mouth and the centre of the oolichan fishery, studying the plants, buildings, and materials associated with the fishery. We talked with the owners of oolichan fishing camps and obtained their permission to photograph the sites. In Nisga'a tradition, we brought food, buckets, and other necessary supplies to the camp. Some of this research was undertaken as part of Dr. Mackin's Ph D. research at the University of British Columbia (*cf* Mackin 2004).

Structures and processing employed in oolichan fishing

In longer retellings of Nisga'a oral histories about Txeemsim and the oolichan, buildings and implements employed in the procurement of resources are often described in great detail. The dip-net is one of the implements described in stories and shared across generations during the fishery. Although methods of fabricating nets varied with different parts of the coast, Nass Valley oolichan nets were traditionally made from twine fabricated with stinging nettle (*Urtica dioica*, Nisga'a name *sdatx*) (Figure 1). The stem fibers were softened and then twisted or rolled together to make a fine cord (Turner

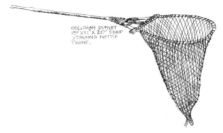

Figure 1: Stinging nettle twine has traditionally been employed in the fabrication of oolichan fishing nets.
Photo: Nancy Mackin.

1998: 175). Since the technology was entwined with peoples' ability to obtain food, the connection between nettle and nets became part of the landscape through language and display. As described below by Nisga'a leader and Elder Dr. Joseph Gosnell, stinging nettle twine and nets are featured in place names from the distant past and in museum displays intended for the present and future.

The oolichan net was made from...the stinging nettle. Within our family's hunting area which starts just a little north of this community and continues up, is Lax Ansdatx which means the rotten stinging nettle. Its fibres were used to make the nets in the early days. The fibres were softened and then twisted together; they look like modern day twine. It was fine enough to be able to be used for nets. In fact I think a stinging nettle net from the Nass may be in the Canadian Museum of Civilization in Hull (Joe Gosnell, Hleek interview 2003).

The twine from twisted fibers extracted from the stems of stinging nettle was first made into mesh, and then fashioned into nets. To impart power to the net in order to enable it to catch oolichan, however, the Nisga'a people would prepare their bodies with cleansing ceremonies before making the net (Alison Nyce, National Museum of the American Indian 2006). The power infused into the oolichan net benefited everyone, since the oolichan were a common resource – shared by all the Nisga'a people (ibid). Interestingly, the Txeemsim story describes the same cleansing ceremonies associated with gambling rituals. Maintaining the ceremonies is a part of the spiritual aspect of resource management and is therefore remembered in numerous ways including as part of games that are enjoyed by children as well as elders. The net builders' codifications are recalled through countless generations and remembered by stories, place names, games, plants, and the procurement of foods.

As described by Dr. Gosnell and Alison Nyce, oolichan dip-nets connect people across generations: they embody a technology that remains important for symbolic as well as practical value. Other structures associated with the oolichan also connect across generations and speak to changes in the climate and landscape that occur across time. For example, fishing through the ice for oolichan necessitated the construction of platforms and retaining systems that helped the fishers work safely on the cold, water-slicked ice surface. At Fishery Bay, the river was, until recent times, still frozen over in early March when the oolichan arrived. Oolichan dip nets were dipped into rectangular troughs cut into the ice, an ancient practice that is still employed among Nisga'a fishers of the Nass River. The ice provided a solid platform for securing the nets. Fishers worked from a platform which was secured by a pole and beam structure constructed around the perimeter of the ice-free rectangle cut into the frozen river.

Harry Nyce Senior, Director of Fish and Wildlife for the Nisga'a Nation, described gradual changes in oolichan harvesting methods (2003 interview):

So as the fishery took on a different mode, techniques started to change. They used canoes before for harvesting, and then they used powerboats. When the whole river would ice up solid, by springtime the water would start to melt the water underneath, and sometimes we would fish on the ice. The ice would be five feet thick. There were different methods of harvesting: using the modern method, they would use horse and sleigh. They used big boxes, about four or five feet by six feet long cedar boxes, and you would fill them up with oolichan on the ice. The horseman would drive the cedar boxes back to the bin, where it would be unloaded. Later on after the horses we would use skidoos.

More recently, global climate change has further altered oolichan fishing methods. Since 1989, the Nass River at Fishery Bay has rarely provided a solid footing of ice in March. In recent years, fishers of the Nass catch oolichan with nets cast from boats. As the later Nisga'a elder Dr. Bertram McKay explained to Nancy Mackin in 2003:

We have very very heavy north winds blowing here, especially the month of February. Until this El Niño animal appeared: this winter we didn't see any north winds, very little if there was any at all. February is known as the month when the north wind never ceases. Buxwlaḵs means all the leaves on the north side of the tree are blown down to the ground. That's what Buxwlaḵs means, and that's how they refer to February. Now we haven't seen that for a long time, so the weather is getting warmer in winter, and the winter months are getting shorter. Like right now, in the olden days, there would be a lot of snow on the ground. But there is none.

Processing Oolichan

Oolichan is processed in a variety of different ways to get the most value from this first fish of the spring. Oolichan are smoke-dried for storage, and many are rendered to obtain the rich, nutritious oil. Both dried fish and oil are stored for later use and for trade. Some oolichan are strung on cedarbark lines and sun dried on tall frames called ganee'e; sun-dried oolichan may be boiled for later use. Some of the fish are half-smoked, using alder wood (*luux* in Nisga'a) or fully smoked and dried for feasts in early winter. However, because of their high oil content, smoked oolichan do not keep as well as smoked salmon and so were not often stored for mid- or late-winter use.

The structures employed for different stages in oolichan production and the plant materials employed to construct the buildings and fittings that are employed for each stage tell a story about the dynamics of food production within Nisga'a culture. Some technologies are ancient and remain nearly unchanged over countless generations in the Nass Valley, while others have responded to changes in materials available, climate, or lifestyles. Within many of the construction technologies, a balance between constancy and change is evident. Tradition and innovation twine together to form the solution best suited to cultural, economic, and ecological conditions.

Structures and Plant Resources for Rendering Oolichan Grease

Before oolichan can be rendered into the much-valued oil, (Figure 2) they must be aged or ripened for seven to ten days under carefully controlled conditions. It is important for the fish to ripen before they are rendered, partly because the slime is difficult to remove from the fish, and partly because the decomposition process increases the amount of iodine and trace vitamins that are in the oil produced and therefore improves the taste. The quality of oil influences the value of the goods that can be received in trade for oolichan grease, so careful ripening has economic benefits for both the people of the Nass Valley and their trading partners.

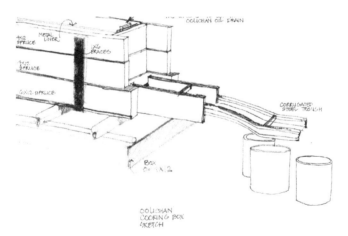

Figure 2: An oolichan rendering bin employed at Johnnie Robinson's camp at Fishery Bay. Drawing: Nancy Mackin

Constructing bins for the controlled ripening of oolichan requires specific materials and construction conventions. Because the oolichan run is so large, the containers for ripening oolichan must have a large capacity and a resistance to collapsing under the weight of the fish. Oolichan bins are constructed from Sitka spruce (*Picea sitchensis*, Nisga'a *seeks*) planks, which have the necessary structural strength and are long enough to construct the walls of the oolichan bins. Western hemlock (*Tsuga heterophylla, giikw* in Nisga'a) boughs line the bins, absorbing excess liquids and keeping the bins fresh as the fish ripen. The oolichan were stored in these bins from the time immediately after they are caught until a week or two later, when they will be processed for oil.

After the fish are left to sit so the exuded oolichan mucus can begin to break down, the grease is extracted by boiling the oolichan in vats of hot water. Thick spruce planks are used to construct the oolichan oil cooking vat, because of its durable nature to withstand the heat and pressure of the cooking process. In the past, the water was heated by adding hot rocks. Nowadays, however, the underside of the vats are metal so a fire can be safely used to heat the water directly in the vats (Figure 3). Present-day cooking vats are also lined with metal to keep the vats water-tight. Before metal sheets were available in the Nass Valley, the cottony material from the airborne seeds of *Populus balsamifera* (black cottonwood) was collected, mixed with grease and used as caulking to seal any gaps in the spruce-wood cooking vat.

Figure 3: Metal-lined oolichan cooking vat.
Photo: Nancy Mackin

In the vat, the oil separates from the fish and rises to the top of the water. When all oil is judged to have come out of the fish, the mixture is cooled and the oil is skimmed off the top (traditionally a wooden skimmer would be used for this, although other implements such as ladles are sometimes used). This process is undertaken within oolichan rendering houses constructed as permanent buildings in Fishery Bay and other centres for oolichan fishing and production. The rendering houses were traditionally made of Sitka spruce or western red-cedar (*Thuja plicata, simgan* in Nisga'a) and had removable panels on the side that enabled the fishers to control temperatures within a building that would contain numerous cooking fires. An ingenious device was employed by the Nisga'a people to ensure that the fires (and the people!) within the building would have adequate oxygen and that carbon dioxide and other gases from the fires were exhausted to the out-of-doors. Rectangular openings were designed into the roof, and these openings had a slightly larger rectangular covering made of several planks fastened together and positioned like a lever on a fulcrum (Nisga'a *ala*). The *ala* enabled people to adjust the direction or degree of openness of the roof opening by using a rope that hung into the interior of the space. Modern rendering houses have metal fresh-air intake and exhaust vents to accomplish a similar level of environmental controls.

Oolichan rendering houses now use plywood panels instead of traditional cedar planks to cool temperatures inside the building while the fires are burning and to permit transfer of materials from the water side into the building. (The main entry is on the side of the building opposite the water, so without the removable panels people would have to carry the fish around the building).

After the oolichan are rendered, *Salix hookeriana* (Hooker's willow) branches were woven to make baskets for straining the boiled oolichan, which were poured into the baskets from the cooking vat when they were still warm (Emma Nyce pers. comm. 2012). Nowadays, metal troughs and strainers separate the valued parts in the oolichan grease-making process.

Fresh, Half-smoked and Smoked Oolichan

Oolichan are sometimes eaten fresh, either roasted over the fire as in former times or, nowadays, fried. More recently, fresh or half-smoked oolichan are often frozen for later use (McNeary 1994: 38). Much of the oolichan catch is traditionally preserved by smoking, drying or rendering. The fish are pierced by sticks threaded through the mouth and gills and then hung on drying racks. When the weather is dry, it could take as little as five days for the fish to cure. Some communities also salt oolichan to preserve them. There is little waste, since about 85 percent of an oolichan is edible. Each part of the fish contributes to a healthy diet; for example, the fat in oolichan is an excellent source of iodine (see also Kuhnlein et al. 1996).

Throughout the Nass Valley, oolichan continue to be smoked in a traditional Nisga'a smokehouse. The smokehouse is itself an indigenous work of technology, since the materials, form, height, and fabrication (Figure 4) details all contribute to a structure that does not get too hot (the smokehouse and its contents would burn down and the fish would cook and spoil!) and so that the amount of smoke is exactly correct for taste and preservation of the fish. Red-cedar planks were used in the construction of the walls and roof of the smokehouse as shown in the drawings and photograph (Figure 5). The vertical planks were spaced slightly apart so that the air could circulate within the smokehouse.

Numerous additional plants employed in the preparation of fresh or smoked oolichan. For example, red alder (***luux***) (*Alnus rubra*) was employed as the fuel for preparing half smoked oolichan (Dr. Bertram McKay 2003 personal communication) and continues to be used in the same way today among Nisga'a people.

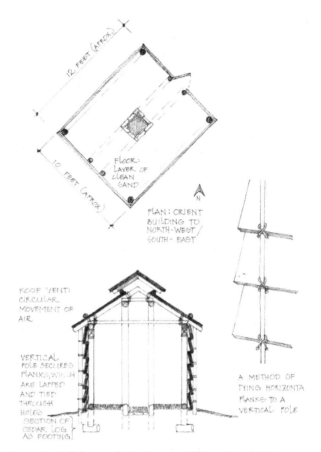

Figure 4: Oolichan rendering house at Fishery Bay. 2007.
Sketch: Nancy Mackin

Figure 5: Nisga'a Elder Lawrence Adams in his smokehouse, built with his grandfather in a way similar to the drawing (Figure 4) but with vertical planks on the sides. Photo: Nancy Turner

In the Nass Valley today, oolichan oil or grease still has numerous culinary uses, some of which are based on long tradition. Oolichan oil was traditionally added to berries as a condiment and to help preserve them. Nowadays, the valued oil is added to berries to enhance their taste and make a favorite dessert. "Lava berries" (*t'ipyees)* (the fleshy leaves of *Sedum divergens*), which appear in early May on the rocky lava plains of the Nass Valley, are enjoyed eaten with oolichan grease (Figure 6), as are salmonberries (*miik'ooks*) (*Rubus spectabilis*), blueberries (*maaý im gilix*) (*Vaccinium alaskaense, V. ovalifolium*), western dock (*tl'ok'ats*) (*Rumex occidentalis*), riceroot (*gaasg*) (*Fritillaria camschatsensis*) and other plants indigenous to the Nass Valley region.

Figure 6: "Lavaberries" (leaves of *Sedum divergens*) and other berries were traditionally mixed with a portion of oolichan grease and stored for winter use. Photo: Nancy Mackin, 2006

Sun-drying Oolichans on a *Ganee'e* (frame for drying oolichans)

The oolichan drying structure, or *ganee'e*, has been in use from the time immemorial in the Nass Valley. The three-poled rack is efficient, essential, ecologically responsive, sturdy, timeless—a work of architectural mastery, as described by Harry and Deanna Nyce in the interview below.

Harry Nyce: *The ganee'e structure is made of one of two kinds of cedar trees: on the coast they used yellow cedar* [Cupressus nootkatensis], *and now we use the red-cedar* [Thuja plicata]. *Historically, they were made of cedar poles about ten inches in diameter at the base, and what we use now is eight to six inches in diameter. The older models that were used at Fishery Bay were taller: they were between fifteen to twenty feet high, and they were supported of course at* the bottom. As they got higher, there were platforms made. The supports were permanent; on certain levels, about every four to five feet, there were permanent supports for the platform. They used planks, and they were moved up to hang the oolichan to sun-dry, and they would move up and of course move down as needed.

What we are using now is a smaller version. We have one now in Ottawa, a ganee'e that my fish crew put together. It is a lot smaller. It would fit inside a room like this—about ten feet tall. We shipped over oolichans, but the oolichans kept dripping; they didn't know how to stop the oolichan from dripping. It is just symbolic of a ganee'e that we sent over to Ottawa for the purposes of an exhibit.

The ganee'e themselves are used for two things: one is the oolichan, and some of the fishers when they were drying other fishes used the ganee'e for drying other fishes as well. If it were sunny, they would dry k'ayukws, which are strips of fish drying. They would cold smoke it then hang it on the ganee'e using crosspieces so the air gets through the ganee'e. That was the two uses. My grandmother told the story of that, but I never did see it: it was quite some time ago.

So different methods came into play with the harvesting. And so likewise with the ganee'e, the younger people are starting to use different things. I'm sure you'll probably see aluminum ganee'e made some day, because the structures are very similar, because you use different material. For the most part, what you see in the valley are the smaller ganee'e made from cedar—cedar poles.

Deanna Nyce: *There was spirituality about them as well, when they raise it.*

Harry Nyce: *Certainly because of their use of the food, it was very much important that the area be kept safe for one, and sacred for the other. All the people were very much environmental at heart, so some of the practices were very much spiritual.*

Nancy Mackin: *Would the ganee'e be constructed a certain way or a certain time that the ganee'e would be built in time of year? Were they built every year, or were they left to stand?*

Harry: *Some of them were left to stand. The ganee'e at Fishery Bay was left to stand, because that is where the major work for the fishing operation was carried out. In the communities: I don't recall, in the early days when I was growing up in Gitwinksihlkw, I saw only one that was built and left up, but there could have been more. They were taken down. The major reason was at Fishery Bay, because people moved away, it was just a camp for that, and so people moved away. When you put them in the other communities other things can happen, like children could start climbing and those sorts of things, so they were taken down.*

The three poles were supported by smaller part of the tree, the part and the top is small, or we would harvest a small tree. They were fastened. There would be a triangle that at one point would be about five or six feet, depending on the individual that was making it. It wouldn't come together as a teepee: it was more or less straight. It remained straight up and down.

Nancy: *What tools were used?*

Harry: *Mostly shale rock was used. My understanding is that there was a lot of inheritance of sharp rocks that were made into heads. Those were the main ones. We have some. You can see the handle on them, and it is tapered. They just hit them towards the person that is working (Harry shows downward cutting motions). They just notch out of the tree enough to add into the pole itself.*

Nancy: *Why three poles instead of four: would there be a reason?*

Deanna: *It's to do with how the air flows through three is very different from how the air flows through four, and it's an efficient way of drawing air. It is incredible technology, and a very efficient form.*

Harry: *Rather than having the four, where one side would cover up the other side, with a triangle the wind could go through in various directions, there wasn't a preferred direction because it was a triangle. To anchor, you would go down about four feet into the ground. They would probably last about six to ten years, depending on the tree itself, the weather, and the saturation on the ground. They would test them, and rock them. Historically a lot of them used spruce roots to fasten them, and those that weaved would weave rope out of cedar and roots as well, to make the ganee'e stronger. It would wrap around the pole so that it is secure. Of course it is notched in there, so it is really sitting in there, all you are doing is fastening it to the pole itself. There is not much movement, you are not rocking it. All you are there for is to place the oolichans on the ganee'e* (Deanna and Harry Nyce interview 2003).

In several ways, the ganee'e is emblematic of the highly complex economic and social fabric of Nisga'a and other northerly peoples of the Pacific Northwest Coast (Mitchell and Donald 1988). Construction personnel had to be related in a specific way to the person stringing oolichan on a ganee'e. As Gitwinksihlkw elder Grace Azak explains, "The person who erects the drying rack is the woman's wilksi . itkw" (Ayuukhl Nisga'a Volume IV: 177). Because wilksi .•itkw, the woman's male relatives on her father's side, such as her paternal uncles and nephews, are of prime economic and cultural importance in the matrilineal Nisga'a society, they must be builders of the ganee'e, a structure keyed to economic and cultural strength. The ganee'e is also an architectural expression of gender divisions of labour that still characterize some traditional economic activities. Along with weaving nets, mats, fabrics, and rope, gathering plants and shellfish, and manufacturing clothing and baskets, women were mostly in charge of processing and storing fish and meat. Within traditional social structure and when possible, men hunted and fished, made fishing and hunting equipment (with the nets and rope made by women)—and provided all woodworking, including making ganee'e.

They start putting the rope around [the ganee'e] and start tightening it up— you need the help of a young tough teenager... 'til it's tight enough.... You start putting it on the strings from the top and you work down. The ganee'e is left and they put sticks and branches on top...so the crows or whatever don't sit there...[and] start eating it from the top...They're going to come back in May to pick it up, real dry (Hleek—James Gosnell, in Ayuukhl Nisga'a Volume IV 1993:176).

Grace Azak (Ayuukhl Nisga'a Volume IV 1993: 177) notes that pole spacing is determined by the length of the cedar bark strips strung between poles:

Cedar bark is used. The inner bark of the cedar is made into narrow strips about ¼", [and then] strung through the gills and mouth of the oolichans. The length of the cedar bark strings is one arm and one half...It is the men who make the drying racks using three cedar poles set upright in holes in the ground and nailed together at distance far enough apart to hang the oolichans between the poles.

Structures for Trade of Oolichan Products

Since oolichan grease was a necessity of life on the Northwest Coast and the adjacent interior, Native peoples frequently traveled long distances to secure or deliver a supply of the valued grease. Many historic "Grease trails," named from the oolichan grease transported over them, extend between communities and from coastal regions into the interior all along the major rivers of northwestern North America. There is, for example a famous Grease trail from the Upper Nass to the Upper Skeena River. Along this trail, numerous bridges were constructed across the many fast-moving streams, rivers, or chasms that could not be traversed by raft or canoe. Ranging from simple log structures to complex suspension bridges made of poles, branches and red-cedar ropes, these bridges provided links for trade as well as social connections that enabled people to exchange ideas and technologies (Guernsey 2006). Some bridges and trails were maintained for hundreds of years or longer, as explained by Reverend William Collison in the early twentieth century.

We picked up [the trail] at the head of canoe navigation on the Nass, and followed it for 20 miles to the junction of the Cranberry River with the Nass. Here we crossed the Cranberry on an ancient Indian bridge, and branched off on an old hunting trail along the Nass to Meziadin, leaving the Grease Trail, which took a north-easterly direction to Kitwancool Lake, and thence to its terminus on the Upper Skeena. This trail has a long and interesting history, for over it Indians from the far interior have travelled for centuries. I only regret that I did not take more careful stock of this historic highway (Collison 1915).

Recent Significance of Architecture for Oolichan

Structures associated with oolichan production – bridges, *ganee'e* (three-poled racks for drying oolichan), rendering houses, smokehouses and bridges – have been employed in the Nass Valley for thousands of years (Ayuukhl Nisga'a Volume 4). The woods and other plant materials employed in building these structures – Pacific crabapple (*sk'an-milks*) (*Malus fusca*), amabilis fir (*ho'oks*) (*Abies amabilis*), subalpine fir (*alda) (Abies lasocarpa),* Pacific yew (*haxwdakw)* (*Taxus brevifolia*), white birch (*haawak'*) (*Betula papyrifera*), juniper (*ts'ex*) (*Juniperis communis*), red alder (*luux*) (*Alnus rubra*), and lodgepole pine (*sginist*) (*Pinus contorta*) as well as western red-cedar (*simgan*) (*Thuja plicata*), yellow cedar (*sgwinee*) (*Chamaecyparis nootkaensis*), Sitka spruce (*seeks*) (*Picea sitchensis*), Rocky Mountain maple (*k'ookst*) (*Acer glabrum*), and western hemlock (*giikw*) (*Tsuga heterophylla*) – have more than three thousand years of use by Tsimshian-speaking peoples (including Nisga'a, Gitxsan, and Coast Tsimshian), according to archeological evidence (Macdonald 1976; Cybulski 1992; Macdonald and Cybulski 2001; Cybulski 2002). More recently, structures and places associated with oolichan have had significance in the British Columbian treaty process. It was at Fishery Bay, during oolichan season, that the Nisga'a people gathered with other Northwest Coastal peoples to pursue their claim that their Aboriginal title to land had never been extinguished.

The oolichan is a resource that is unique to this area. The Nisga'a really were the central figures for the oolichan fishing. Even though they do have it in some of our neighbouring groups, this was the largest gathering site, traditionally. We had up to five thousand people who would gather at Fishery Bay each year to trade and to fish for the oolichan. In that, it was a sort of a forum for everybody, all Nations, it was a forum. And so when contact was made and then there was starting to be an encroachment on the land, this became the forum for the discussion of the land claim, the oolichan time, because it was the one time when everybody would be together...And so it was at the time of the oolichan that they had their meetings about the land to decide what to do: what are your people going to do, what are the Haida going to do, what are the Gitxsan going to do? And so it was the time for the Nisga'a to meet as well. And that was when it was decided that we were going to have one claim, for all the families; not that each family would be separate, but they would be all together. That was the Common Bowl idea because of the oolichan, which is a common resource. You cannot deny that resource to anybody. The resource is shared—that's the whole concept of sharing. That was around the late 1800's. We had the petition in 1913, but they started working on it, I think it was 1897. There was an actual meeting of all of the chiefs of the Nass Valley, and quite a discussion on what to do, what could we do. There was a meeting towards Laxgalts'ap (sometimes called Greenville) of all of the chiefs, and they all placed their name in a bowl, saying okay, "I submit my name and my title to this land for this claim". It was a unanimous decision, to move forward (Allison Nyce Interview 2003).

The oolichan fishery, then, was a focus of assembly that influenced Pacific Northwest Coastal Indigenous peoples' efforts towards recognition of title to their traditional lands. Fishery Bay was especially significant. The structures of Fishery Bay became meeting places. The summer dwellings brought generations together. The Grease Trail network and complex of bridges invited peoples from distant places to add their support. Since Indigenous peoples' land in most of what is now British Columbia was confiscated without treaties, people had to act. With the oolichan fishery as a catalyst, and the places and technologies of the fishery as symbols, peoples of the coast collected funds, hired legal counsel, sent petitions to centres of government in England and Ottawa. Finally, in May 2000, after 113 years of work and effort, the Nisga'a people, in conjunction with Canada and the province of British Columbia, ratified the first modern treaty in the province (*cf* Aboriginal Affairs and Northern Development Canada Fact sheet: the Nisga'a Treaty). Within the treaty were expressed provisions for Nisga'a control of the Nass River oolichan fishery.

Architectural and industrial design works of the Nisga'a oolichan fishery benefited more than politics, people, and cultures. Ecosystems have also benefited from the Indigenous knowledge embedded within oolichan fishery technologies. Since the 1940's, Canada's First Nations asked the Minister of Fisheries not to grant oolichan licenses to any non-aboriginal individuals or companies. The Minister complied. The ban extended to all oolichan-producing areas except the Fraser River, where commercial fishing of oolichan continued – and where oolichan are now nearly extirpated. Now, the Nisga'a Nation manages the Nass River oolichan fishery, which is one of the few places in the world where oolichan remain abundant. The Nisga'a people have won awards for their exemplary – and traditional – systems of monitoring and procuring oolichan. Essential to the ongoing traditional fishing practices are the structures and implements at Fishery Bay. With the assistance of the Elders they remind successive generations of the traditions around oolichan and traditional food production.

However, Nisga'a Elders and leaders worry about the future of the oolichan fishery. Elder and fisher Horace Stevens, for example, expressed his concerns: "The young people don't work long enough these days to get enough oolichan. The old way is to work through the night and have four or five loads to cook. The young people are happy to fill one boat, but then they don't have enough to cook" (Elders' workshop Oct. 2007). He worried that the smaller catch meant that not enough oolichan were being prepared to last through the winter. The Nisga'a knowledge of the land extends to understanding how to plan and prepare for the future and how to conserve resources for future generations. The buildings and implements are also based on these same ideals.

The ideals could have been lost when Indigenous children of Canada were sent away from their homelands to distant residential schools starting in 1920. However, the places of tradition, such as Fishery Bay, remained standing as reminders of philosophies, spirituality, and practical knowledge gained over countless generations. Elders were there in those places of tradition (Figure 7), ready to teach the children when they returned home from residential school, even for short periods of time. When Aboriginal residential schools closed in the 1950's and the Indigenous children of British Columbia returned home, the aging buildings of Fishery Bay and other oolichan producing areas stood as reminders of ancient knowledge. The fishing practices remained alive. The buildings were repaired or rebuilt, employing the same core ideas as structures of the past. Today, the buildings and places and plant-derived technologies encode valued information. Still, Horace Stevens worried: "We [who know the fishing practices] are so few now. We are lucky we have our Elders." As in the past, the Elders have their works of architecture that will outlive them. "The buildings tell the stories", says Nisga'a llder Emma Nyce (2003).

Figure 7: Fishers on the Nass River March 2007. Photo: Robert Mackin-Lang

The Oolichan fishery and Change

The Nisga'a fishing structures make visible the respect held for resources. The complexity of visible and aurally communicated traditions comprise a methodology that has remained unchanged for countless generations. Technology has changed some aspects of the fishery, as boats propelled by motors are now used and the Nisga'a have graduated from bentwood box cooking to spruce vats enhanced by metal bases and liners complete with ventilation systems and airtight lidded buckets. Importantly, however, the transmission of knowledge remains the same, as elders teach young citizens. Fishery Bay camp leaders (owners) also teach younger people how to catch, store and cook oolichans along with the cultural rules that go along with the process.

Other changes pose threats to the oolichan fishery, however. Like the weather, people's behavior can be seen as volatile towards the sustainably of the resource. Nisga'a leadership worries about over fishing, especially by those who do not have a vested interest in the renewal of the resource. For example, in the spring of 2008 an Indigenous skipper brought his seine boat from a southern community, fished, and took a huge haul down to Prince Rupert to sell (a practice which is culturally forbidden). He did not even ask for permission. This type of activity threatens the stock. Another example involved behavior of commercial shrimp fishers, working at the mouth of the Nass and other rivers, who often scrape the bottom of the sea destroying countless numbers of oolichan and other fish (Harry Nyce pers. comm. 2003) Dealing with these types of tensions has frustrated the Nisga'a since the late 1800's in their attempts to preserve the fishery. Today, with the Treaty, Nisga'a government can (and likely will, if this type of behavior continues) apply sanctions to restrict activities that contravene traditional knowledge practices.

Canada-wide and Global Implications of Nisga'a Oolichan Traditions

In Canada, the oolichan fishery practices and associated technologies and structures are increasingly being recognized as evidence of the importance and value of oral history. A recent Supreme Court decision, Tsilhqot'in Nation *vs.* British Columbia (2007), recognized oral tradition evidence as equal to that contained within written historical documents. The decision went on to describe how oral traditions interweave with dynamic food gathering systems and with the structures that are constructed to support the seasonal resource management activities. "Examples of oral history include evidence of when a witness participated in seasonal rounds and how he or she learned from parents or grandparents about how to construct and live in traditional shelters: How the community interprets those [oral] traditions is an unfolding process, based on their environment and culture" (Tsilhqot'in Nation *vs.* British Columbia 2007: 43-44).

Around the world, the scientific community is also coming to recognize that Nisga'a and other Indigenous resource management systems are based on carefully maintained oral histories that date back hundreds or even thousands of years. Since academic science typically can observe changes over just a fraction of the time invested by Indigenous knowledge, some scientists are now searching for ways to respectfully bring their knowledge together with that of Indigenous peoples in an effort to ensure longevity and health of ecosystems and the species within them. The Nisga'a fishery has been able to overcome many of the hurdles that are often experienced when academic scientists and Indigenous knowledge holders collaborate on policy. Working closely with LGL Limited Environmental Research Associates, a team of scientists based on Vancouver Island, British Columbia, Nisga'a Fisheries have been able to consistently and sustainably harvest oolichan in the Nass River Valley. By contrast, other oolichan stocks, such as those in Washington and Oregon, have since May 2010 been listed as threatened under the Endangered Species Act (NOAA National Marine Fisheries Service 2012). The Nass River oolichan fishery sets an example for global policies of co-management. Further, oolichan fishing traditions are emblematic of Indigenous food-gathering knowledge that is dynamic, practice-based, and responsive to environmental signals – and therefore highly relevant to global food policies and practices.

Endnote

[1] Alternative spellings include oulachen, eulachon, ooligan. The fish is also known to some as "candlefish". It is believed that the state of Oregon in the United States represents a pronunciation of "oolichan" by Cree traders (Byram and David 2001).

Acknowledgements

Deepest appreciation is directed to the Nisga'a Elders and knowledge-holders who have offered their knowledge and wisdom to this research, including Horace Stevens, Emma Nyce, Harry Nyce, Allison Nyce, Charlie Robinson, Grace Azak, Alice Azak, and Dr. Joseph Gosnell. We would also like to thank Dr. Nancy Turner for her generous offering of knowledge and photographs. Funding for this research was provided by the Social Sciences and Humanities Research Council (Canada) (2005 grant "Textbook of Traditional Ecological Knowledge in Northern British Columbia". Nancy Mackin PI and 2004 grant "Dynamics of Traditional Ecological Knowledge Acquisition and Transmission". Nancy Turner PI.and the University of British Columbia

Graduate Fund. Robert Mackin-Lang and Robert Lang have graciously donated their time, support, and photographs.

References

Ayuukhl Nisga'a (1993) Four volumes of oral history published by Wilp Wilxo'oskwhl Nisga'a.
Aboriginal Affairs and Northern Development Canada fact sheet: the Nisga'a Treaty. Accessed on-line Nov. 12 2012 at http://www.aadnc-aandc.gc.ca/eng/1100100016428/1100100016429
Byram S, David G L (2001) Ourigan. Wealth of the Northwest Coast. *Oregon Historical Quarterly,* 102(2):126-157
Collison WH (1915). The Oolichan Fishery. *In the Wake of the War Canoe.* London: Seeley, Service, and Col Ltd. 65-74
Cybulski JS (1992) A greenville burial ground: Human remains and mortuary elements in British Columbia Coast History. Hull: Canadian Museum of Civilization
Cybulski JS (2001) Human Biological Relationships for the Northern Northwest Coast. *In: Perspectives on Northwest Coast Prehistory,* 107-144
Guernsey B (2006) Aboriginal bridges of Northwestern British Columbia. Royal British Columbia Museum Living landscapes series accessed on-line at http://www.livinglandscapes.bc.ca/northwest/bridges/index.html
Hirch M (2003) Trading across time and space: Culture along the North American "Grease Trails" from a European perspective. Paper presented to the Canadian Studies International Interdisciplinary Conference held at the University College of the Cariboo, Kamloops Sept 12-14, 2003. Accessed on-line May 3, 2007 at http://www.cwis.org/fwdp/pdf/Oolichan%20Paper%20(Mirjam).pdf
Kuhnlein HV, Yeboah F, Sedgemore M, Sedgemore S, Chan HM (1996) Nutritional qualities of Ooligan grease: A traditional food fat of British Columbia First Nations. *Journal of Food Composition and Data Analysis,* 9:18-31
MacDonald, George F (1976) Prehistoric Art of the Northern Northwest Coast. *In: Indian Art Traditions of the Northwest Coast,* Roy L. Carlson ed. Proceedings from Symposium entitled The Prehistory of Northwest Coast Indian Art held at Simon Fraser University in 1976: 99-129
MacDonald, George F, Cybulski JS (2001) Introduction: The Prince Rupert Harbour Project. *In:* Jerome S. Cybulski, ed., *Perspectives on Northern Northwest Coast Prehistory* (Hull: Canadian Museum of Civilization): 1-24
Mackin N (2004) Nisga'a Landscape and Architectural Histories: Ecological Wisdom and Community-led Design. PhD dissertation, Vancouver: University of British Columbia
McNeary, Stephen A (1994) Where Fire Came Down: Social and Economic Life of the Nisga'a. New Aiyansh: Wilp Wilxo'oskwhl Nisga'a Publications?
Mitchell D, Donald L (1988) Archeology and the Study of Northwest Coast Economies. *In: Prehistoric Economies of the Pacific Northwest Coast.* London: JAI Press: 293-351
Moller, Henrik, Fikret B, Phillip O'Brian L, Mina K (2004) Combining Science and Traditional Ecological Knowledge: Monitoring populations for Co-Management. Ecology and Society 9(3): http://www.ecologyandsociety.org/vol9/iss3/art2.
Nisga'a Tribal Council (1993) The Land and Resources: Traditional Nisga'a Systems of Land Use and Ownership. Ayuukhl Nisga'a Study Volume IV. Wilp Wilxo'oskwhl Nisga'a Publications
NOAA National Marine Fisheries Service. (2012) Accessed on-line Dec. 1 2012 at http://www.nwr.noaa.gov/Other-Marine-Species/Eulachon.cfm

Nyce A (2006) National Museum of the American Indian oolichan net display. Commentary on-line at http://www.nmai.si.edu/mp3/listening/06_nisgaanet.mp3

REF B S, David G L (2001) Ourigan. Wealth of the Northwest Coast. *Oregon Historical Quarterly,* 102 (2): 126-157

Royal British Columbia Museum (2007) Living Landscapes: Grease Trails. Accessed on-line Oct. 1 2007 at http://www.livinglandscapes.bc.ca/northwest/oolichan_history/grease_trails.htm

Stewart H (1977) Indian Fishing: Early methods on the Northwest Coast. Vancouver, BC: J.J. Douglas Ltd

Tsilhqot'in Nation v. British Columbia (2007) Supreme Court of British Columbia Decision Nov. 2007. Accessed on-line at http://www.courts.gov.bc.ca/jdb-txt/sc/07/17/2007bcsc1700.pdf

Turner NJ (1998) Plant Technology of British Columbia First Peoples. Vancouver: UBC Press.

Turner NJ (2005) The Earth's Blanket: Traditional teachings for sustainable living. Vancouver, BC: Douglas and McIntyre

Chapter – 11

Salmon Food Webs: SAANICH First Nation Peoples' Intrinsic Interconnectedness to Salmon Fishing and Conservation on Southern Vancouver Island, British Columbia, Canada

Roxanne Paul*

Abstract

The Saanich (Coast Salish) are an Indigenous people of southern Vancouver Island on the southwest coast of Canada who for many centuries have relied heavily on anadromous Pacific salmon (*Oncorhynchus* spp.) as a major food source. Over time, Saanich fishers have become experts on many aspects of these salmon, including knowledge of the different species, their life cycles, feeding ecology, migration routes, and population dynamics. Much of this knowledge is centered at Saanich Inlet and Goldstream River, a major salmon spawning stream flowing into this inlet. Some Saanich and other Coast Salish people still fish for salmon at Goldstream using traditional gaff hooks and spears, but they have also adopted more modern gear in recent times. Pacific coho (*O. kisutch*) and chum (*O. keta*) salmon stocks in particular have fluctuated over the decades, evidently due to impacts of commercial fishing, pollution from oil leakage and domestic sewage in Saanich Inlet. Saanich fishers have practiced their own conservation protocols, as described in this chapter, and co-management opportunities have provided them with a broader voice in salmon conservation.

---

*roxannep100@hotmail.com

Keywords: Saanich First Nation; coast Salish Indigenous People; pacific salmon, Co-management, fisheries conservation, traditional ecological knowledge.

Introduction: The Salmon and the People

"We always bring along our elders who were there first, and before we even start to fish (*Goldstream River*), we'll say a prayer together to honour the fish, and we thank for our food." (John Elliott, Elder and Fisher, Tsartlip First Nation 2002).

Three species of Pacific salmon – coho (*Oncorhynchus kisutch*), chinook (*O. tshawytscha*) and chum (*O. keta*) – migrate through Saanich Inlet into Goldstream River on the southern coast of Vancouver Island to spawn (Figure 1). These waters are traditional fishing sites of the Saanich First Nation (Coast Salish) communities. Goldstream River salmon are genetically distinct

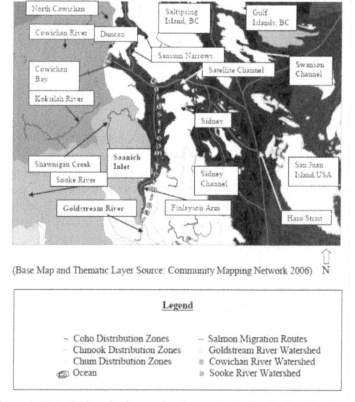

Figure 1: Watersheds and salmon migration routes leading to Sannich inlet.

from any other anadromous salmon population and are a keystone species[1] (Endnotes). These salmon also have immense cultural, nutritional, economic and ecological value for the Saanich people, who have depended upon them from time immemorial, probably millennia.

The purpose of this research (part of my master's program at the University of Victoria 2001-2006) was to examine how Saanich Indigenous fishers' traditional ecological knowledge (TEK) about salmon – including their traditional fishing methods – can contribute effectively to protecting and enhancing the runs of wild coho, chinook, and chum salmon at Goldstream River and Saanich Inlet. In the following sections, the salmon of Saanich Inlet and Goldstream are described, including the timing of their spawning runs, recent declines in their populations, and efforts to manage and enhance the fishery by the federal Department of Fisheries and Oceans as well as its character as a co-management model. The traditional knowledge and practices of Indigenous fishers of Pacific North America are described in general, followed by an outline of the methods used in this study to document the TEK of Saanich fishers regarding use and conservation of the Goldstream salmon. My results, including details of the Saanich fishers' knowledge and practice around the use of salmon – especially the Goldstream salmon – are presented. Finally, I propose recommendations for using TEK and Western Scientific Knowledge together to develop an effective co-management strategy for the Goldstream salmon fishery.

The long-term viability of the Goldstream salmon is of ongoing concern, with unexplained fluctuations in their populations. Hatchery enhancement of the commercially valuable coho and chinook salmon stocks, including incubation of fertilized eggs, feeding and release of fry into the river, has been in effect at Goldstream River since 1982 (Appendix A) and artificial spawning channels were created in an effort to increase the stocks. In addition, restrictions have been placed on the numbers of fish the Saanich are able to take. The declines in salmon populations and the resulting restrictions placed on the fishery have impacted the Saanich people's ability to secure enough fish for their needs over the year.

The salmon of saanich inlet and goldstream river

All Pacific salmon species are anadromous, hatching in fresh water, migrating to the ocean where they spend most of their adult lives, and returning to freshwater to spawn. The adults die after spawning and the fertilized eggs remain in the gravel spawning beds until they hatch and the young fry salmon emerge into the river. They live in fresh water, feeding on small aquatic life until partly grown. Then, as smolts, they swim downstream to the estuary and out into the ocean, where they live out their adult lives until they are ready to spawn. Three species of salmon (coho, chinook and chum) spawn in Goldstream River (see Appendix B).

Saanich Inlet is a temperate marine fjord located on the west side of the Saanich Peninsula on the southeastern tip of Vancouver Island, British Columbia (BC), extending over 20 kilometres north of the Goldstream estuary. The salmon returning to Goldstream River pass through the mouth of the river (at Reach 1), and either remain there or migrate further upstream to spawn in Reaches 2 and 3. All three reaches are assessed as primary salmon habitat (Bocking et al. 1998). The migratory route of the Goldstream salmon, extending through Saanich Inlet and up Goldstream River to the spawning grounds, is called the 'Goldstream terminal'.

The entire Goldstream region is situated within the Coastal Douglas-fir Biogeoclimatic Zone, characterized by immense stands of mainly coniferous trees, especially Douglas-fir (*Pseudotsuga menziesii*), grand fir (*Abies grandis*) and western redcedar (*Thuja plicata*) (Pojar et al. 2004). The floodplain vegetation is comprised of giant western redcedars, Douglas-firs, and grand firs, as well as large deciduous trees: bigleaf maple (*Acer glabrum*), red alder (*Alnus rubra*) and black cottonwood (*Populus balsamifera* ssp. *trichocarpa*). There is a dense shrub understory in the floodplain of salmonberry (*Rubus spectabilis*), gray currant (*Ribes bracteosum*) and thimbleberry (*Rubus parviflorus*), as well as sword fern (*Polystichum munitum*), lady fern (*Athyrium filix-femina*), skunk cabbage (*Lysichiton americanus*) and other temperate rainforest species. Beyond the estuary and adjacent alluvial floodplain, Goldstream River is lined on both sides by hills and sharp cliffs, with mainly coniferous forest. The shrub layer is dominated by salal (*Gaultheria shallon*), oceanspray (*Holodiscus discolor*), dull Oregon-grape (*Mahonia nervosa*) and trailing blackberry (*Rubus ursinus*), as well as numerous herbaceous and bryophyte species (Green and Klinka 1994).

Since the arrival of Europeans in the region, the Goldstream area has been altered in significant ways. In the mid and late 1800s gold mining, road building and construction of a dam and reservoir impacted the watershed. The major north-south running highway on the island is constructed through several kilometers of narrow canyon right alongside the spawning areas of the river. (In April 2011, a fuel tanker trunk crashed and overturned along the highway, spilling some 42,000 litres of gasoline and 700 litres of diesel fuel, and killing thousands of young salmon fry in Goldstream river). Logging took place in the vicinity of the river from 1938 to 1995. Goldstream Provincial Park, established in 1958 from a gift of land by the Victoria Water Board to the province of British Columbia, encompasses the lower salmon-bearing portion of the river and has had a picnic ground built along one of the spawning areas, as well as hiking trails and a nature house at the mouth of the river. Further up the river is a large public campground. In 1995, the salmon spawning reaches of the river and the surrounding riparian zone were designated as an Ecological Reserve (Figure 2), and logging was terminated in the area. There is an increasing realization that maintaining the ecological integrity of Goldstream River habitat is critical to the perpetuation of healthy Goldstream salmon populations and to the overall security and well-being of the Saanich Peoples.

Recent Declines in Goldstream Salmon Stocks

Goldstream salmon fall within the BC South Coast, West Coast Vancouver Island (WCVI) and southern Strait of Georgia salmon stock categories assigned by Fisheries and Oceans Canada (herein DFO), the national governmental body responsible for fisheries management in Canada (DFO 1999b; 1999; 2001d; 2002d). According to DFO stock assessments, South Coast BC coho stocks, Southern Strait of Georgia coho and chinook stocks and WCVI chinook stocks dropped to "seriously low" population levels from the 1970s to the 1990s and at the beginning of the 21st century. By 1999, coho salmon populations had decreased below long-term averages more drastically than other BC salmon species (Baxter 2000; DFO 1999c; 1999d). These stocks remained at low abundances in 2005 but were projected to increase slightly in 2006 (DFO 2005c).[2]

Due to high mortality rates at sea over the 25-year period leading up to 1999, WCVI and southern Strait of Georgia coho and chinook stocks were subjected to intensive conservation measures (Copes 1998; DFO 1999d). Efforts to restore these declining coho and chinook populations include enhancement programs implemented under DFO management plans such as the hatchery coho and chinook stock enhancement initiative in place at Goldstream River.

Goldstream River Fisheries Management

A summary of life histories of the Goldstream salmon, including the amount of time they spend in the river prior to their migration to saline waters, is provided in (Table 1). There are currently many more hatchery-raised than wild or naturally spawned coho and chinook in Goldstream River (Bocking et al. 1998; DFO 2002b), with a ratio of hatchery-raised to wild stocks of approximately 9:1 (DFO 2002c; McCully, pers comm. 2002). Some of these hatchery fish are released at Goldstream and other sites where salmon enhancement programs are in place (e.g. Craigflower, Noble, Tod and Colquitz Creeks) (Goldstream Volunteer Salmonid Enhancement Association [GVSEA] 2001; Till 2005). The remaining 10% wild brood stocks of coho and chinook returning to Goldstream are at high risk of being extirpated by the domestically raised, hatchery stocks returning to this spawning site.

Table 1: Life History Characteristics of Goldstream River Salmon.

Salmon species	Average spawning age	Spawning season	Average adult fork length[a]	Average adult weight	#Eggs laid/female average age of fry at ocean migration/smolt stage
Coho	3 yrs old	Nov-Jan	55 cm	4 kg	> 5000 12 months
Chinook	3 yrs old	Oct-Dec	80 cm	16 kg	< 4000 - >14,000 < 3 months
Chum	5 yrs old	Sept-Dec	65 cm	5 kg	2,000 - 4,000 < 1 week[b]

Sources: Baxter 2000; Candy and Quinn 1996; Harvey and MacDuffee 2002
[a] Tip of nose to fork of tail fin.
[b] Chum fry swim to estuary immediately after emerging from gravel and migrate from estuary to sea after a few weeks.

Notably, DFO officers identify hatchery salmon as "wild" stocks after the second generation of hatchery raised salmon spawn at the river (DFO 2002b, 2002c). Goldstream chum, on the other hand, have retained 100% of their natural genetic lineage (GVSEA 2005; Mc.Cully P. pers comm. 2002).

Joint stewardship and applied resource management practices built on First Nations' rights to fish (and to harvest other natural resources) are commonly referred to as cooperative management or "co-management" (Berkes and Henley 1997; Berkes et al. 1991; Notzke 1995; Pinkerton 1989). Co-management as defined by the Royal Commission on Aboriginal Peoples involves "institutional arrangements whereby governments and Aboriginal [and sometimes other parties] enter into formal agreements specifying their respective rights, powers, and obligations with reference to the management and allocation of resources in a particular area" (RCAP 2:666 In: Berkes and Henley 1997:29).

Numerous case studies highlight strengths and weaknesses, successes, trials and problems resulting from co-management efforts between federal (or state-level) and First Nations (or local-level) authorities (Weinstein 1995; Notzke 1995; Berkes 1999; Pinkerton and Robinson 2001; DFO 2001b; First Nations Panel on Fisheries 2004; Confederacy of Nations 2004; 2005a; Gitxsan Chief's Office 1998; 2006). [For a comprehensive, case-by-case listing, refer to Notzke (1995)]. Schrieber (2001) emphasized that lack of attention to local concerns is the dominant contributor to biological and social crises in fisheries, and illustrated that the flow of social and economic benefits from the fishery back into the community is integral to power-sharing and meaningful co-management arrangements with government fisheries managers.

The Selective Surplus Salmon Fishery of the Gitxsan (a First Nation of the Skeena River of northern British Columbia), similar to that of the Saanich First Nation Excess Salmon to Spawning Requirement (ESSR) fishery, is a legalized, commercial sale fishery entered into by agreement between and co-managed by DFO fisheries managers and the First Nation band councilors (Notzke 1995; Pinkerton and Weinstein 1995; First Nation Panel on Fisheries 2004: 28; DFO 2005a). The joint agreement enables and permits harvest and sale of fish, in addition to the Gitxsan's rights to fish for Food, Social and Ceremonial (FSC) purposes, as is the case with the Saanich Peoples' fishing rights. These agreements (for both the Gitxsan and Saanich First Nations) were developed in part as fishers began exploring options for conducting more selective fishing methods in order to target enhanced or plentiful stocks, and rebuild depressed wild stocks that had been over-harvested in mixed-stock fisheries (Pinkerton and Weinstein 1995:66; Gitxsan Wet'suwet'en Watershed Authority 1998; Taylor 2003).

Traditional Knowledge and Practices of Indigenous Fishers of Pacific North America

The United Nations Environment Programme's (UNEP) signatories to the 1997 Convention on Biological Diversity called for recognition, protection, and promotion of indigenous knowledge (UNEP 1997). The application of indigenous ecological knowledge (herein termed traditional ecological knowledge or TEK) in biodiversity conservation initiatives is gradually gaining more widespread acceptance as it is becoming increasingly recognized that science alone has proven insufficient in alleviating loss of biodiversity and other issues of environmental degradation facing society today (Flett et al. 1996; Turner 1997; Turner et al. 2000; Brodnig and Mayer-Schoenberger 2000; Garvin 2001; Mackinson 2001; Nigel et al. 2003). A prominent definition of TEK referred to by many researchers on the subject (Notzke 1995; Huntington 2000; Kimmerer 2000; 2002) has evolved from the work of Fikret Berkes (1993; 1995; 1999): Traditional ecological knowledge is "...a cumulative body of knowledge, practice, and belief, evolving by adaptive processes and handed down through generations by cultural transmission, about the relationship of living things (including humans) with one another and with their environment." (Berkes 1999:8). Research on integrating fishers' ecological knowledge in fish biology and fisheries management suggests that one of the main failures of former fisheries management systems has been the exclusion of the dynamics or behaviour of the fishers (e.g. frequency, location and target of fishing effort) as an essential consideration of the system (Hilborn et al. 1995; Freire and Garcia-Allut 1999).

Most fishers know a lot about fish distribution and behaviour. Much of this knowledge is based not only on individual observations and experiences, but also those of parents, grandparents and others they have been fishing with (Makinson and Nottestad 1998). Interviews with fishers can elicit important ecological knowledge about fish behaviour and fishing practices (Johannes 1993; cf. Gottesfeld and Johnson 1994). Reports from the oral history of the Nuu-Chah-Nulth First Nations who reside nearby on the west coast of Vancouver Island further illustrate the importance of TEK contributions to our understanding of past BC ecosystems. Though previously unknown to ecologists or historians, past BC ecosystems supported bluefin tuna (*Thunnus thynnus*) whose populations have subsequently been extirpated in this region (Haggan 2000). This oral testimony was subsequently, further supported by archaeological evidence (Crockford 1994; 1997), which confirmed the former existence of a bluefin tuna fishery along BC's Pacific coastline.

Prior to the 1900s, First Nations used fishnets of various types (e.g. seine, gillnets, reefnets, and hand-held dipnets) made from spun fibres harvested from willow bark (*Salix* spp.) or stinging nettle plants (*Urtica dioica*) to catch salmon in streams and at sea (cf. Claxton and Elliott 1994). Traditional fishing technologies included adjusting fish net mesh sizes, using hand carved, steam bent gaff and bentwood hooks (made of yew wood (*Taxus brevifolia*), deer bone barb and cedar wood (*Thuja plicata*) lashing) and spears carved from yew and other woods. Fishing technologies would have also included considerations and methods for targeting the species, run, size and gender of the fish in accordance to what they believed would sustain future fish populations (Berkes 1999).

The Saanich Salmon Fishery

Saanich Peninsula and the area around Saanich Inlet are within the traditional territory of the Saanich First Nation (Jenness 1938; Claxton and Elliott 1994; Mos et al. 2004). For many generations, the families of the North Saanich (Tseycum and Pauquachin), South Saanich (Tsartlip and Tsawout) and the Malahat (who live at Mill Bay on the west shore of Saanich Inlet) have fished coho, chinook and chum stocks in Goldstream River, the waters of Saanich Inlet, and adjacent straits. Chum, being the most abundant salmon species returning to Goldstream River, were and are a major food resource, harvested each year from mid or late October to early December. Saanich Tribal Fisheries is a Saanich First Nation organization that acts as the administrative and management body on fisheries matters including those regarding Goldstream River and Saanich Inlet salmon stocks. Both the inlet and the river are important traditional fishing localities for the Saanich (Figure 2 and 2.1).

The Saanich Nation comprises the communities of Tsartlip, Tseycum and Pauquachin on the west side of the Saanich Peninsula and the Tsawout on the east side as well as the Malahat on the west bank of Saanich Inlet (Figure 2). The total on-reserve population of these communities is approximately 2250 people (Mos et al. 2004; Aboriginal Canada Portal 2006; Department of Indian Affairs and Northern Development 2006).

Saanich First Nation people have fished salmon from Goldstream River and Saanich Inlet since pre-European settlement times (Bocking et al. 1998:3), and have continued to rely on fresh and dried salmon as a staple source of protein year round (Elliott 1990; Simonsen et al. 1995: Mos et al. 2004). Saanich people continue to fish salmon at Goldstream River and Saanich Inlet. Saanich Peoples' rights "to fish as formerly" are set out in the 1852 Douglas Treaties which were signed by Saanich First Nations referred to then as the South Saanich tribes (which are the bands now referred to as Tsartlip and Tsawout and the

Figure 2: Goldstream river estuary and riparian zone.

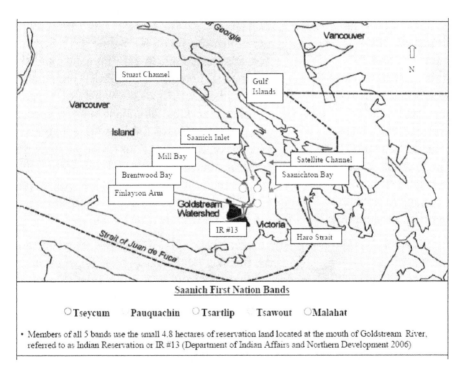

Figure 2.1: Location of saanich first nation bands (Base Map from: Bocking et al. 1998).

Malahat band who are descendents of the South Saanich Tribes) and North Saanich tribes (now called the Pauquachin and Tseycum bands). The Douglas Treaties stated that Saanich people would "retain their hunting and fishing rights on unoccupied lands" (Madill 1981: 9; Province of BC 2000; Union of BC Indian Chiefs 2003).

In the case of the Saanich indigenous salmon fishery, and as will be shown in my results, the Saanich tribal chiefs, elders and fishers hold longstanding and extensive knowledge about freshwater and marine ecosystems, salmon ecology, methods that protect and respect salmon and salmon habitat, salmon fishing practices, storage and preparation of salmon, their traditional food source and wild salmon conservation. The application and integration of TEK with available scientific data about the salmon populations is an urgent requirement in the midst of our uncertainty about fish stocks (Paul 2006).

The next section outlines the methods used in this study to document the TEK of Saanich fishers regarding use and conservation of the Goldstream salmon.

Research Methodology

Interviews

Traditional ecological knowledge about salmon was documented through collaboration with Saanich First Nation fishers. Saanich fishers were asked to share their knowledge and perspectives about the Goldstream and Saanich Inlet fishery in semi-directed interviews that ranged in duration from one to four hours. A survey research instrument was created for conducting these interviews. The overarching goal of the interview schedule was to document knowledge and life experiences of Saanich Peoples who have fished salmon in these waters for subsistence purposes over several years.

I applied to the University of Victoria Human Research Ethics Committee to undertake interviews for this research and received approval. Interviewees were selected on a chain peer-referral basis which served to identify key, experienced informants rather than a random sampling of the community. This is a recommended approach for undertaking research integrating TEK (Huntington 2000). A Saanich fisher who fished Goldstream River and Saanich Inlet was referred to me by my master's program supervisor who had worked with him previously. That fisher participated in a one-on-one interview with me, using open-ended questions and discussions, and referred one of his peers to me. All fishers referred were interviewed until no new names came up. Seven fishers were interviewed in total, including six from Tsartlip and one from Tsawout (Appendix D). All participants provided me with their informed consent and permission to tape-record our sessions and use the information for my studies. The interviews were transcribed *ver batim* and major themes were then obtained by thematic analysis.

Thematic Analysis

Thematic analysis is commonly used to categorize qualitative information from interviews. It is a process of grouping together similar responses by using a series of codes or flags such as particular themes. A theme is a pattern found in interview responses that will allow description and organization of the possible observations and may be extended to interpret aspects of a phenomenon. Boyatzis (1998:4) described thematic analysis as a useful tool for "…systematically observing a person, an interaction, a situation, an organization, or a culture." It can translate methods and results into forms accessible to readers from different fields, orientations or traditions of enquiry (Silverman 1993). It is a good methodology to use with interviews seeking TEK within an Aboriginal community, as it is useful for analyzing individual and shared responses to questions within the questionnaire.

Thematic analysis of interview transcripts was undertaken in order to encode qualitative information and document dominant, emerging themes regarding Saanich Peoples' past and present salmon fishing experiences. If a recurring theme was identified in the responses of most participants, it was considered a major theme. "Summaries of fishers' thoughts about salmon conservation and ecology relevant to this study are provided independently of tabulated results found in Tables 2 and 3". Discrete and unique topics relayed by Saanich First Nation fishers follow.

Results

The Saanich First Nation Salmon Fishery

Saanich Fishers' Insights Regarding the Goldstream Salmon

The Saanich fishers interviewed conveyed valuable TEK including descriptive, biological information about morphology, physical characteristics, food web, feeding behaviour and migration patterns of Goldstream salmon as well as alterations to salmon habitat the fishers had observed and their own conservation and restoration efforts at Goldstream River and Saanich Inlet over time. They also relayed knowledge about their former and current fishing practices and technologies used to fish salmon. Some observations of size and abundance of salmon caught, and accounts of relatives' fishing experiences, were recorded during the interviews.

In general, according to those interviewed, Saanich salmon fishing opportunities have markedly declined over the past 150 years or so. Some of the reasons cited for this are eradication of many small, local salmon-bearing streams due to road construction, pollution, and overfishing of salmon and shrimp (a primary food source for salmon).

Saanich Inlet tributaries are important habitat for anadromous salmon (Simonsen et al. 1995: online; Simonsen et al. 1997; Elliott J. pers. comm. 2002). "The coho spawn in little streams...they just seem like ditches to some people" (Elliott J. pers. comm. 2002). The elimination of numerous small streams through East and West Saanich on southern Vancouver Island has disrupted the Saanich Inlet ecosystem by cutting off, fragmenting and altering the direction of surficial and groundwater flow to local creeks and streams. "So when they put in the...West Saanich Road, it started to knock off all those streams change their direction" (Elliott J. pers. comm. 2002). Elliott observed that these roads acted as physical barriers, forcing coho salmon to reroute their migration path and subsequently attempt to dig redds (spawning nests) and lay eggs in inhospitable aquatic environments where spawning and rearing of juveniles could not occur (e.g. the banks of the inlet). The roads and culverts also blocked nutrients originating from those streams from flowing into Saanich Inlet to nourish the shrimp and other marine organisms that subsisted upon them. "...Those are the streams that are feeding our Inlet and they're feeding the coho. That's what they are there for. They eat those [*nutrients*] and whatever comes out of those streams, that is their food. And, it is also feeding the young salmon, the ones that are just coming out of the stream..." (Elliott J. pers. comm. 2002).

John Elliott, Earl Claxton Sr. and other Saanich fishers recalled their elders teaching them that Saanich Inlet was a nursery and important feeding ground for wild salmon migrating through and feeding on the then plentiful food sources within it (Claxton Sr. E. pers. comm. 2002; Elliott J. pers. comm. 2002). "...My dad always said that this, Saanich Inlet is a nursery for salmon. That's the way he described it... They come here, they are born, the little [*salmon smolts*] feed, they grow up here and then they leave" (Elliott J. pers. comm. 2002).

The Shrimp of the Inlet

A notable decrease in shrimp has taken place in Saanich Inlet since the roads were built throughout the Tsawout and other Saanich First Nation reservations (Elliott J. pers. comm. 2002). John Elliott and Joe Bartleman (pers comm. 2002) expressed that the decrease of shrimp in the region is likely due to increased commercial shrimp fishing and pollution in the inlet. The continuance of the commercial shrimp (and prawn) fishery that was established as a test fishery in Saanich Inlet in 1999 is increasing the demand for shrimp and causing further decline to the shrimp population (Bartleman J. pers. comm. 2002). Shrimp are a major prey for salmon so a declining shrimp population may be a contributing factor to the subsequent decrease in size and abundance of salmon and other fish stocks that used to pass through Saanich Inlet in the thousands (Claxton Sr. E. pers. comm. 2002; Elliott J. pers. comm. 2002). Elder John Elliott (pers. comm. 2002) dissected the stomachs of recently deceased salmon at Goldstream River to determine what they had been eating. He found mostly shrimp in the stomachs of these salmon, indicating that shrimp is a main part of the diet of the salmon feeding in Saanich Inlet prior to migrating upriver (to Goldstream River spawning grounds). The declines in shrimp have thus negatively affected the livelihoods of Saanich fishers, both directly, and indirectly as a major food of salmon (Elliott J. pers. comm. 2002).

Conservation Practices of Saanich First Nation Fishers

The Saanich were aware of the necessity to control and limit their fish catch. Earl Claxton Sr. (pers. comm. 2002), for example, was careful to teach his grandchildren about conserving the fish: "…but I always say don't get anymore than about 30, even 30's too much." Earl Claxton led field trips for children and youth attending kindergarten to grade eight classes at the Saanich Tribal School and showed them how to fish Goldstream River salmon in the least wasteful ways. He also taught them not to catch more than was needed for their families' winter supply of chum salmon.

Joe Bartleman (pers. comm. 2002) and his family (of the Tsartlip band) had fished Goldstream River and Saanich Inlet for over 200 years. They especially liked to fish grilse coho salmon, which were caught in the inlet when they were approximately 30 cm long. The coho grilse were a main part of the Saanich Peoples' food fish but have subsequently been eliminated from their diet due to coho fishing restrictions over the past 15 to 20 years. "I must have been in my early 20s last time I ever went out for those [coho], because they closed it completely, and when it reopened again, we could still take them at 12 inches, and I resisted" (Bartleman J. pers comm. 2002). Joe Bartleman had witnessed a decline of coho returning to Saanich Inlet feeding grounds first-hand over the twenty years previous to my interview with him. His testimony illustrated that even during times when Fisheries and Oceans Canada (DFO) reopened coho fishing in Saanich Inlet, he exercised restraint from his traditional coho fishing practice in the interest of protecting future generations of coho salmon. A similar conservation ethic was evident from the interview with [then] Tsartlip Elected Chief, and lifelong fisher, Simon Smith (pers comm. 2002). "…For the chum, our people believe these are ready to spawn so they don't want to catch them... eggs are something our people like too and we do believe in conservation, we won't fish them if we believe there isn't enough in the river." Accounts by Earl Claxton, Sr., Joe Bartleman and Simon Smith's of their families' fishing practices illustrated that Saanich fishers are cautious about the quantity of salmon that they fish and that conservation is a well-ingrained and intrinsic part of their ongoing fishing practices.

Changes to Saanich Peoples' Salmon Fishing Efforts

In the past (prior to the 1950s), traditional, rotational fishing practices maintained the general productivity and abundance of Goldstream salmon stocks. Joe Bartleman spoke of his parents' experiences with rotation fishing of chum at Goldstream River (between 50 and 70 years ago).

I think we had more of a demand on the resource at that time, as there were always different families in that stream doing their fishing at different times. It was never all of us there at once. We couldn't possibly leave them enough room. But, because there was all that rotation, and with all these people taking fish out of the stream, we could allow more than 17,000 (*chum*) because we were taking them as they come in. (Bartleman J. pers. comm. 2002)

Over the past 50 to 70 years, there have been some changes in fishing patterns such as the transition from traditional, rotational family fishing practices which entailed catching more salmon at the river to a modern commercial seine boat (chum) salmon fishery taking place in Saanich Inlet[3].

Changes to the Goldstream River Salmon Stock Fishery

Six out of seven fishers interviewed confirmed that Goldstream has historically been a predominantly chum salmon-bearing stream. This is consistent with DFO's historical records of abundance of salmon that returned to the river (escapements) from 1932 to 2005 (DFO 1932-2005). Today, Goldstream River is a mixed stock fishery providing spawning habitat for enhanced coho and chinook as well as naturally spawning chum salmon. Five of the seven fishers expressed concern that hatchery enhancement of coho and chinook to higher levels than the stream's natural, historical populations could negatively impact its wild chum populations. In addition, two out of seven fishers relayed that Goldstream coho and chum were smaller in length, girth and weight than they were when they fished them in the past. John Elliott (pers comm. 2002), for example, observed that the coho he caught in Saanich Inlet in the past two years were about the same length but a lot thinner (~2 lbs/1 kg lighter) than the ones he had caught in the previous 40 years.

Saanich Tribal Fisheries

Salmon are still caught by Saanich fishers at Goldstream River, primarily to feed themselves and their families. Saanich Tribal Fisheries consists of three community leaders voted in by the Saanich First Nation to work on important matters pertaining to fisheries. Saanich Tribal Fisheries has regular meetings with elders and people who fish for the four Saanich bands (Bartleman J. pers. comm. 2002).

In the past, Saanich fishers would travel by dugout canoe to sell or trade salmon they caught in Saanich Inlet for tea, sugar, flour, bread and other products in coastal communities of British Columbia and the United States (See Table 2 for timeline). Current federal and international trade, fisheries and immigration laws now delimit the range in which Saanich fishers can travel in order to catch, offer, trade or sell salmon. Salmon caught in Saanich Inlet was sometimes brought as gifts or for trade when people were visiting relatives or other members of local tribes. Saanich fishers also used to sell salmon caught in the inlet to the market in the town of Sidney or along the side of West Saanich Road. The money from the sale of the salmon was used to buy fishing gear, food, and anything else they needed for the next fishing trip (Claxton Sr. E. pers. comm. 2002).

Table 2: Traditional and contemporary fishing methods of Saanich First Nations.

Fishing practices and technologies (pre-European[1] settlement of 1800s to the 1960s)
Practices specific to Goldstream River
  i. Families set up fishing and smoking camps near riverbeds (Oct.- Dec.).
  ii. Started fishing at river mouth and moved upriver (so as not to take too many from one place).
  iii. Fished at night, shared and passed on histories of the river, salmon and ceremonies.
  iv. Fishing implements (wooden shafts, gaff hooks, spears, nettle fibre fishing nets, dip-nets, stakes and weirs) were handmade from wood, tree sap and twine. (Smith S. pers. comm. 2002). Wooden stakes and stinging nettle weirs were used to lure salmon to one side of stream for harvesting (Bartleman J. pers. comm. 2002).
  v. Chum salmon caught were cured (cut, salted and laid out in containers to dry overnight) or staked and hung to dry under a canopy in the sun and wind. They were also smoked by hanging over burning alder wood at the campsite by the river.[2] Salmon were then piled in layers and bundled with rope, or strung together on a rope line. Children were taught how to fish on site.

Practices specific to Saanich inlet
  i. Families fished near the mouth of Goldstream River and in the inlet waters throughout the Saanich Peninsula.
  ii. Families fished in canoes carved from cedar trees all day and at night, sharing stories about how to read the wind and tides to find where the salmon were, passing on traditional ecological knowledge of special fishing places, salmon, the inlet, salmon food ceremonies and special words to speak upon catching salmon.

iii. Canoes or wooden boats using hook and line or net fishing in the inlet. Beach seine net fishing, rod and line as well as spear fishing salmon from the shoreline. Gillnet, seine and troller (hook and line) boat fishing of coho, chinook and chum (for family cultural and communal Food, Social and Ceremonial as well as trade and small scale commercial sale).
iv. Salmon were stored in wooden crates on the boats. Once freezers were kept at home (~ the 1930s), fishers kept salmon they caught on ice in boxes on board the boat.
v. Salmon stored in crates on the beach overnight and carried or transported to homes by horse and buggy up to the mid 1900s then by fuel operated vehicles thereafter.
vi. Caught chum, coho and chinook salmon in the inlet.

Recent fishing practices and technologies used (1960s-present)
Practices specific to Goldstream River
i. Gaff hooks, spears, dip-nets, weirs, (for stakes luring), used for salmon fishing during day or nighttime fishing trips. Salmon then strung together on a rope, pulled downriver and transported home where they smoke or airdry the fish.
ii. Bring home raw chum caught from the river (in cedar boxes or other containers) and freeze them; or bring them to a local smokehouse where they are cured (cut, salted, laid out to dry overnight, staked, hung, then cold or hard-smoked by cooking them over burning alderwood).

Practices specific to Saanich inlet
There are currently 1-4 seine boats (contracted by Saanich Tribal Fisheries) fishing chum during a 10-day fishing season (Oct); fish are loaded onto a truck, and delivered to the home residences of Saanich First Nations people living in the Saanich First Nation villages.

Changes in salmon fishing practices (1940s to present)
Practices specific to Goldstream river
Nets, lines and twine (used for binding gaff hooks to fishing poles) were woven from stinging nettle plant stems in the past. Hooks, weights, poles, spears, knives and axes, (to chop cedar, alder or other trees for fishing poles and firewood), used to be made of carved stone and cedar. Since the 1940s, synthetic fibre fishing nets, line, hooks, weights, rods and knives have been the main fishing implements used and most are now purchased at department stores (not handmade).

Practices specific to Saanich inlet
Seine fishing used to be done by families from cedar canoes by drifting cotton nets with mesh sizes adjusted to the size of salmon they wanted to catch. Since ~1940, metal and fiberglass hook and line seine boats with synthetic nets and gasoline powered motors have been used to fish chum. People used to catch many more chum in Saanich Inlet by using seine net, troll, canoes mounted with line or net, and by spear fishing in the past than they do now.

Saanich Peoples' ongoing spiritual ceremonies and beliefs about Goldstream salmon
Practices specific to Goldstream river
Saanich people bring their children to Goldstream to teach them about their heritage and ancient connection to the river so that the younger generation will feel connected to the stream and learn and pass on the spiritual beliefs their parents and elders learned (e.g. the salmon have a species name and a prayer name that Saanich People use to honour and respect the salmon and which reflects their beliefs of gathering food; and, how to fish without needlessly wasting or injuring the salmon; and, how to fish in a way that will protect the next generation of salmon).

Practices specific to Saanich inlet
Chum and a few coho (1 male and 1 female for [unspecified] annual longhouse ceremony) harvested by Saanich Tribal Fisheries boat for cultural and communal celebrations and longhouse ceremonies (e.g. potlatch, naming and blessing ceremonies) (Smith S. pers. comm. 2002).

It was tradition for heads of families to safeguard traditional ecological knowledge about river, estuarine and ocean stewardship, including the presence (or absence and relative abundance) of salmon and other environmental considerations pertinent to Goldstream River and Saanich Inlet waters. This history (like an orally conveyed almanac) is passed on verbally during social gatherings and may take on religious and spiritual contexts important among families of the four Saanich villages. Such information is sometimes used for predicting when a good salmon run will occur and may be passed on when one or more families take fishing trips together. Joe Bartleman (pers. comm. 2002) relayed that: "Before immigration laws and fisheries restrictions, Saanich fishers used to rotate fishing areas much more frequently."

Traditional Ecological Knowledge of the Goldstream Salmon Run

By listening to their elders, some Saanich fishers have learned to recognize environmental cues that signal the timing and abundance of the upcoming season's Goldstream River salmon run. Earl Claxton Sr. (pers. comm. 2002) recalled his mother (Elsie Claxton) teaching him that seasonal stages of local plants indicate the timing of the (Goldstream) salmon run: "My mother said that when the Spiraea or the Oceanspray (*Holodiscus discolor*) is in full blossom, it is time for you to go out there and get ready for the salmon run.... When the Spiraea is just starting to turn brown, it is time to set up your (*fishing*) camps." Earl's mother also passed on to him the knowledge that extremely low tides occur when the moon is full.[4]

Emmanuel Cooper (pers. comm. 2002) also spoke of environmental indicators about abundance and timing of the salmon run which he learned from his ancestors: "If there are a lot of fir cones on the (Douglas-fir) fir tree in the Springtime, you know there is a big salmon run coming.... When blackberries are ripe [probably trailing blackberry bush, {Rubus ursinus}, which is native to the area], you know the salmon run is coming down." These ecological indicators guided the timing of Saanich peoples' salmon fishing activities over many generations.

A little known fact is that there are two runs of chum salmon at Goldstream River. "The first run, in October, are longer, skinnier. They are leaner. The fat is burned off them. The lean ones are really good for smoking. The ones from the second run around November have more fat (Bartleman J. pers. comm. 2002).

The following section describes the themes expressed by all seven fishers interviewed (major themes), presented alongside their associated subthemes (with subheadings indicating information that is specific to Goldstream River

Table 3: Goldstream salmon ecology and conservation themes from interviews with Saanich first nation fishers.

Theme	A. Findings specific to Goldstream river	B. Findings specific to Saanich inlet
1. Traditional knowledge of salmon ecology and biology	i. Sockeye and pink salmon are not endemic to Goldstream. Salmon are sensitive to pollution and the smell of theriver. ii. Observed changes in salmon population dynamics over time (fewer salmon returning to the river now than in the past).	i. Observed changes in salmon populations over time (fewer coho, chinook and chum to be caught in Saanich Inlet) ii. Observed changes in salmon migration behaviour over time - Feeding habitat and migration paths have changed (i.e. coho are no longer using their former migration routes to spawn in local creeks; naturally spawned juveniles no longer feed in local creeks in the abundance that they used to; fewer salmon return to Saanich Inlet from local creeks or remain in the inlet after migrating from Goldstream River (probably due to pollution in the inlet in recent years) though the inlet was commonly used by salmon as nursery and feeding grounds. - Overfishing of all salmon–especially offshore coho and chinook -Shrimp is a main food of salmon.- Salmon are staying at greater depths in the water column. - Adult salmon returning to Saanich Inlet are smaller. - Increased pollution from oil leaks from boats and increased disposal of fecal coliform in the Inlet and overfishing of salmon (especially coho and chinook) at sea/offshore - Some Saanich People getting stomach sickness from eating seafood in Saanich Inlet-this is thought to be due to pollution

(Contd.)

Contd. Table 3

Theme	A. Findings specific to Goldstream river	B. Findings specific to Saanich inlet
2. Salmon conservation, restoration and stewardship efforts by Saanich Peoples	i. Refraining from harvesting river stocks (including adult coho, coho grilse and all species of salmon eggs/roe). ii. Restoring riparian zone by: planting native vegetation along streambanks to improve salmon spawning and rearing habitat and to decrease erosion processes in wetlands, along streambanks and in creek beds; removing invasive vegetation such as Himalayan Blackberry, English Holly and English Ivy. (Wetland stewardship viewed as a part of fishery conservation efforts- Saanich people partake in streamkeepers and wetland keeper's projects in Saanich.) iii. On-going in-stream salmon population surveying.	i. Refraining from fishing coho. ii. Fishing excess chum in Saanich Inlet to reduce effects of overspawning in the river. iii. Keeping log of abundance of chum caught in Saanich Inlet for the chum fishery and report total annual catch to DFO officers.

and Saanich Inlet). Subthemes are descriptions provided by the majority (4 or more) of fishers interviewed. These descriptors effectively characterize the major theme. Some of these thematic results such as that of 'traditional knowledge of salmon ecology and biology' (Table 3) included fishers' observations as well as their accounts of TEK passed down from elders, relatives or other Saanich fishers. This section begins with a summary of Saanich Peoples' traditional fishing practices (Table 2).

The results reported above are an illustration of the breadth of knowledge Saanich Indigenous Peoples have retained about Goldstream River, Saanich Inlet, the abundance, characteristics, ecology and conservation of Goldstream salmon, and of their traditional and contemporary fishing practices, patterns, preparation, storage and ceremonial purpose for these respected fish at these sites. The importance of salmon to Saanich Peoples' culture is also evident in the symbolism of their paintings, carvings and crafts (Figure 3).

Figure 3: Roxanne's Muledeer hide Salmon drum crafted by a traditional hunter and fisher from Tsartlip, Saanich (anonymity requested).

Discussion

An Effective Application of a Co-Management Strategy for the Goldstream Salmon Fishery

Collectively, the voices of the Saanich fishers interviewed for this project provided a distinctive contribution of TEK in the form of oral testimonies rich in history, and honour for the land, people and resources as well as the cultural, social and spiritual traditions. Fishers respectfully shared their knowledge of Goldstream salmon, the salmon habitat and the fishery, as well as traditional and modern fishing methods. Saanich fishers made direct observations of cumulative impacts upon coastal and marine ecosystems caused by changes to the landscape and marine biodiversity (e.g. loss of streams and lowered abundance of shrimp or other sources of salmon nourishment, and redirection of water flow of Goldstream River).

The descriptive results from interviews with Saanich fishers provided some insights about how and why changes in salmon abundance and conservation efforts came about over the past 130 years or so (since road construction began in East Saanich in the 1870s) (BC Archives 2003). The major changes to the landscape, physiography, quality of the river waters, the inlet waters, salmon abundance, enhancement and migration patterns, the salmon food web and adjustments to Saanich fishers' fishing methods and conservation practices relayed in my results are important local observations which are not immediately apparent to those engaged solely with population assessment, monitoring, enhancement and other salmon conservation initiatives at local and regional scales. These environmental indicators are not discovered or monitored by natural scientists on a frequent basis. Researchers engaged with salmon population monitoring at Goldstream River, Saanich Inlet and beyond may find it useful to learn how historic changes to the local landscape and aquatic environments measure up alongside historic changes in salmon abundance at the river (Figure 4).

Figure 4: View of Saanich Inlet from Goldstream river estuary.
Photo: Roxanne Paul, October 2003

The results reported in this chapter indicate that the retention of local, subsistence practices (e.g. harvesting wild salmon in Saanich Inlet and Goldstream River) and of the passing on of traditional ecological knowledge from one family and generation to the next is an important part of First Nation cultures. Indeed, the perpetuation of Saanich First Nations' knowledge from generation to generation is directly linked to the ecosystem integrity and biodiversity of coastal and marine habitat encompassing the Saanich Peoples' homelands. In effect, Saanich First Nations' salmon fishing customs, patterns and dynamics are an inclusive part of the natural system (here, the Goldstream terminal salmon migration route).

Prior to the onset of increased commercial fishing in Saanich Inlet, the Goldstream salmon stocks were closely safeguarded by the heads of Saanich First Nation families who knew the inlet, the river, the fish and the coastal lands intimately. The ecological knowledge relating to these places would be passed on to other members of the community during harvesting and at social and cultural gatherings and events. These people would in turn become stewards of the river, the inlet and the fishery. One of the main changes in the fishery is that in the past, there was a lot more rotation of fishing, so that all the families would have enough room, enough time and enough salmon to fish. The knowledge from experienced Saanich Indigenous people who were managing the fishery was preserved and passed on in a more localized manner in the past when more families were fishing together in the inlet or camping at the river in groups.

Today, Saanich Tribal Fisheries is a Saanich First Nation council that addresses local fisheries matters. Saanich First Nation families do not fish the inlet or river as often as they used to. The smokehouse that was formerly used by Saanich people at Goldstream River was torn down around 1992 (Morris Sr. I. pers.comm. 2002). Prior to that time, Saanich fishers would invite visitors to Goldstream Park to their smokehouse to demonstrate their traditional salmon fishing, cutting and smoking methods. It is undetermined as to whether a new smokehouse will be built in its place. In 2005, however the Minister of Fisheries and Oceans Canada announced the need to focus on First Nations' fishing opportunities and Goldstream River was identified as one of the places where this should occur (DFO 2005). Perhaps plans for increasing First Nations' fishing opportunities at Goldstream River could include reconstruction of a new smokehouse for Saanich fishers.

Given that Saanich Indigenous peoples have relied on salmon of Saanich Inlet and Goldstream River since time immemorial, probably thousands of years, and given that they have rights to fish these salmon as defined in the Canadian Constitution, they need to be even more closely involved as participants in monitoring, management and decision-making protocols affecting the salmon and their fisheries. This is important because the salmon and the fishery are integral components of the cultural fabric of the Saanich Peoples lands, waters, resources and homeplace (Figure 5).

Figure 5: Mount Finlayson, Goldstream Park.
Photo: Roxanne Paul, October 2006

Conclusion

Some traditional communities, such as those of the Saanich First Nations, retain a wealth of information about traditional ecological knowledge pertaining to local wildlife such as their local salmon resources as well as protocols for preserving the ecosystem integrity of the land, water and resources. In the case of the Saanich People, this knowledge includes deep-rooted understandings about local creeks as well as freshwater and marine species which salmon are dependent upon to survive. The knowledge and awareness of these subtle yet eternally persevering interdependent relationships and their role in the web of life is directly linked to the Saanich Peoples' physical and spiritual attachment to their home place and the perpetuation of traditional ecological knowledge acquired from their elders, relatives, community members and peers.

What would happen if the global community thought holistically about the interconnectedness of local ecosystems that support their food resources? Would our local resource harvesting and consumption activities change? Could we become more physically and spiritually connected with the food we consume and the communities where these foods are harvested? I believe this book compilation about biocultural diversity can open up a good forum for further developing and harnessing consciousness and conscientiousness about our place and role within local food systems.

Endnotes

[1]Keystone species is defined by Power et al. (1996:609) as "a species whose effect is large, and disproportionately large relative to its abundance."
[2]Though conservation efforts are in place, WCVI coho and chinook stocks are not listed as species at risk under Canada's Species at Risk Act (Baxter 2000; DFO 1999; Environment Canada 2005).

³The Saanich Tribal Fisheries fishing vessel(s) begin harvesting the fall chum salmon run in Saanich Inlet between September and November solely in years when 15,000 chum (the number considered by Fisheries and Oceans Canada to be the 'carrying capacity' for the chum population at Goldstream River) 'escapements' return to the river.

⁴Salmon tend to gather near estuaries to feed during spring tides when river and ocean floor sediment is disrupted due to upwelling caused by the moon's magnetic pull on the water when flood and ebb tidal currents are strong. This phenomenon occurs for two to three days before and after the full (or new) moon because the disruption of gravel and sediment uncovers and transports smaller organisms out of the river to the estuary and nearshore saltwater environments where salmon swim in to feed upon them (DFO 1996; Stowe 1996).

Acknowledgements

This research study was made possible by the Saanich fishers who gave selflessly and openly of their time, knowledge and experience. The Saanich People I would like to thank are: Joe Bartleman, Earl Claxton Sr., Emmanuel Cooper, John Elliott Sr., Ivan Morris, Sandy Morris and Simon Smith. Thank you to Dr. Nancy J. Turner, Dr. Michael C.R. Edgell, Dr. Tom E. Reimchen, and Dr. Rosemary Ommer, my academic supervisory committee from the University of Victoria who were all very attentive to my interests and supportive of my work from the outset and throughout my graduate studies. I am also very appreciative of Dr. Ranjay K. Singh for his role in recruiting for, editing and compiling this important collective work about traditional foods, local knowledge and biodiversity conservation. My sincere thanks to the Department of Geography and the School of Environmental Studies at the University of Victoria; the Sierra Club of British Columbia; the Edward Bassett Family Foundation; and the Lorene Kennedy Environmental Studies bursaries for the funds and other resources provided that helped to sponsor this study. Thank you also to my mother, father, brother-in-law, grandmother, and my late grandfather for their unconditional support and a special thanks to my sister for her undying encouragement, devotion and guidance during my academic pursuits at the University of Victoria and beyond.

References

Alcorn JB (1993) Indigenous Peoples and conservation. *Conservation Biology*, 7(2):424-427
BC Archives (1873) Saanich settlers' position regarding East Saanich Road. GR-0868
Berkes F (1999) Sacred ecology: Traditional ecological knowledge and resource management. Philadelphia: Taylor and Francis
Berkes F, Henley T (1997) Co-management and traditional knowledge: Threat or opportunity? Policy Options, March: 29-35

Berkes F, Folke C, Gadgil M (1995) Traditional ecological knowledge, biodiversity, resilience and sustainability in: Biodiversity Conservation: Problems and Policies. *In:* Perrigs, C.A. et al. (ed). Dordrecht (The Netherlands): Kluwer Academic. pps. 281-289

Berkes F, George P, Preston RJ (1991) Co-management: The evolution in theory and practice of the joint administration of living resources. *Alternatives Journal*, 18(2):12-18

Bocking R, Firth R, Ferguson J, Yazvenko L (1998) LGL Limited Environmental Research Associates. 1998. Goldstream River overview and level 1 fish habitat assessment and rehabilitation opportunities (including riparian). Sidney: Te'mexw Treaty Association, and the Ministry of Environment, Lands and Parks

Boyatzis RE (1998) Transforming qualitative information: Thematic analysis and code development. Thousand Oaks: Sage Publications

British Columbia Department of Lands, Works. Originals, 1871-1883, 44 cm. Letters inward to the Chief Commissioner of Lands and Works. Province of BC. 2001. [online] URL: www.search.bcarchives.gov.bc.ca/sn1A91DF1/bsearch/TextualRecords#form Accessed: 5 April 2003

Brodnig G, Mayer-Schoenberger V (2000) Bridging the gap: The role of spatial information technologies in the integration of traditional environmental knowledge and western science. The Electronic J Inform Syst in Develo Countries 1(1):1-15. [online] URL: http://www.is.cityu.edu.hk/research/ejisdc/vol1.htm, Accessed: 24 January 2005

Claxton E Sr, Elliott Sr J (1994) Reef net technology of the Saltwater people. Brentwood ay: Saanich Indian School Board

Confederacy of Nations (2004) Resolution No. 9. Nuu-chah-nulth fisheries litigation. Saskatoon, Saskatchewan. Assembly of First Nations. May 18, 19 and 20, 2004

Crockford S (1994) New archaeological and ethnographic evidence of an extinct fishery for giant blue fin tuna on the Pacific North Coast of North America. Fish Exploitation in the Past. Proceedings of the 7th meeting of the ICAZ Fish Remains Working Group. *In:* Van Neer W (ed). Annales du Muse Royal de l'Afrique Central.Sciences Zoologiques. No. 274. Tervuren

Department of Indian Affairs and Northern Development (DIAND) (2006) First Nation profiles. Registered population of Malahat First Nation as of May 2006. [Online]. URL: http://pse2-esd2.aincinac.gc.ca/FNProfiles/FNProfiles_ GeneralInformation. asp? BAND_NUMBER=647&BAND_NAME= Malahat+First+Nation. Accessed: 27 June 2006

Dwyer PD (1994) Modern conservation and indigenous peoples: in search of wisdom. *Pacific Conservation Biology*, 1:91-97

Elliott D Sr (1990) Janet Poth ed. Saltwater People: A resource book for the Saanich Native Studies program. School District # 63 (Saanich)

First Nation Panel on Fisheries (2004) Our place at the table: First Nations in the BC Fishery. Canada. Budget Printing

Fisheries and Oceans Canada (1932-2005) DFO BC 16 Reports. Annual reports of salmon stream and spawning grounds for Goldstream River flowing into Finlayson Arm/Saanich Inlet. Victoria: Fisheries and Oceans Canada

Fisheries and Oceans Canada (2001b) Excess Salmon to Spawning Requirements (ESSR) License 2001-2009 for Saanich Tribal Fisheries-Area 19. Victoria: Fisheries and Canada Pacific Region

Fisheries and Oceans Canada (2005a) First Nations-North Coast-Commercial salmon. [online] URL: www.pac.dfo-mpo.gc.ca/northcoast/commercl/default.htm Accessed: 24 February 2005

Fisheries and Oceans Canada (2005b) Personal communication with Gerry Kelly. South Coast Area Fisheries Manager. Telephone interview Re: ESSR chum fishery and Goldstream Salmon. 29 March 2005. Pacific Biological Station. Nanaimo, BC

Flett L, Bill L, Crozier J, Surrendi D (1996) A report of wisdom synthesized from the traditional knowledge component studies. Northern River Basins Study Synthesis Report No. 12.

Freire J, Garcia-Allut A (1999) Integration of fisher's ecological knowledge in fisheries biology and management. A proposal for the case of the artisanal coastal fisheries of Galicia (NW Spain). International Council for the Exploration of the Sea (ICES). Session S: Evaluation of complete fisheries systems. Economic, social and ecological analyses. C.M.1999/S:07.17 pp.

Gadgil M, Berkes F, Folke C (1993) Indigenous knowledge for biodiversity conservation. *AMBIO*, 22(2):151-156

Garvin T, Nelson S, Ellehoj E, Redmond B (2001) A guide to conducting a traditional knowledge and land use study. Edmonton: Natural Resources Canada. Canadian Forest Service

George EM (2003) Living on the edge: Nuu-Chah_Nulth history from an Ahousaht chief's perspective. Winlaw: Sono Nis Press

Gitxsan Chief's Office (2006) Treaty talks. [online] URL: www.gitxsan.com/html/treaty.htm. Acccessed: 22 Feb 06

Gitxsan Wet'suwet'en Watershed Authority (1998) Year of crisis for Pacific fisheries: Is there a future for salmon and for commercial fishing? BC Aboriginal Fisheries Commission. *British Columbia Aboriginal Fisheries Journal*, 4(2):4

Gottesfeld L, Johnson M (1994) Conservation, territory, and traditional beliefs: An analysis of Gitksan and Wet'suwet'en subsistence, Northwest British Columbia, Canada. *Human Ecology*, 22(4):443-464

Green RN, Klinka K (1994) A field guide to site identification and interpretation for the Vancouver Forest Region. BC Ministry of Forests. Victoria. BC land management handbook No. 28

Haggan N, Pitcher T, Rashid S (2003) Back to the future in the Strait of Georgia. University of British Columbia Fisheries Centre. Proceedings from the Georgia Basin/Puget Sound Research Conference. Puget Sound Action Team. Westin Bayshore, Vancouver, BC

Healy C (1993) The significance and application of TEK: Traditional ecological knowledge (wisdom for sustainable development). Centre for Resource and Environmental Studies. Canberra: Australian National University. Pps. 21-27

Hilborn R, Walters CJ, Ludwig D (1995) Sustainable exploitation of renewable resources. *Annual Review of Ecology and Systematics*. 26:45-67

Hilborn RC, Walters CJ (1992) Quantitative fisheries stock assessment: Choice, dynamics and uncertainty. London: Chapman and Hall

Huntington HP (2000) Using traditional ecological knowledge in science: Methods and applications. *Ecological Application*, 10(5):1270-1274

Jenness D (1938) The Saanich Indians of Vancouver Island. Unpublished Manuscript, available in type in Special Collections, UBC and Provincial Archives, Victoria

Johannes RE (1993) Traditional ecological knowledge of fishers and marine hunters in: Traditional ecological knowledge (Wisdom for Sustainable Development). Centre for Resource and Environmental Studies. Canberra: Australian National University. pp. 144-146

Kimmerer RW (2000) Native knowledge for native ecosystems. *Journal Forestry*, August:4-9

Mackinson S (2001) Integrating local and scientific knowledge: An example in fisheries management. *Journal of Environmental Management*, 27(4):533-545

Mackinson S, Nottestad L (1998) Combining local and scientific knowledge. *Revise of Fish Biology and Fisheries*, 8:481-490

Madill D (1981) British Columbia Indian Treaties in Historical Perspective Ottawa: Treaties and Historical Research Centre. Department of Indian Affairs and Northern Development

Mos L, Janel J, Cullon D, Montour L, Alleyne C, Ross PS (2004) The importance of marine foods to a near-urban First Nation community in coastal British Columbia, Canada: Toward a risk-benefit assessment. *Journal of Toxicology and Environmental Health Part A*, 67: 791-808.

Newell D (1993) Tangled webs of history: Indians and the law in Canada's Pacific coast fisheries. University of Toronto Press, Toronto, Ontario, Canada

Notzke C (1995) A New Perspective in Aboriginal Natural Resource Management: Co-management. *Geoforum Journal*, 26(2): 187-209

Paul R (2006) Counting on their migration home: an examination of monitoring protocols and Saanich First Nations' perspectives of coho (*Oncorhynchus kisutch*), chinook (*O. Tshawytscha*) and chum (*O. keta*) Pacific salmon at Goldstream River and Saanich Inlet, Southern Vancouver Island, British Columbia. Master's Thesis. University of Victoria

Pinkerton E (1989) Co-operative management of local fisheries: New directions for improved management and community development. Vancouver: University of British Columbia Press

Pinkerton E, Weinstein M (1995) Fisheries that work: Sustainability through community-based management. Vancouver: The David Suzuki Foundation

Pojar J, MacKinnon A (1994) Plants of Coastal British Columbia including Washington, Oregon and Alaska. Vancouver. BC Ministry of Forests & Lone Pine Publishing

Pojar J, Flynn S, Cadrin C (2004) Douglas-Fir/Dull Oregon Grape Plant Community Information. Accounts and measures for managing identified wildlife. Accounts V. Province of BC

Power ME, Tilman D, Estes JA, Menge BA, Bond WJ, Mills LS, Daily G, Castilla JC, Lubchenco J, Paine RT (1996) Challenges in the quest for keystones. *Bioscience*, 46:609-620

Province of British Columbia (BC) 2001 References. Douglas Treaty payments. [online] URL: www.gov.bc.ca/tno/history/payment.Htm Accessed: 5 April 2003. Government of British Columbia Treaty Negotiations Office

Robinson C (2001) Working towards regional agreements: Recent developments in co- operartive resource management in Canada's British Columbia. *Australian Geographic Studies*, 39(2):183-197

Schreiber DK (2001) Co-Management without involvement: the plight of fishing Communities. *Fish and Fisheries*, 4(2):376-384

Silverman D (1993) Interpreting qualitative data: methods for analyzing talk, text and interaction. London: Sage Publishing

Simonsen BO, Davis A, Haggarty J (1995) Saanich inlet study report on First Nations consultation. BC Ministry of Environment, Lands and Parks. June 1995. [online] URL: http://wlapwww.gov.bc.ca/wat/wq/saanich/sisrofnc.html Accessed: 29 October, 2003

Stowe K (1996) Exploring ocean science. 2nd Ed. New York: John Wiley and Sons Inc

Taylor G (2003) Perspective of the commercial salmon fishery in: Solutions for salmon conservation-Proceedings from the World Summit on Salmon. 10-13 June 2003. Vancouver : Simon Fraser University

Turner NJ (1997) Traditional ecological knowledge in: The rainforests of home: Profile of a North American bioregion. Washington DC: Island Press. Pp. 275-298

Turner NJ, Ignace MB, Ignace R (2000) Traditional ecological knowledge and wisdom of Aboriginal Peoples in British Columbia. *Journal of Applied Ecology*, 10 (5):1275-1287

Union of BC Indian Chiefs (2003) The Douglas Reserve policy. Vancouver: Union of BC Indian Chiefs (UBCIC). [online] URL: www.ubcic.bc.ca/douglas.htm Accessed: 5 April 2003.

United Nations Environment Programme (UNEP) (1997) Ad hoc open-ended inter sessional working group on article 8(j) and related provisions on the United Nations Convention on Biological Diversity. 5th meeting. Oct. 15-19 2007. Item 5 of the provisional agenda. Montreal, Canada

Appendix A: Glossary of specialized terms used in this chapter

1. Terms Relating to Salmon

Carrying Capacity: The maximal population size of a given species that an area can support without reducing its ability to support the same species in the future.

Escapement: The quantity of sexually mature adult salmon (typically measured by number or biomass) that successfully passes through a fishery to reach the spawning grounds.

Enhancement: The application of biological and technical knowledge and capabilities to increase the productivity of fish stocks. It may be achieved by altering habitat attributes (e.g., habitat restoration) or by using fish culture techniques (e.g., hatcheries, production spawning channels).

Hatchery Salmon: A salmon whose parents were born in a hatchery or a salmon that has spent a portion of its life cycle in an artificial environment; Any salmon incubated or reared under artificial conditions for a part of its life. This definition does not distinguish between a salmon one generation removed from the wild and a salmon whose parents were highly domesticated products of the hatchery (Figure 6).

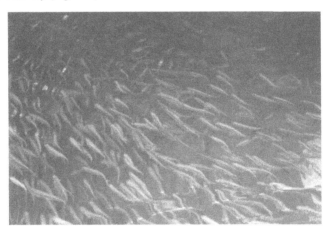

Figure 6: Hatchery Raised Coho Fry, Goldstream River Hatchery. Photo: Roxanne Paul, October 2006.

Natural Salmon: Any salmon produced in the natural environment as a result of natural reproduction. A natural salmon could be wild (see definition below) or it could be the progeny of hatchery parents that spawned in the natural environment. It is impossible to distinguish a natural and wild salmon by field observation alone.

Population: Group of interbreeding salmon that is sufficiently isolated from other populations so that there will be persistent adaptations to the local habitat.

Stock: A group of salmon spawning in a specific stream at a specific season, which do not interbreed to a substantial degree with any other group of salmon. Several stocks linked by a low level of straying may constitute a metapopulation.

Wild Salmon: A salmon whose parents have spawned in the wild and has spent its life in the natural environment; any naturally, spawned salmon belonging to an indigenous population.

Indigenous means a population whose lineage can be traced back to 1800 in the same geographical area or that resulted from natural colonization from another indigenous population. This is a difficult definition to apply since we do not have continuous records of salmon populations going back to the 1800s. Its application is more appropriate in defining what is not wild rather than what is wild. For example, a hatchery salmon population introduced recently by humans to a creek would not be considered wild. Where there is doubt, a population should be considered wild unless there is clear proof that it is not.

Appendix B: Common, scientific and saanich language names for BC Salmon species (Baxter 2000; Claxton and Elliott 1994).

Common name	Vernacular names	Scientific name	Saanich language name*
Coho	Bluebacks, Silver	*Oncorhynchus kisutch*	fÁ,WEN
Chinook	Spring, King, Tyee	*O. tshawytscha*	STOKI
Chum	Dog, Calico	*O. keta*	QOLEW
Sockeye	Sukkai, Red	*O. nerka*	fEKI
Pink	Humpback	*O. gorbuscha*	HENNEN

*Saanich language salmon species names sourced from: Claxton, Earl Sr. and Elliott, J. Sr. 1994.
Reef Net Technology of the Saltwater People. Brentwood Bay: Saanich Indian School Board.

Appendix C: Summary of habitat descriptions of reaches 1 to 3 Goldstream river, Salmon escapement enumeration survey area (modified from Bocking et al. 1999).

Habitat Descriptors	Reach 1	Reach 2	Reach 3	Total Reaches 1-3
Total Area (Reach length x Avg. bankfull width)	16,275 m^2	20,828 m^2	9,261 m^2	Range = 46,364 – 45,846 m^2 = 518 m^2; Mean = 46,105 m^2
Reach Length	930 m	1,270 m	630 m	2,830 m
Gradient	1.2%	1.4%	1.7%	Average = 1.4%
Salmon Habitat Length	735 m	1,327 m	564 m	2,626 m
Habitat Types	71.8% glide, 23.7% riffle, 4.5% pool	38.6 % glide, 54.8 % riffle, 6.6% pool	36.5% glide, 52.2% riffle,	Average = 49 % glide, 43.6 % riffle 6.4% pool, 1% cascade
Average Bankfull Width Area (Habitat length x Bankfull width)	17.5 m	16.4 m	14.7 m	Average = 16.2 m
Average Wetted Width Salmon Rearing Habitat	12.7 m 12,255 m^2	11.7 m 21,763 m^2	13.4 m 9,261 m^2	Average = 12.6 m Range = 45,846 m^2 - 43,279 m^2 = 2,567 m^2; Mean = 44,563 m^2
Average Water Depth	0.5 m	0.7 m	0.6 m	Average = 0.6 m
Volume of Salmon Habitat Waters (Salmon Habitat Length x Avg Bankfull Width x Avg. Water Depth	6,431 m^3 = 6,431,000 L	15,234 m^3 = 15,234,000 L	4,975 m^3 = 4,975,000 L	Range = 26,640 – 25,525 m^3 = 1,115 m^3; Mean = 26,083 m^3 = 26,083,000 L
Overall Spawning Gravel Quality	Good -9% low, 64% medium, 27% high quality	Fair – 31% low, 59% medium, 10% high quality	Fair – 38 % low, 56 % medium, 6 % high quality	Fair/Good- 26 % low, 60% medium, 14 % high quality

Appendix D: Saanich First Nation fisher interviewees.

Bartleman, Joe. Tsartlip Elder. Goldstream River and Saanich Inlet fisher. Personal Interview. Standardized Survey Instrument. *Saanich Tribal Fishery Interview Questionnaire Transcript: Historical and Current Fishing Methods and Practices (at Goldstream River and Saanich Inlet)*. 10 December, 2002. 848 Stelly's Cross-Road, Brentwood Bay, Saanich, BC.

Dr. Claxton, Earl, Sr.. 2002. Tsawout. Cultural Historian and Researcher, Saanich Indian School Board. Goldstream River and Saanich Inlet fisher. Personal Interview. Standardized Survey Instrument. *Saanich Tribal Fishery Interview Questionnaire Transcript: Historical and Current Fishing Methods and Practices (at Goldstream River and Saanich Inlet)*. 22 July, 2002. Language Centre – 7449 West Saanich Road, Brentwood Bay, Saanich, BC.

Cooper, Emmanuel. 2002. Tsartlip Elder. Goldstream River and Saanich Inlet fisher. Personal Interview. Standardized Survey Instrument. *Saanich Tribal Fishery Interview Questionnaire Transcript: Historical and Current Fishing Methods and Practices (at Goldstream River and Saanich Inlet)*. 29 August, 2002. 7543 West Saanich Road, Brentwood Bay, Saanich BC

Dr. Elliott, John. 2002. Tsartlip. Linguist and Cultural Historian, Saanich Indian School Board. Goldstream River and Saanich Inlet fisher. Personal Interview. Standardized Survey Instrument. *Saanich Tribal Fishery Interview Questionnaire Transcript: Historical and Current Fishing Methods and Practices (at Goldstream River and Saanich Inlet)*. 18 Sept., 2002. Language Centre. 7449 West, Saanich Road, Brentwood Bay, Saanich, BC.

Morris, Ivan Sr. 2002. Tsartlip Elder. Goldsteam River and Saanich Inlet fisher. Personal Interview. *Standardized Survey Instrument. Saanich Tribal Fishery Interview Questionnaire Transcript: Historical and Current Fishing Methods and Practices (at Goldstream River and Saanich Inlet)*. 7 Aug., 2002. 45 Tsartlip Drive, Brentwood Bay, Saanich, BC.

Morris, Sandy. 2002. Tsartlip Elder. Goldstream River and Saanich Inlet fisher.Personal Interview. Standardized Survey Instrument. *Saanich Tribal Fishery Interview Questionnaire Transcript: Historical and Current Fishing Methods and Practices (at Goldstream River and Saanich Inlet)*. 7 August 2002. Church Road, Brentwood Bay, Saanich.

Smith, Simon. 2002. Tsartlip Elected Band Chief. Goldstream River and Saanich Inlet fisher. Personal Interview. Standardized Survey Instrument. *Saanich Tribal Fishery Interview Questionnaire Transcript: Historical and Current Fishing Methods and Practices (at Goldstream River and Saanich Inlet)*. 30 October, 2002. Tsartlip Band Office. Stelly's Cross-Road, Brentwood Bay, Saanich, BC.

Chapter – 12

Tsampa of Ladakh: Adaptation of a Traditional Food at Higher Altitude and Emergent Changes

Konchok Targais, Dorjey Angchok*, Tsering Stobdan, R.B. Srivastava and Ranjay K. Singh

Abstract

Ladakh, comprising of Leh and Kargil district is truly described as a high altitude cold-arid desert region of J&K state, India, where local people grow barley *(Hordeum vulgare)* and prepare various food products out of it. The roasted barley flour, locally known as *tsampa* is one of them, which goes as ingredients in a variety of local dishes and beverages. It is eaten either in raw or in cooked form and the traditional method of *tsampa* preparation is through the *Ran-tak* (traditional water mill), for which efforts have been made to describe it in detail. Like other parts of the world, this remote area in the trans-Himalaya has also witnessed exchange with the outside industrial world which has brought with it many changes (both desirable and undesirable) and the traditional foods were also not left behind. One major cause for decline in the traditional *tsampa* preparation has been the availability of cheap subsidized produce from the lowland India, on the other hand the increase in information exchange with the outside world has also helped local people to realize the need and importance to conserve their own locally grown crops and integrate into the mainstream fooding system. However, *tsampa* prepared

---
*achuk_iari@rediffmail.com

using the traditional method is still being preferred by the local inhabitants and there is need for preservation of the traditional method of *tsampa* preparation and variety of foods prepared from it. Therefore this chapter is an effort to document in detail the traditional method of *tsampa* preparation through *Ran-tak*, as practiced in Leh district of Ladakh region.

Keywords: Barley, traditional food, Himalaya, Ladakh, *tsampa*, *Ran-tak*.

Introduction

Globally, the native and indigenous communities living in fragile ecosystems and far flung areas have developed location specific ecological knowledge and practices to sustain their life support system. Similarly, for centuries, Ladakh enjoyed a stable and self-reliant agricultural economy based on growing barley, wheat and peas and keeping livestock which are compatible with social-ecological systems. The growing season is only a few months long every year, similar to the northern countries of the world. Animals are scarce and water is in short supply. The Ladakhis developed a small-scale farming system adapted to this unique environment. The land is irrigated by a system of channels which funnel water from the ice and snow of the mountains. The principal crops are barley and wheat. Rice was previously a luxury in the Ladakhi diet, but, subsidized by the government, has now become a cheap staple (Rizvi 1996). Ladakh region in the trans-Himalaya comprises of two districts viz. Leh and Kargil. District Leh (study site) is situated roughly between 32 to 36 degree north latitude and 75 to 80 degree East longitude and altitude ranging from 2300 mtrs to 5000 mtrs above sea level. The district is bounded by Pakistan in the west and China in the north and eastern part and Himachal Pradesh in the south east (Figure 1). Topographically, the whole of the district is mountainous with three parallel ranges of the Himalayas, the Zanskar, the Ladakh and the Karakoram range. It lies on the rain shadow side of the Himalayan, where dry monsoon winds reaches Leh after being robbed of its moisture in plains and the Himalayas mountain the district combines the condition of both arctic and desert climate. Therefore Ladakh is often called 'cold desert', where intensive sunlight, high evaporation rate, strong winds and fluctuating temperature characterize the general climate (Angchok et al. 2009). It is generally said that a man sitting in the sun with his feet in the shade can have sunstroke and frostbite at the same time. Because of high mountains all round and heavy snowfall during winter, the area remains landlocked to the outside world for nearly five months in a year.

Under these harsh conditions, one of the major reasons behind human habitation is the ingenuity of local people, who has devised new and sustainable way of living. And, one major product of this ingenuity is the traditional foods, evolved over time and established in the fooding system of Ladakhi people. And, among the various traditional foods of Ladakh, *tsampa* (and other foods where *tsampa* is the major ingredients) holds a special place for. They are easy to make, simple, least fuel consuming, ingredients available locally and preferred by the local inhabitants, which is a true representative of the region. William Moorcroft, a veterinary surgeon traveling in the 1820's, was the first Englishman to give a detailed account of Ladakh (Reifenberg 1998) where he describes *tsampa* as one of the major food of Ladakh.

Figure 1: Ladakh on map.

Staple diet of the region is barley, or grim, which is roasted and ground for use as flour, commonly known as *tsampa* or *namphey* (in whole of this text it referred as *tsampa*). The area sown under barley in Leh district has been stagnant and there has not been a significant variation for many decades, which in the year 2010-11 was 4421 hectares (Figure 2). On the other hand there is an increase in the population of Leh district with a decadal growth of +25.48% (for detail

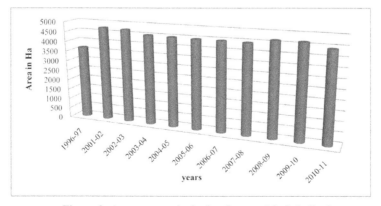

Figure 2: Area sown under barley (hectares) in Leh district.

see Table 1). Therefore, it is argued that despite barley (more specifically *tsampa*) being an important ingredient in *Ladakhi* food and its importance in socio-cultural life of the people, there is a sharp decline of its use. With the advent of power driven mechanical method of milling, the traditional method of *tsampa* preparation through *Ran-tak* (locally owned water mill) system is also facing the peril of decadence. However, *tsampa* prepared using the traditional method is still being preferred due to its longer shelf life and palatability. Therefore, it is essential to understand and document this ancient traditional method of *tsampa* preparation and the related traditional foods and beverages prevalent in this part of the world, and the present study was undertaken in this direction.

Table 1: Statistics of Leh district.

Parameters	Description	Reference year
Population	68380	1981
	117232	2001
	147107	2011 (provisional)
Decadal Growth of Population (%)	+25.48	2001-11
Cultivators	14305 (65.48%)	1971
	17415 (58.49%)	1981
	22041 (37.92%)	2001

Source: Census department, Leh.

Objectives

Considering the above statements, it was endeavored to take up the following objectives for the present study:

1. To understand and document the traditional method of *tsampa* preparation,
2. To document various traditional foods and beverages related to *tsampa*, and
3. To understand the status and present scenario of *tsampa* as a traditional food.

Research Methodology

The present study is designed to explore and understand the prevailing traditional system of *tsampa* preparation in the trans-Himalayan Leh district of India, where average altitude of human habitation is over 3000 meters amsl and the mean maximum and minimum temperature is 18.9°C and -5.8°C respectively observed during the last decade. Average annual precipitation is less than 200 mm of which more than 70% is in the form of snowfall (as per DIHAR, Leh). In this tras-Himalayan region, the survey was conducted in two adjoining villages to Leh town (Gangles and Skara village) and two villages outskirt (aprox 100 kms) to Leh town (Nurla and Saspol village). These four villages were purposively selected to make the study more inclusive by integrating urbanization factor and prevalence of *tsampa* preparation through *Ran-tak* (traditional water mill) in the selected villages. From each of the selected villages, ten households were selected randomly among those who prepare *tsampa* through the traditional method.

Before the survey, a pilot study was undertaken to gain background experience about the local communities and the resource available, which further helped the researchers to establish rapport with the local people. Based on the information collected during the pilot study, and knowledge gained from literature reviews, *tsampa* as a traditional food was identified for the present study. The selection was based on relevance to the whole community and the input given by local people, and various literatures arguing the importance of such traditional food, in high altitude area of Ladakh.

The study was carried out in an interactive mode involving local people at every step of the research. Locals were encouraged and help share their knowledge about the *tsampa* preparation through *ran-tak* and the prevalent food diversity related to *tsampa*. The detailed information was collected through a combination of methods comprising community observation (participatory and distant observation), personal interviews (PI) and focused group discussions (FGD). For the PI, whole household was taken as a unit instead of single individual. The reason being, at many occasions the key informant was aided by his/her other family members in providing information. Moreover, it acted like a semi-group discussion in itself where members of different age groups, sex and orientation contributed to make the information more comprehensive. One FGD in each village was organized, apart from the personal interview. It helped to cross check the information gathered during personal interviews. Typically the interview methodology focused on informal and open dialogues, group discussions as well as semi structured set of questions were used. All the interviews included an initial visit to the village and a series of follow up meetings. Wherever possible the information was noted down, and in many occasions where it was not feasible to note down, the narratives were audio taped with their permission. At each step of data collection, prior informed consents (PIC) from each respondents were obtained and their consent were taken for publication of recorded information. Furthermore, there has been request from many respondents to share back with them the pooled published information, which will be taken care of after the publication.

Results and Discussion

The findings of the study is broadly categorized into three groups viz. (i) traditional method of *tsampa* preparation through the use of *ran-tak*, (ii) prevalent *tsampa* related traditional foods and (iii) changes in this traditional system.

Tsampa preparation

A step wise (Figure 3) traditional method of *tsampa* preparation is as follows:

a. *Khanng* (Washing/Cleaning): A horizontal water channel (6-7 ft long and 2 ft wide) with a sharp 2-4 feet fall at the outlet is selected. A goat or yak's fur woven sheet locally known as *chali* is spread uniformly in the water channel on which barley grain is poured in small quantity (4-5 Kg) depending upon the width of the channel. Flow of water is maintained in such a manner that it initially carries dust and lighter particles/wastes leaving behind the grains and other heavier particles. The grains are stirred slowly so that lighter particles are washed away and collected at the end of the water channel. Once the grains are free from the lighter debris, it is stirred more forcefully so that the force of water flow carries the grain leaving behind the heavier impurities such as coarse sand, pebbles etc.. There is no fixed rule for maintaining flow of water and the force with which grains are stirred. The washed grains are collected at the end of the water channel with a sharp water fall outlet in a Yak or goat's fur sack called *Khulu*. The sharp water fall at the end of water channel makes collection of the washed grain simple and easy. The whole process demands high dexterity and experience due to which only the older people are engaged in the process.

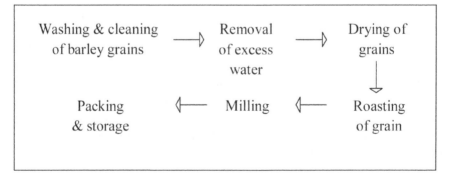

Figure 3: Flow chart of traditional method of *tsampa* preparation.

b. Removal of excess water: The cleaned grain is kept in the *Khulu* for few hours till the excess water gets removed. Delay in removing the grains from *Khulu* causes sprouting of grains and results in poor quality *tsampa*.

c. Drying: After removing the excess water barley grains are spread uniformly in shade or under open sun for 1-2 days depending upon the ambient temperature. The grains are occasionally turned upside down with a wooden spreader known as *Pan-kha* till an optimum moisture level is reached.

d. Roasting: Roasting is traditionally done outdoor on a circular iron pan known as *Lha-na*, which is about 28 inch in diameter. A wooden triangle shaped spreader with a long handle, locally known as *Yug-du* is used for mixing and spreading of grains on the hot *Lha-na*. The *Lha-na* is kept heated on a *Chula* (fire or stove) approximately 1 foot high and flame is kept low to ensure uniform roasting of grains. Twigs of willow, poplar or seabuckthorn are generally used as firewood. To roast a jug full of washed grain, it takes approximately 45 to 60 seconds. The roasted grain is removed from hot *Lha-na* when almost 50 percent of grains get popped. If the roasted grains are meant for direct consumption as snacks locally called *Yo-za*, then it is roasted a little longer till 80-90 percent grains get popped. After roasting, the lot is again cleaned off unwanted particles by a combination of sieving, winnowing and hand picking of unwanted materials. This process again demands experience due to which skilled individuals (mostly ladies) are engaged.

e. Milling: Traditionally, roasted barley is grind in self operated water mill called *Ran-tak* (Figure 4 & 5a,b). The term *Ran-tak* comprises of two words, '*Rang*' meaning 'self' and '*thag*' stands for 'grind' as it runs on sheer gravitational force of water without much human intervention during the milling process. During on-spot obse7rvation of the milling processes, it was observed that power driven mechanical mill runs at much faster speed i.e 3000 to 3500 revolution

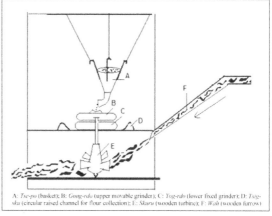

A: *Tse-po* (basket); B: *Gong-rdo* (upper movable grinder); C: *Yog-rdo* (lower fixed grinder); D: *Tsug-sku* (circular raised channel for flour collection); E: *Skuru* (wooden turbine); F: *Wah* (wooden furrow)

Figure 4: *Ran-tak* – traditional water mill (schematic diagram not to scale).

per minute (rpm) as compare to 150-180 rpm in *Ran-tak* resulting in heating of the product, which may be the reason for difference in quality. Temperature difference of 15°C was observed in the finished product obtained by traditional milling process as that of the modern day process. Various components of a *Ran-tak* are described below:

Figure 5: Inside **(a)** and outside **(b)** view: *Ran-tak* (traditional water mill).
Photo: Dorjey Angchok

Tse-po (basket): The basket locally known as *Tse-po* is made up of a hard straw with capacity varying from 25 to 50 kilogram. The basket is of conical shape with a small opening at the bottom. The opening is connected with a wooden screw-regulated furrow known as *rHog*. The downward angled wooden outlet i.e., '*rhog*' is connected to a wooden j-shaped wooden piece called *thak-tharak* made of apricot or walnut tree, which vibrates when it comes in contact with the rotating grinder. This rhythmic agitation of this *thak-tharak* regulates the dropping of grains into the grinder which lies beneath.

Grinder

The grinders comprises of two circular stone pieces approximately 24 to 30 inches in diameter. The upper rotating part is called *Gong-rdo* while the lower fixed base is called the *Yog-rdo* which has a small hole (*Za-kung*) in the center. Generally bluish color stone are preferred for making grinder since it is considered hard enough for this purpose. The surface of the rotating and fixed stone which comes in contact for grinding are not completely smooth. The rough surface allows proper grinding of the roasted grain. The surface of the upper grinder which comes in contact with the lower fixed stone is slightly concave while that of the lower grinder coming in contact with the upper stone is convex. Hence the actual grinding of the roasted grains takes place at 2-4 inches periphery of the two circular stones. The grinded material gets collected in a circular channel called *Tsig-sku* around the grinder and a brush made up of well-massaged sheep skin with wool called as *phyab-yak* is used for wiping and collecting the flour.

Wooden turbine

The wooden turbine is called *Skuru* and it comprises of a bell shaped central axis (3-3.5 feet height) tapering towards the upper end. A 1-1.5 ft flat iron piece is inserted in the upper end of the turbine; the other end is attached to *Gong-rdo*, the upper movable stone grinder. Depending upon the diameter of the axis, 9 to 11 fins are attached to the central rotar keeping 5-6 inches distance between each fins. Length of each fin is about 1 foot. The central rotating axis is generally made op of Juniper wood while the fins are carved from willow or poplar wood. The base of the turbine is fitted with a cone shaped white stone (marble) which rotates on another flat marble stone (*pakor*) fitted in apricot/walnut log. Fine tuning height of the turbine base is done to get different size of flour powder from fine to coarse.

Water head

It is a wooden furrow called as *Wah* which gushed the water into the turbine from a height of 7-8 ft head at an angle of 45-50p , which actually rotates the turbine. The furrow tapper toward the lower end which actually aids in maximizing force of the water before it hits the turbine. The slightly carved periphery of the fins helps in direct hit of water flow at a point. The furrow is 1 foot width at the head and 11- 12 feet long. Approximate 500 litres of water flow per minute from the furrow to rotate the turbine. There is provision for ventilator inside the mill chamber which is used to view and check the turbine and for making fine adjustment. During peak winters, controlled fire is burn near the ventilator to avoid ice formation on turbine fins.

Housing

The *Ran-tak* is housed in a two storey stone built structure. The lower structure house the turbine while the upper storey houses the grinder and other accessories for grinding and collection of the finished products. The ceiling of the lower house structure is made up of oleaster or mulberry logs due to its resistance to water and durability.

Packing

The finished product i.e *tsampa* is shifted from *Ran-tak* to the household store room (called as *Zot*) and kept in wooden boxes each of 50-200 kilogram capacity with a small opening at the base for taking out their daily requirement of *tsampa*.

Food and beverages made with tsampa

Tsampa is consumed, either in raw or cooked form. A wooden bowl, locally called *pheykor,* filled with *tsampa* is kept on the table of the living-cum-kitchen room throughout so that it can be consume by the family members and the guest without the need of offering and consent by the housewife. *Tsampa* forms a part of sociocultural life of the people of Ladakh and no social activity is complete without it. It forms a key ingredient for offerings made during religious rituals, festive occasions and on New Year ceremony. It is used as the most convenient ration for traders, pilgrims and travelers. Traditional doctors, *Amchis,* often prescribe patients to take *tsampa* based food as an effective treatment against variety of ailments. Every household prepare *tsampa* twice a year, once during onset of spring which is meant for consumption till autumn. The second lot is prepared after crop harvesting in autumn and is consumed till onset of spring. Each household prepares around 200 kilogram *tsampa* in one lot, however, it varies depending on family strength. For preparing *tsampa*, barley is measured in terms of *Nas-khal* and one *Nas-khal* is approximately 14-15 kilogram.

A variety of dishes and beverages are being prepared from *tsampa* (or it being an ingredient). Some of the popular and important *tsampa* based dishes and beverages are described below (adapted from Angchok et al. 2009):

i. **Kholak**: It is the most commonly used food that involves no cooking (Figure 6). The dish is the ultimate quick and easy breakfast, lunch, trek or anytime food. *Kholak* is made by adding *tsampa* to any liquid or semi-liquid and bought to a consistency, where it does not stick to the hand through proper mixing and accordingly e.g. **Tsiri Kholak**: *tsampa* mixed with diluted *chhang* (fermented barley beverage); **Sbangphe**: the *chhang* residue is dried and ground together with *tsampa*. It is then mixed with butter tea, or sometimes water, and left overnight to be eaten the next morning; **Chuu kholak:** *chuu* means water. *tsampa* is added to water with a pinch of salt or sugar. Mostly used by hunters or travelers when

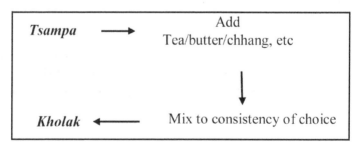

Figure 6: Flow chart of *Kholak* preparation.

nothing else is available; **Cha Kholak:** *cha* means tea. *Kholak* made by adding *tsampa* to tea (made from green tea leaves and salt in place of sugar); **Der Kholak:** *Derba* means butter milk. This is made by adding *tsampa* to butter milk; **Chhang Kholak:** *Chhang* means local barley made fermented drink. *tsampa* is added to *chhang*. It has a sour taste and is mostly preferred by travelers; **Phemar** (*Kholak* for the sweet-tooth): *Phemar* is usually served only to guests or on special occasions such as weddings. *tsampa* is added to salt tea with a lump of butter. Sugar can be added depending upon individual's choice; **Chubtsos** (one of the major ingredients of *phemar*): For *chubtsos*, wheat grain is boiled and made soft. Then it is strained, sun dried and roasted in a big pan, it is then mixed half and half with *tsampa* and kept specifically for this dish (*Kholak*). To make *kholak*, *chubtsos* is added to tea with a large lump of butter and sugar to taste. It taste better than the rest of the *kholaks*, and is mostly served on special occasions only; **Kushi Phey Kholak** (crushed dried apple with *tsampa*): This is made in villages for wedding and the parties to celebrate a baby's first year. **Chuli Phe Kholak** (Powdered dried apricot with *tsampa*): This is a refreshing food mostly used by travelers and is mildly laxating.

ii. **Paba**: The tsampa forms the key ingredient in preparing the popular dish called *paba*, which is prepared by boiling the flour in salted water till a tough consistency is obtained. The flour used for *paba* is called as *Yotches*, which is a mixture of flour of barley and local pea (*Nak-rhan*) in the ratio of 4: 1. *Paba* is taken as lunch during agricultural practices. The cooled *paba*, locally known as *zan-khang*, is also taken with either vegetable *curry* or simple grounded chilly (*tha-ner*). Pure barley flour can also be used for making *Paba* which is generally consumed by people with poor digestion.

iii. **sNham-thuk**: This soup can be eaten at any meal, generally at breakfast, when it makes a cheering start, especially in winter. It is given whenever somebody caught a cold (quite a frequent occurrence in a *Ladakhi* winter even if otherwise healthy). To make it more nutritious small pieces of meat, local pea and *chhurphe* (cottage cheese) is also added. The *tsampa* is often mixed with powdered pea, apricot or apple in different proportions as per need and affordability.

iv. **Cha Shrul** (Butter tea with *tsampa*): This is butter tea with enough *tsampa* mixed in to make it soup like.

v. **Chhang shrul**: Mostly, liberal dollops of *tsampa* will be put on top of the poured *chhang* and this is called *chhang shrul*. As a sigh of honour, guests are served *chhang* with a small piece of butter on their cup.

Changes in the traditional system

The traditional foods are no longer a source of pride for local *Ladakhis* (Norberg 1992), and in the haste to appear modern (so called), people are rejecting their own culture one woman rightly quotes.

Now when I'm a guest in a village, people apologize if they serve the traditional tsampa, instead of rice or instant noodles etc.

Rice, which had previously been a luxury in the *Ladakhi* diet, now has replaced the traditional barley and wheat based traditional foods. Ladakh's local economy, which has provided enough food for its people is now being invaded by produce from industrial farms located on the other side of the Himalayas (Angchok et al. 2009), and these produce are readily available at subsidized rates through the well established Public Distribution System (PDS) in majority of the villages. Due to which for many *Ladakhis*, it is no longer worthwhile to continue farming and grow local crops.

In context of *tsampa* preparation, use of *Ran-tak* for *tsampa* preparation is fast declining especially in and around Leh town. It was observed that around 30 years back there were approximately 100 fully functional *Ran-tak* along 5-7 kms glacier fed stream stretching from Gangles to Skara in Leh town. However, only ten *Ran-tak* are presently functional. The main reason behind the decline is introduction of conventional flour mills in the region and a shift from traditional to cosmopolitan food. However, people who still prefer the traditional food opt for *tsampa* prepared by the traditional milling process. The flavor, texture, shelf life and possibly the nutritional composition of *tsampa* prepared by traditional method is much better as compare to the one obtained through the modern day power driven mechanical mills.

On the other hand, there has been desirable changes in attitude towards the local foods and its importance due to influx of a cosmopolitan culture as a result of boost in tourism, and better connectivity with the outside world with improving communication infrastructure (virtual and real). The exchange of information has made the locals realize that the organically grown (due to lack of availability of chemical fertilizers and pesticide and most importantly due to the cold and dry environment incidence of pest is negligible in Ladakh) local crops are much better than the produce from the industrial world. Therefore, to help local farmers grow local produce and make the system economically viable efforts are being made by the local governing body (Ladakh Autonomous Hill Development Council) in collaboration with various NGOs and government departments. The campaign has already begun and it is hoped that the local foods will not be restricted to traditional ceremonies, but will once again be a part of fooding system in Ladakh.

Conclusion

Tsampa play an important role in the life of local population in their diet, customs, rituals, etc from centuries. However, with the opening of the region to modern world and introduction of modern technologies, a sharp decline in use of *tsampa* has been observed. The method of *tsampa* preparation is undergoing changes. There is a need to preserve the art of *tsampa* preparation which yield product with better flavor, texture and shelf life. The role of *tsampa* in everyday life of a Ladakhi and the variety of dishes being prepared from it needs to be documented and pass on to the younger generation.

Acknowledgements

We are immensely grateful to all the local respondents and experienced villagers of Gangles, Skara, Nurla and Saspol for sparing time to share with us their knowledge on the subject.The logistic supports in preparing this article obtained from the institutions to which first to fourth authors belong is acknowledged.

References

Anonymous (2013) Statistical hand book of Leh, Ladakh Autonomous Hill Development Council, Leh, Government of Jammu & Kashmir, India. http://leh.nic.in. Cited 20 Nov 2013

Anonymous (2013) Weather Records of Defence Institute of High Altitude Research, Leh. Accessed 10 Nov 2013

Angchok D, Dwivedi SK, Ahmed Z (2009) Traditional foods and beverages of Ladakh. *Indian Journal of Traditional Knowledge,* 8(4): 551-558

Norberg H (1992) Ancient futures: Learning from Ladakh. Oxford University Press, New Delhi.

Reifenberg G (1998) Ladakhi kitchen. Melong Publication of Ladakh, Leh

Rizvi J (1996) *Ladakh - Crossroads of High Asia.* Oxford University Press

Targais K, Stobdan T, Mundra S, Zulfikar A, Yadav A, Girish K, Singh SB (2012) *Chhang* - A barley based alcoholic beverage of Ladakh. *Indian Journal of Traditional Knowledge,* 11(2012): 190-193

Chapter – 13

Bioculturally Important Indigenous Fruit Tree *Mahua* (*Madhuca* spp.; Sapotaceae): It's' Role in Community-Based Adaptive Management

Anshuman Singh*, Ranjay K. Singh, Sarvesh Tripathy and BS Dwivedi

Abstract

Mahua (*Madhuca* spp.) is an important tree in India, with high socio-economic, environmental and spiritual values. We carried out this study among the local communities of Azamgarh district (eastern Uttar Pradesh) and with the *Gond* and *Baiga* tribes from Raisen and Dindori districts, Madhya Pradesh. Among the five *mahua* species found in Indian subcontinent, *Madhuca latifolia* is widely distributed in the north and central plains, the focal area of our study. *Mahua* has been vital to the survival of these communities, particularly the *Gond* and *Baiga* tribes, from the time immemorial. It has been a major source of food energy in the form of sugar, seed oil and alcohol. As well, different parts of the tree are used as ethnomedicine. The tribal people's high regard for *mahua* is reflected by its sustainable and regulated use. In this study, we present an in-depth analysis of conservation, management and sustainable use of *mahua*, and also suggest its role in socio-economic adaptation, with particular reference to climate change.

Keywords: Adaptation, biocultural values, community-based conservation, *Mahua*, *Madhuca* spp., traditional knowledge.

---

*anshumaniari@gmail.com

Introduction

There is a growing realization of the rampant degradation and continued reduction of natural resources, placing key food production systems around the globe at risk. It is argued that the world is facing a profound challenge to feed a population expected to reach 9 billion people by 2050. About 25 percent of our life supporting systems, including soil and water resources and biodiversity in all its forms, are degraded (FAO 2011). Added to this, ongoing global climate change endangers global food production (IPCC 2007) and the developing countries are likely to be the most greatly affected (Rosenzweig and Parry 1994). This state of affairs compels us to look into the possible reasons and accordingly design more effective systems for sustainable food production and natural resource management and enhancement. Issues such as rapidly changing global scenario, environmental risks, food shortages, need for empowerment of the weaker sections of society and concerns of developing and underdeveloped countries are also to be taken into account in developing location specific solutions to the present crises. Over the last few decades, the concept of sustainable development has emerged as a powerful source and it has marked a paradigm shift in understanding the intricate human-environment relationships. Thanks to this shift, the traditional view that the environment is merely a commodity has changed, and many humans have realized the incredible worth of "Mother Earth" and the resources she provides (Hopwood et al. 2005). Sustainable agriculture requires the continued availability of key natural resources: healthy soil, fresh water, and diverse, productive species. The conservation and sustained use of these precious resources cannot be realized without integrating trees and forests in planning and development schemes at the grassroots, national and regional levels (Prinsley 1992). From a socio-economic, environmental and spiritual standpoint, trees are of great importance to humankind. They sustain human life both directly and indirectly by providing a wide range of products and services for our livelihoods. Trees represent one of the important components of virtually every after terrestrial ecosystem (Seth 2004).

Mahua (*Madhuca* spp.), belonging to the family Sapotaceae, is such a tree. It has been a traditional source of livelihood for the poor and disadvantaged people, mainly tribal communities, living in the different parts of India. The genus *Madhuca* comprises of five economically important species: *Madhuca longifolia, M. latifolia, M. butyracea, M. neriifolia and M. bourdillonii*. These species are well known for their wide variety of uses since ancient times. Among these five species, *Madhuca latifolia* is widely distributed in the north and central Indian plains, the locations of our study. To avoid confusion in taxonomy and nomenclature, throughout this article we have used the common name *mahua* – derived from the Sanskrit word '*madhu*' meaning honey – to refer to this tree. *Mahua* flowers, which are exceptionally rich in sugars (66-72 per cent of dry weight), are used by the local communities as food, in ethnomedicine and for the preparation of an alcoholic beverage. The economic importance of *mahua* is also apparent by its mention in ancient Indian literature and travelogues of foreign travelers. *Mahua* trees are deciduous in nature and are found in abundance in the tropical rain forests of Asian and Australian Continents (Awasthi et al. 1975). It is a multipurpose forest tree which provides an answer for the three major "F's": food, fodder and fuel (Patel et al. 2011). The whole tree, its flowers and seeds have been very useful in the Indian economy for a long time. The corolla (the whorl of flower petals), an important bioresource, is rich in fermentable sugars and can be utilized as a substrate for bio-ethanol production (The Wealth of India 1962). They are edible and used as a sweetener in preparation of many local dishes, including *halwa, kheer, poori* and *burfi* (Patel and Naik 2008). In view of the immense significance of *mahua* as a key species in the livelihoods and the cultural heritage of the local and tribal communities, the present study was carried out with the following objectives: (i) to identify the key roles of *mahua* in sustainable development, with emphasis on environmental restoration and ecological processes, (ii) to understand the role of the *mahua* tree and its different products in the livelihoods of different communities, and (iii) to highlight the present state of affairs and the future prospects of this biocultural resource from a natural resource management perspective.

Research Methodology

Study sites

We conducted three case studies among the local communities of Azamgarh district (eastern Uttar Pradesh) and *Gond* and *Baiga* tribes from Raisen and Dindori districts, Madhya Pradesh, India to complement the secondary data on the biocultural dimensions of the *mahua* tree. The district Azamgarh, located between at 25° 383' and 26° 27' N latitudes and 82° 403' and 83° 52' E longitudes, falls in the eastern Gangetic plains and has a humid sub-tropical climate. Although rich in natural resources, Azamgarh is identified as one of the country's 250 least developed districts where about 91% of the population lives in villages. More than 85.0% of the farmers posses small and marginal land holdings and have subsistence agricultural economy with integrated farming systems (Singh et al. 2009). Community members speak the Bhojpuri dialect of Hindi. Their food resources are accessed from both terrestrial and aquatic ecosystems, imparting a dynamic and integrative character to the community-ecosystem interactions. The traditional cultural ethos and social taboos have compelled the local community to respect the nature. Over generations, the local people have developed a deep understanding of human-environment interactions and associated natural processes, and this has been helpful in the development of climate resilient socioecological systems (Singh et al. 2013).

The Raisen district lies in the central part of Madhya Pradesh state and is located between the 22° 47' and 23° 33'N latitudes and the 77°21' and 78° 49' E longitudes. It encompasses a total area of 8,395 square kilometers. It is a landlocked district and shares boundaries with the districts of Vidisha in the north, Sagar in the northeast, Narsinghpur in the southeast, Hoshangabad in the south and Bhopal in the west (Ahmad et al. 2012). The district is rich in forest resources with a total forest cover of about 3, 33,670 hectares. The tribal communities depend mainly on the forests for their livelihoods. The district comes under agro-ecological subregion of the central highlands (Malwa and Bundelkhand), characterized by hot sub-humid climate and rainfed agriculture. The *Gonds* (the largest tribe in Madhya Pradesh) have good population in this district. This tribe has a subsistence farming base, along with collection of forest products for their livelihood. The Dindori district of Madhya Pradesh, India lies in the Narmada valley and is surrounded by the Kachar and Satpura mountain ranges, rich in natural beauty and wild resources. It is a tribal dominated district with *Baigas* being one of the dominant tribes. This economically poor district is surrounded by Jabalpur in the northeast, Shahdol in the northwest, Bilaspur in the southwest, Rajnandgaon and Balaghat in the south and Mandla in the west. It also touches the boundary of Chattisgarh state (Sharma and Dwivedi 2006). The selected tribes (*Gond* and *Baiga*) speak the *Baigani* dialect of the Chhattisgarhi language.

Sampling and data collection

The village Sonapur, from Jahanganj Block of Azamgarh district (Uttar Pradesh), was selected purposively. A list of 65 key knowledge holders (KKHs: 40 men and 25 women), rich in biocultural knowledge on *mahua* tree, was prepared with the help of village elders and the Gram Panchayat (village institution) officials. The data pertaining to the *Gond* tribe were collected from 5 KKHs (50-70 years in age) from the Sipluri village, Sanchi Block, district Raisen, Madhya Pradesh. The data on the *Baiga* tribe were collected from the villages of Dhanua Sagar, Kohka and Modaki, of Dindori district, Madhya Pradesh. A total of 12 KKHs (4 from each village) of varying age (45-70 years) were selected randomly and were studied in 6 groups (2 groups in each village). The methodological framework of data collection was the same for all the three communities. The villages and tribes were sampled purposively, based on the objectives of the study. Before our data collection, rapport building was done with the help of key communicators.

Most of the observations were taken under the actual conditions in *mahua* groves during transect walks. Combinations of personal interviews (with open-ended questions) and focus group discussions were used to collect the data. Before the final application, an interview schedule with open-ended questions was pilot tested and any questions with ambiguous meaning were refined and/ or omitted. The villagers and key respondents from all the three study sites were asked questions related to three different aspects of *mahua*. The first aspect was the system(s) of management and the role of *mahua* in sustaining natural resources, supporting biodiversity and maintaining healthy ecological processes. The second aspect related to the significance of this tree as a biocultural and livelihood resource. The third aspect pertained to the present state of affairs of this bioculturally valued tree, including an investigation of the future needs for its sustainable management. Prior informed consent was obtained from each community to publish the results.

Results and Discussion

Systems of management and environmental roles of mahua

We found that clan- and community-based conservation and management of *mahua* plantations have been an integral part of traditional sociocultural and livelihood adaptation in the study regions (Figure 1). In the village Sonapur of the Azamgarh district, the *mahua* groves used to be managed by a particular clan consisting of an average 5 households. In certain cases, individual households having joint family systems had an exclusive right to manage and access the plantations for collection of the flowers and fruits. In the case of joint management by a clan, a rotation system prevailed wherein a particular household would be given periodic (weekly or fortnightly) access to pick the fallen flowers and collect other plant parts. In exceptional cases, grove management rested with the entire village community.

Figure 1: A clan managed *mahua* grove.
Photo: Ranjay K. Singh

As far these clan and community-based management systems prevailed until the late 1970s. Thereafter, in the following decades, management of this tree and its groves largely became a private affair. The reasons for this shift include the weakening of social bonds, fragmentation of joint family and land holdings, and changing land tenure systems. Previously, there existed a women's institution called *sajhiya* (at the clan level) with the task of collecting *mahua* flowers in small groups during March-April. These women's groups used to visit the *mahua* groves in the early morning and, while collecting the flowers from beneath the trees, they sang folk songs related to nature and their society. In the recent past, however, due to the socio-political and economic changes, this institution has almost disappeared. Presently, hired labourers are employed to pick the *mahua* flowers, and they are allotted either 1/4th or 1/3rd of the produce they collect.

Several clans together formerly planted the mosaics of *mahua* trees and developed sustainable harvesting protocols for them. An outsider to the clan needed prior permission to access the *mahua* trees and collect their flowers. In the case of unauthorized access and/or harvesting without the clan's consent, a clan meeting was held and the culprit was socially criticized. Sometimes, such acts also attracted fines, either in cash or kind. During study, one of the respondents (Mr. Lachhiram Singh, 70 years, Sonapur village, Azamgarh district) narrated an interesting system of social punishment for the unauthorized grazing of farm animals in *mahua* groves. The accused was asked to present her/his views in a social gathering and, based upon the magnitude of offense, s/he was censured. If such persons failed to appear before the social *panchyat* (customary institution) within 24 hours, her/his animal would be sent to the nearby *maveshi-ghar* (government cattle house). Then s/he would pay a fine to get the animal released.

The *Gond* and *Baiga* tribes of the studied villages practice subsistence agriculture and depend considerably on forest resources. Most of the forests in this region are under the management of the State Forest Department. The local tribal communities are given permission to access and use the forest resources for their livelihoods. It is said that *mahua* is the backbone of almost every tribal community of Madhya Pradesh. This was found to be true in the case of the *Gond* and *Baiga* tribes in our study. The *mahua* groves, as well as the scattered trees, form an integral part of the livelihood adaptations of the *Gonds* and *Baigas*, ensuring a consistent supply of the food, fuel, fodder, ethnomedicines, timber and inputs for agricultural production. The state *mahua* forest plantations maintained along roadsides and in wastelands are apportioned to different tribal clans based on their consensus to access the trees.

Other than forest plantations, the *Gond* and *Baiga* also plant *mahua* trees on their own lands as well as on shared community lands. Most of the plantations managed at the individual and clan levels are raised on the bunds of agricultural fields. In these tribes, the task of *mahua* flower collection rests largely with the younger women and children. They visit the *mahua* groves early in the morning and collect the fallen flowers. About 2-3 hours after the beginning of the flower collection, the male and female elders of each household will assist them in this work. Traditionally, these tribal people follow strict social norms while collecting the *mahua* flowers. For example, the collection of flowers usually starts after the celebration of tribal *Chait* festival in the month of March (Kala 2011a). In most cases, the flowers are collected for household consumption and only the excess amounts are sold for cash. This also ensures the sustainability of *mahua* by preventing its overexploitation. The tribal people are afraid of collecting flowers and seeds from sacred *mahua* groves. There is a general belief that since sacred groves were pure, the gatherers might be punished by spirits and deities for any unauthorized collection (Kala 2011b).

Sacred groves refer to the tracts of natural vegetation dedicated to certain local deities, from which no harvesting of plants or animals is permitted by the local communities. From an ecological perspective, these sacred groves are recognized as a unique cultural institution for conserving biodiversity within highly anthropogenic landscapes (Deb et al. 1997). The sacred forests and vegetation groves, including those of *mahua*, have been instrumental in the conservation of specific landscapes that have immense biocultural and religious significance for a country like India. In most of the cases, community members are aware of their environmental importance. In fact, they reflect a strong tradition of community-based natural resource management (Ormsby and Bhagwat 2010).

Economic development of biological resources often causes deterioration in environmental quality. Significant alterations in natural ecosystems have occurred due to burgeoning human populations, economic exploitation and associated processes, including deforestation and environmental pollution. One study, for example, revealed that opencast coal mining caused massive destruction of the natural vegetation in the Singrauli industrial area of Madhya Pradesh, India which was once covered with dense, tropical deciduous forests in which *mahua* was a common species (Singh and Singh 2006). The respondents from the studied villages revealed that *mahua* groves provide refuge for a diverse number of wild flora and fauna – birds and animals, both domestic and wild. In Sonapur village of Azamgarh district, the green leaves of *mahua* were used as fodder for goats and cattle during the fodder crisis, especially during droughts and famine. As well, the grasses growing between the rows of *mahua* trees were grazed by community members' farm animals.

The *mahua* tree has a large and deep penetrating root system that helps to stabilize soil aggregates and thus prevents soil erosion. *Mahua* plantations can be established on marginal and unproductive soils (Lal 2007). It is identified as a potential tree species for fallow and degraded lands in low rainfall areas (Ramachandra and Kumar 2003). In India, *mahua* trees are commonly planted in less fertile, hard lateritic soils. In Sonapur and other villages of Azamgarh district, *mahua* trees are common in sodic and degraded soils. *Mahua* trees are significantly drought tolerant. However, the saplings and even older trees do not survive waterlogging. Owing to the *mahua* tree's hardy nature, it can even be grown in rocky soils, and the trees are commonly raised in degraded rocky lands, including the salt affected soils (Singh 2001). Added to this, *mahua* plantations have been established on those community lands which otherwise do not support arable crops. Such observations point to the fact that even poor and degraded soils can be put to productive use by planting the *mahua* trees. Although no direct evidence is available, it may be assumed that *mahua* plantations could improve the fertility of unproductive soils by increasing their physico-chemical properties (Shukla and Misra 1993), raising the nutrient status and improving the water balance (Joffre and Rambal 1993).

Mahua as a biocultural and livelihood resource

Traditional communities worldwide, including those in India, practice different forms of nature worship. One such significant tradition of nature worship is that of providing protection to patches of forests dedicated to deities or ancestral spirits under the belief that to keep them in a relatively undisturbed state is expressive of an important relationship of humans with the divine or with nature. These vegetation patches are designated as sacred groves (Malhotra et al. 2001). From time immemorial, different Indian tribes have revered the *mahua* trees and plantations as the vehicles of Gods and Goddesses and have worshipped them. For instance, the Gond tribes of Madhya Pradesh consider *mahua* groves as the abode of deities and they pay devotion to *mahua* for the multiple benefits obtained, as also to worship the Goddess Khedapati (Kala 2011). The Mundas and Santhals of Bihar also worship *mahua* trees (Sinha 1995b). This tribal belief has been highly effective in the conservation of 'sacred *mahua* groves' in their original forms (Sinha 1995). Although such sacred groves often cover a miniscule proportion of the total area of the country, the number of such groves is estimated to be between 100,000 and 150,000 (Malhotra 1998). For about 600,000 villages in the Indian countryside, sacred groves form an integral part of the rural landscape. In Kodagu district of Karnataka state in the Western Ghats of India, sacred groves have relict populations of *Madhuca neriifolia* (Thw.) H.J. Lam, a threatened, wild relative of *mahua* (Bhagwat et al. 2005).

In the study regions, most of the *mahua* plantations are of trees that have been raised from seed and they exhibit a great variability. The respondents in our study revealed that *mahua* trees grow well in light black soils rich in organic matter and humus. The trees have a long gestation period and begin bearing flowers and fruits 12-15 years after planting. The flower (with cream- coloured, fleshy corolla) is the most economically important part of the tree. When the mature tree is in bloom, the ground beneath it is cleared of weeds. The falling of the corollas continues for about 20-30 days and this period generally extends from March 15th to April 15th. The productive economic life of the trees is about 60 years. In the first year of production, the *mahua* tree yield on average 75-150 kg of fresh flowers. There is a gradual increase in flower yield up to about 40 years of age, when the trees reach full bearing potential, producing 0.4-0.6 t of fresh flowers. On an average, 15-20 kg *mahua* flowers are collected every day by a household during the flowering season. A single *mahua* tree, on average, yields about 0.26 t flowers within a span of 20 days. In beginning, for the first 8 days, about 1.5 to 2 *kuri* (one *kuri*= 5 kg) flowers are collected from a fully bearing tree. This period is considered the most important from a flower yield standpoint as flower yield continues to decrease subsequently (Figure 2). The *mahua* flowers have great significance to the *Gond* tribe, and thus these people have adopted this tree as the backbone of their livelihood. Each member of a *Gond* family collects about 2-3 kg *mahua* flowers per day, and the average collection per family (averaging 5 members) per day would be 12-15 kg. In one season, each *Gond* household collects about 6-7 q of *mahua* flowers. We found that from one tree (40-50 years old), about 2.0 to 2.5 quintal of *mahua* flowers can be collected in one season. The women use to dry the flowers and process them for sale as well as for their own family's consumption. On average, the dried *mahua* flowers are sold @ Rs. 10-12 per kg. The bark of *mahua* tree is also collected and sold (Rs. 10-15 per kg) to local traders who sell it for use in treating coughs, asthma and indigestion.

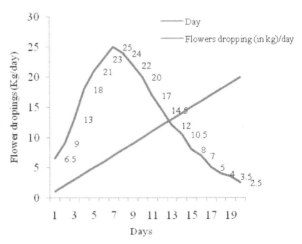

Figure 2: Average flower yield of *mahua* per kg per day in one season as perceived by the users.

In the village of Sonapur in Azamgarh district, we observed a number of ethnobotanical uses of the *mahua* flower. A local, energy rich food called *latta* was prepared from a paste of the dried flowers (corollas, without anthers). The farmers, after consuming *latta*, worked in the fields with greater endurance. During summers, a similar ethnic food, called *thokwa*, was prepared by mixing the fresh flower juice with wheat flour. The dried flowers, after being boiled with goat's milk in an earthen pot, were given to patients suffering from joint pain. The dried *mahua* flowers were also given to cows and buffalo to bring them into puberty. Older women respondents revealed that these practices are gradually disappearing, due in large part to modern ways of living, changing food habits and alienation of people from their traditional values.

The flowers of *mahua* are of great significance to the household economy of the *Gond* and *Baiga* tribes. The dried flowers after soaking into water are mixed with wheat flour to make energy rich chapatti (*mauhar*). During droughts, the flowers are mixed with maize flour and eaten as an emergency food. This practice helps minimize the demand for wheat as the economically poor *Gond* community has fewer land holdings to produce wheat for year round consumption. The dry flower petals, after cleaning, are made into paste and added with *tilli* (*Sesamum*) seeds to make an ethnic food called *laddu*. After extracting the juice from dry *mahau* flowers, the residue is boiled in water and mixed with wheat flour along with some dry fruits to make an energetic ethnic food called *kanchi*. *Gond* women use the flowers as a cosmetic for the face as well as to treat dermatological disorders. The flowers are also applied to heal body wounds. The extract from fresh flowers is used to cure ear-aches. The dry flowers, given to milch animals, enhance milk production.

Most of the *Gond* and *Baiga* community members in the studied villages ferment the dry flowers to prepare an alcoholic beverage locally called "*Daranga*". The dry flowers are put in a big earthen pot containing water and are left for about 10 days in winter and 5-6 days in summer to ferment. When the liquor is ready, it is used almost daily and a certain amount is also kept for celebrating different festivals and social events. The tribal people enjoy the delicacy of this alcoholic beverage on almost every cultural occasion, starting from childbirth to use at the death of a person. Two parts of *mahua* liquor and one part of honey is mixed, and given to those suffering from asthma. Children are also given about one and half teaspoons of the liquor with hot water to avoid cold symptoms. A massage of the liquor gives relief from muscle cramps. The initial 15 minutes of fermentation yields stronger liquor and is often given to mothers for abortion at one or two month's pregnancy. Persons suffering from paralysis are advised to take the *mahua* liquor after eating pigeon meat. Wound infections in animals, as well as in humans, are also cured by using the liquor.

The animals are given liquor to overcome the problems of indigestion and loose motion.

In Azamgarh district, the unripe fruits were as a vegetable during the summer, a time characterized by a shortage of green vegetables. The ripe fruits were eaten and the seeds were collected by children and women to extract the oil and make oilcakes for application to the potato crop. The oil extracted from *mahua* seeds was used in massaging to alleviate waist pain in women. The seed oil was also used to fry an ethnic wheat flour based food called *poori*. During the turbulent monsoon, characterized by blowing winds and cold temperatures, the dried flowers were given to bullocks to reduce the effect of pounding rains. The dry leaves were collected during the month of March and piled up for use as bedding. The leaves, after mixing with dung and animal urine, were used as compost. At the time of preparing jaggery during the winter season, the leaves of *mahua* were used as fuel. Among the *Baiga* and *Gond* tribes, the dry flowers after collection are sold to local traders @ Rs. 10-15 to generate an income, which is used for procuring foods and other needs of a household. The ripe fruit (called *Gulainda*) is eaten, while *gulli* (seed) is collected for use as a source of extracting food oil. The oil is used for food preparation and in massaging joints and waist pain (among women). The oil cake is burned to repel the snakes and other poisonous insects.

Dona (a kind of local organic utensil for serving food) is made from the green leaves of *mahua* and sold in the local market to generate income. These tribes, whose agricultural fields are situated near the *mahua* groves used to direct the irrigation channels and drains during the rainy season to channel the *mahua* leaf litter into the lower areas of land where indigenous varieties of rice (*lochai, patharchatti*, etc.) are to be planted. During March and April the dry leaves are collected and piled up in home gardens to use as fuel, as well as for mulch and making compost by mixing with cow dung and urine. The small twigs and branches are used as fuel for cooking. Use of green leaves in *mandap* [a shaded shelter prepared using sugarcane (*Saccharum officinarum*) plants] is integral to the marriage ceremony among the *Gond* and *Baiga* tribes. On the occasion of *harchhath puja* (a local festival), *mahua* flowers are offered to the deities. These festivals and events signify the huge cultural value of *mahua*.

We found instances where manure prepared from the *mahua* leaves was applied to agricultural crops. Again, tribal farmers participating in the study used to burn the dry *mahua* leaves, called *khari*, before planting out rice seedlings. This practice, as revealed by the respondents, helped to control weeds, insect pests and diseases in the rice nurseries and supplemented the nutritional needs of the rice seedlings. During the month of May, the unripe fruit of *mahua* is used as a vegetable. From one tree, about 40-50 kg *mahua* seeds can be harvested, and the seed oil is used in food as well as massaging for joint and waist pain. The oil is also applied to boils and blains. The great importance of *mahua*, which is considered as a holly tree is reflected in its decorative use in marriages as well as in celebrating the *Sharhul* (festival) during the month of April when the collection of flowers starts. At the time of childbirth, a liquor prepared from *mahua* flowers is used as an anticeptic wash to clean both the baby and the mother. At delivery, the mother is given a little quantity of *mahua* liquor for alleviate pain and the baby is also given in small quantity.

During the month of May, the unripe fruit is used in preparing vegetables. From one tree, about 40-50 kg seeds of *mahua* can be harvested and the seed oil is used in preparing food as well as massaging in for joint and waist pain. The oil is also applied on boils and blains. Looking to the great importance, *mahua* tree is considered a holly tree and its significance are valued in marriages as well as celebrating the festival (*Sharhul*) during the month of April.

Present Status of Mahua Conservation

Our study throws light on the present state of affairs regarding the management and use of *mahua* as a natural and biocultural resource. As far as tree management and conservation are concerned, the *mahua* groves seem to be in a state of neglect. This situation is more pronounced in Azamgarh district as compared with its tribal dominated counterparts of Raisen and Dindori districts in Madhya Pradesh state. We identified a set of factors, applicable to all the study sites, which might explain this situation. First and foremost, the local community in Azamgarh seems to have turned away from nature. The weakening of social bonds, growing dependency on modern technology, and the forces of globalization are gradually eroding the deep-rooted cultural ethos maintained through close, day-to-day interactions with the natural world. The tribal communities which have succeeded in maintaining their cultural heritage and are heavily dependent on the forest resources seem to have a better understanding of the importance of natural resources. If these precious resources are not to be degraded or lost, it will be necessary to devote attention to their sustainable management through decentralized and participatory projects such as social and community forestry and joint forest management.

Policy planning and development schemes should recognize the critically important role of communities and local knowledge holders and practitioners in resource use and conservation. Our argument is in line with that of other scholars who have championed the role of community in natural resources management (Agrawal and Gibson 1999; Leach et al. 1999; Kellert et al. 2000; Berkes 2012). The role of communities in the conservation and sustainable use of natural resources is based on the firm commitment of local people and institutions towards this cause. It is not a new approach *per se*. Instead, it seeks to revive and renew traditionally established socio-cultural norms and institutions and mechanisms for managing and conserving the natural environment (Kellert et al. 2000). In this regard, revival of such rural institutions, as the *sajhiya* system (collective management of trees to minimize risks, and the drudgery involved in the collection of flowers and subsequent processing) of the Azamgarh women should be given top priority. Environmental education and ecoliteracy tools are widely accepted as effective approaches for sustainable development. These tools have the promise of offering lasting solutions to environmental challenges by providing information on the use of locally available resources. They offer economic benefits, food security, ethnomedicinal solutions, and cultural continuity to both the local communities and nations (Pilgrim et al. 2007). The most important outcome of ecoliteracy and environmental education is the range of sustainable resource management practices that have evolved from them. The importance of these practices in biodiversity conservation strategies has also been highlighted in India (Singh 2010).

The villagers in Azamgarh district reported that in the past few years, *mahua* trees have started exhibiting erratic reproductive behaviour and the flowering time has been postponed. Some respondents revealed that normal flowering, which formerly started around mid-March, now occurs in the first or second week of April (Figure 3). This might adversely interfere with cycles of pollination, interactions with birds, and other ecological processes resulting in reduced flower and fruit yield. The irregularity in flowering was also noted from our research in the *Gond* villages. The tribal respondents reported that flowering times were either earlier or later than in previous times. In some years, flowering starts as early as middle of February and sometimes flowers appear as late as April. Cold spells, high wind velocities and thunderstorms, now regular phenomena during *mahua* flowering time, have adversely affected flower and fruit yields. The unpredictable climate, together with socioeconomic and ecological changes, has adversely

Figure 3: A late flowering *Mahua* tree. (26th April 2012). Photo: Ranjay K. Singh

impacted the local communities, especially the women, who now need to work harder and longer to collect the *mahua* flowers; flower productivity per tree has been reduced by about 25-30 % as perceived by the respondents. A similar study (Sushant 2013) reported about 25% reduction in the production of *mahua* flowers within the last 20 years. It is interesting to note, however, that the productivity of forest resources, including of *mahua* trees, is relatively less affected as compared with agricultural crop production (Sushant 2013), reflecting the resilient nature and risk bearing capacity of this species and its ecosystems.

In developing the strategies to mitigate greenhouse gas emissions, there has been much interest in forestry-related measures, either to reduce or to offset net carbon emissions. With the Bali Action Plan categorically placing reduced emissions from degradation and deforestation (REDD) activities on the agenda of future international climate change negotiations, there is now a strong possibility for converting the natural and community forests of India into carbon stocks (Singh 2008). However, there are inherent policy flaws and shortcomings in land use planning and land management at village and district levels in India. These flaws need to be addressed and the sociocultural institutions of communities more closely aligned with policy-making, allowing communities to sustain their community forests and sacred groves.

Conclusion and Future Thrust

This study explored the present state of *mahua* conservation: the value of this tree as a natural resource, its different ethnobotanical uses and the emerging issues regarding socioeconomic and climatic changes. Our observations in Sonapur village of Azamgarh district indicate that the once deep-rooted traditional bonds and cultural affinity of the community with *mahua* and its different products have almost disappeared in as this transitional society becomes more integrated into the globalized economy. If immediate and concerted efforts are not made, the community-based management of sacred *mahua* groves in these societies will soon become a part of history. The more remote nature- and forest-dependent communities, such as the *Gond* and *Baiga* tribes participating in this research, seem to have maintained their traditional attachment with this bioresource. Nevertheless, sociopolitical and ecological changes such as those caused by globalization and modernization coupled with climate change have threatened the very existence of the *mahua* tree, which has been historically an integral part of livelihood adaptations for these isolated communities. Based on the key results of this study, we recommend the following policy guidelines to sustain *mahua* and its associated natural resources for the generations to come.

1. Efforts should be made to rejuvenate the old and senile *mahua* plantations. This can be accomplished with technical experts from the fields of horticulture, tree biology and agro-forestry.
2. Although a scientific package of practices is available, these practices are seldom available to or followed by the *mahua* tree owners. Again, such management recommendations lack the 'organic farming approach'. This situation necessitates a fine tuning of the existing package of scientific practices, taking into account the local environmental considerations and feasibility using home grown organic inputs.
3. Efforts should be made for the targeted distribution of high quality planting material of landraces and improved clones for developing new *mahua* orchards, particularly in the drought prone, salt-affected and marginal lands with poor or no productivity. This will also boost the prospects for carbon sequestration in these degraded and fragile ecosystems.
4. A community-based, decentralized and participatory approach driven by local needs and considerations and incorporating formal knowledge and informal practices should be the key to future management programmes on *mahua* and other bioculturally important species.

Acknowledgements

The authors are thankful to all the key knowledge holders who provided the required information. The logistic support provided by Central Soil Salinity Research Institute, Karnal, India is duly acknowledged.

References

Agrawal A, Gibson, CC (1999) Enchantment and disenchantment: the role of community in natural resource conservation. *World Development*, 27(4): 629-649

Ahmad Z, Khan SS, Khan F, Shah F, Wani AA (2012) Studies on the ethnobotanical and ethnomedicinal uses on the plants of the family Euphorbiaceae of Raisen District (MP), India. *Science Secure Journal of Biotechnology*, 1(2): 43-46

Berkes F (2012) Sacred Ecology, Third Edition. Traditional Ecological Knowledge and Resource Management. Taylor & Francis, Philadelphia, USA

Bhagwat SA, Kushalappa CG, Williams PH, Brown ND (2005) A landscape approach to biodiversity conservation of sacred groves in the Western Ghats of India. *Conservation Biology*, 19(6): 1853–1862

Deb D, Deuti K, Malhotra KC (1997) Sacred grove relics as bird refugia. *Current Science*, 73(10):1853-1862

FAO (2011) 'State of the World's land and water resources for food and agriculture (SOLAW)'. Available on www.fao.org/news/story/en/item/95153/icode

Gautam RK, Jyoti R (2005) Primitive tribe in post-modern era with special reference to Baigas of Central India. *Primitive Tribes in Contemporary India*, 1: 189-209

Hopwood B, Mellor M, O'Brien G (2005) Sustainable development: mapping different approaches. *Sustainable development*, 13(1): 38-52

IPCC (2007) Climate Change 2007: Impacts, Adaptation and Vulnerability- Contribution of Working Group II to the IPCC Fourth Assessment, Cambridge University Press, Cambridge.

Joffre R, Rambal S (1993) How tree cover influences the water balance of Mediterranean rangelands. *Ecology*, 74(2):570-582

Kala CP (2011a) Indigenous uses and sustainable harvesting of trees by local people in the Pachmarhi Biosphere Reserve of India. *International Journal of Medicinal and Aromatic Plants*, 1(2):153-161

Kala CP (2011b) Traditional ecological knowledge, sacred groves and conservation of biodiversity in the Pachmarhi Biosphere Reserve of India. *Journal of Environmental Protection*, 2 (7): 967-973

Kellert SR, Mehta JN, Ebbin SA, Lichtenfeld LL (2000) Community natural resource management: promise, rhetoric, and reality. *Society and Natural Resources*, 13(8):705-715

Leach M, Mearns R, Scoones I (1999) Environmental entitlements: dynamics and institutions in community-based natural resource management. *World Development*, 27(2): 225-247

Malhotra KC (1998) Anthropological dimensions of sacred groves in India: an overview. *In:* Ramakrishnan PS, Saxena KG and Chandrashekara UM (Eds), Conserving the sacred for biodiversity management. Oxford and IBH, New Delhi. pp 423–438

Malhotra KC, Gokhale Y, Chatterjee S, Srivastava S (2001) Cultural and ecological dimensions of sacred groves in India. Indian National Science Academy, New Delhi

Ormsby AA, Bhagwat SA (2010) Sacred forests of India: a strong tradition of community-based natural resource management. *Environmental Conservation*, 37(3):320-326

Pilgrim S, Smith D, Pretty J (2007) A cross-regional assessment of the factors affecting ecoliteracy: Implications for policy and practice. *Ecological Applications*, 17(6):1742-1751

Prinsley RT (1992) The role of trees in sustainable agriculture—An overview. *Agroforestry Systems*, 20:87-115

Rosenzweig C, Parry ML (1994) Potential impact of climate change on world food supply, *Nature*, 367:133-138

Shukla AK, Misra PN (1993) Improvement of sodic soil under tree cover. *The Indian Forester*, 119(1): 43-52

Singh AN, Singh JS (2006) Experiments on ecological restoration of coal mine spoil using native trees in a dry tropical environment, India: A synthesis, *New Forests*, 31:25-39

Singh IS (2001) *Mahua*. *In:* Chadha KL (Ed), Handbook of Horticulture. ICAR, New Delhi pp 228-230

Singh RK (2010) Learning the indigenous knowledge and biodiversity through contest: A participatory methodological tool of ecoliteracy. *Indian Journal of Traditional Knowledge*, 9(2):355-360

Singh RK, Singh A, Pandey CB (2013) Agro-biodiversity in rice–wheat-based agroecosystems of eastern Uttar Pradesh, India: Implications for conservation and sustainable management. *International Journal of Sustainable Development & World Ecology*, dx.doi.org/10.1080/13504509.2013.869272

Singh SP, Gangwar B, Singh MP (2009) Economics of farming systems in Uttar Pradesh. *Agricultural Economics Research Review*, 22:129-138

Sinha RK (1995) Biodiversity conservation through faith and tradition in India: Some case studies. *International Journal of Sustainable Development & World Ecology*, 2(4):278-284

Sushant (2012) Impact of climate change in eastern Madhya Pradesh, India. *Tropical Conservation Science*, 6(3):338-364